LONDON MATHEMATICAL SOCIETY LECTURE NOTE SERIES

Managing Editor: Professor J.W.S. Cassels, Department of Pure Mathematics and Mathematical Statistics, University of Cambridge, 16 Mill Lane, Cambridge CB2 1SB, England

The titles below are available from booksellers, or, in case of difficulty, from Cambridge University Press.

London Mathematical Society Lecture Note Series. 240

Stable Groups

Frank O. Wagner
University of Oxford

CAMBRIDGE
UNIVERSITY PRESS

CAMBRIDGE UNIVERSITY PRESS
Cambridge, New York, Melbourne, Madrid, Cape Town, Singapore,
São Paulo, Delhi, Dubai, Tokyo, Mexico City

Cambridge University Press
The Edinburgh Building, Cambridge CB2 8RU, UK

Published in the United States of America by Cambridge University Press, New York

www.cambridge.org
Information on this title: www.cambridge.org/9780521598392

First published 1997

A catalogue record for this publication is available from the British Library

Library of Congress Cataloging in Publication Data

Wagner, Frank O. (Frank Olaf), 1964-
Stable groups / Frank O. Wagner
    p.   cm. - (London Mathematical Society lecture note series; 240)
Includes bibliographical references and index.
ISBN 0 521 598397 (pbk.)
1. Group theory. 2. Model theory. 3. Geometry, Algebraic.
I. Title. II. Series: London Mathematical Society lecture note series; 240.
QA174.2.W34 1997
512'.2-dc21   97-4011 CIP

ISBN 978-0-521-59839-2 Paperback

To Richmod

Groups & Guidance

Table of Contents

Groups & Gist
Preface

The study of stable groups connects model theory, algebraic geometry and group theory. Considering groups on which a certain very general dependence relation (Shelah's notion of "forking") is defined, one tries to derive structural properties, which may be group-theoretic (like nilpotency or solubility), algebro-geometric (identification of a group as an algebraic group, say), or model-theoretic (e.g. a description of the definable sets).

In this book I shall develop the theory of stable groups, bringing together the various extensions of the original finite rank theory under a unified perspective, and aiming to present the results in their proper setting, *viz.* the most general one possible. However, sections 1 and 2 of the noughth chapter do provide the necessary group- and model-theoretic background, and section 3 discusses most of the known examples. I shall concentrate mainly on the group-theoretic structure of stable groups, but also develop a good deal of model theory on the way. In particular, it will emerge in the fourth chapter that stable groups often occur naturally in a model-theoretic analysis, and that the lack of a definable group has strong structural consequences as well.

Groups & Gratitude
Acknowledgements

This book grew out of my DPhil Thesis at the University of Oxford under the supervision of Angus Macintyre, and my Habilitationsschrift at the University of Freiburg under the direction of Martin Ziegler; I am indebted to both of them. Further thanks are due to Wilfrid Hodges, Anand Pillay, and Bruno Poizat for sharing their extensive insight, and also to the logic groups at Oxford and Freiburg for many valuable discussions.

Part of this book was written while I held a research grant of the Deutsche Forschungsgemeinschaft at Freiburg University.

F.O.W.

Chapter 0

Groups & Goals

0.0 Introduction

The study of stable groups began with the classification of uncountably categorical abelian groups [69] and fields [70] by Angus Macintyre. In 1965 Michael Morley [75] had proved his Categoricity Theorem and defined ω-stability on the way, and Saharon Shelah [102, 103] embarked on his study of the models of complete first-order theories using his new notion of "stability"; these logical constraints were now being applied to algebraic structures. Differentially closed fields were studied by Shelah in [104], minimal groups by Joachim Reineke in [99], and finally simple groups by Gregory Cherlin in [34], where he formulated his famous conjecture: a simple group of finite Morley rank is an algebraic group over an algebraically closed field. Like Boris Zil'ber's conjecture from [134], of which it may be considered an analogue for groups, it is more programme than claim: one should try to do algebraic geometry, and in particular to define the Zariski topology, by model-theoretic means. Although the original conjecture was refuted by a counter-example of Ehud Hrushovski [57], the programme itself was completed successfully in [59] and has already led to the first applications outside of logic in Hrushovski's proof of the relative version of the Mordell-Lang Conjecture [56].

Meanwhile people had started to look at more general stability classes: Cherlin and Shelah were studying superstable division rings [33], fields and groups [37], Chantal Berline and Daniel Lascar [16, 17, 18] generalized many results from stable to superstable groups, and John Baldwin and Jan Saxl analysed stable groups in [4]. Ulrich Felgner [41] and Walter Baur, Cherlin and Macintyre [14] proved the nilpotency of $\aleph_0$-categorical stable connected groups (the latter also improved this to commutativity in the ω-stable case). Finally Bruno Poizat in [92] and [93] generalized the method of generic types to stable groups, one of the main tools in the subject besides chain conditions and rank considerations. An important connection between general stable theories and stable groups was discovered by Hrushovski in [51], where conditions for the existence of an interpretable group in a stable theory were given. This group has a natural interpretation as an automorphism group, and may in turn be used to derive further information about the original structure.

Work on the Cherlin-Zil'ber Conjecture, i.e. the study of simple groups of finite Morley rank, was continued in particular by Alexandre Borovik, Ali Nesin and Poizat [95, 20]. Increasingly techniques from finite group theory were used alongside methods from algebraic group theory, especially ideas from the classification of finite simple groups. It is now hoped that the conjecture (under an additional assumption) can be proven along the lines of the proof of the Classification Theorem – fortunately with significant savings on the way!

Thus the interest in stable groups is twofold. On the one hand it is derived from logic, as the results of Hrushovski and their applications, say, in Steven Buechler's partial proof of Vaught's Conjecture [29] show that even in abstract model theory groups may occur in an unexpected context. Furthermore, due to the homogeneity imposed by their multiplicative structure, groups may serve paradigmatically to test new methods and hypotheses. On the other hand, there is the group-theoretic content, as one tries to extract general properties sufficient for a structural analysis from finite or algebraic groups.

Our aim in this book will be to develop the theory of stable groups in full generality, avoiding repetition as far as feasible. However, the rank-free arguments can be quite involved, depending as they do solely on chain conditions and genericity considerations, where in the ω-stable or superstable context one might have used induction on the rank. If one is merely interested in the superstable result, or in order to facilitate a first understanding, one might assume in addition that the group in question is superstable: the result will in most cases be that undefinable or infinitely definable objects become type-definable or even definable, thereby clarifying the argument.

Many of the results will have evolved in various stages, due to various authors; typically a result may have been proven first for groups of finite Morley rank (often by Zil'ber or Nesin), then for superstable groups (of particular importance here are the seminal papers by Berline and Lascar mentioned above), and groups associated with a regular type (p-simple p-connected groups, due to Hrushovski). Of course many more people have contributed to the subject, and I cannot mention more than a few in this introduction, notably John Baldwin, Andreas Baudisch, Oleg Belegradek, Alexandre Borovik, Roger Bryant, Gregory Cherlin, Brian Hartley, James Loveys, Angus Macintyre, Ludomir Newelski, Anand Pillay, Bruno Poizat, Simon Thomas, and John Wilson. (I have included some group theorists, who will figure most prominently in chapter 1, whose content is more algebraic than model-theoretic.) More specific attributions and a few extra remarks will be collected in a section "Historical and Bibliographical Remarks" at the end of each chapter.

Although I have tried to keep the text as self-contained as possible, it does assume a reasonable familiarity with group theory and stability theory. Nevertheless, the next two sections provide a crash course in elementary group theory and basic stability theory and may be skipped by those familiar with the material. Alternatively, Poizat's yellow book *Groupes stables* [95] is a marvellous introduction; conscientious objectors may also try *The Model Theory of Groups* by Nesin and Pillay [80] or, with a particular emphasis on the classification theory, *Groups of Finite Morley Rank* by Borovik and Nesin [20]. I learnt my group theory from Huppert's *Endliche Gruppen* [60] and my stability theory from Poizat's *Cours de théorie des modèles* [94]; more advanced material can be found in the recent book by Pillay, *Geometrical Stability Theory* [88]. Finally, the last substantive section of this introduction will be devoted to some typical examples – although typicality should be taken *cum grano salis:* the known examples of stable groups are all rather well-behaved, and one of the objectives of this enterprise is to see just how typical this behaviour is in the class of all stable groups.

In the first chapter we shall use the stability assumption to deduce various chain conditions. As these may also appear in different contexts (e.g. o-minimal groups), we shall proceed to consider groups with chain conditions in general, without further use of stability. After a section on connected components and definability of certain subgroups (soluble or nilpotent ones, socle, layer) we again widen our scope and consider groups satisfying the maximal condition for centralizers, $\mathfrak{M}_c$, for which we prove various nilpotency results. In particular, we show that the Fitting subgroup of an $\mathfrak{M}_c$-group is nilpotent. We then consider nilpotency in the smaller class of substable groups. Here, we can show the hypercentrality of locally nilpotent substable groups, and prove various other local-global results. Most of them are connected with Engel conditions; while bounded left Engel elements can be shown to lie in the Fitting subgroup, only partial information is available about the hypercentrality of bounded right Engel elements. A final section deals with the Sylow and Baer-Suzuki theorems; variants of these (particularly nice for $p = 2$) are proven in our context.

The second chapter introduces the method of generic types. Extending the approach of Poizat, we consider an infinite subset of some model of a stable theory together with a definable binary function, which is associative on this set and satisfies some cancellation property. We show that this set-up may always be embedded into a definable group. After an aside or two on group operations and fields we investigate generic properties: given a formula $\varphi(\bar{x})$ which holds on generic elements, under what conditions must it be satisfied by the whole group? This question was put forward by Poizat and has its origin in the theory of algebraic groups, where a generically true equation holds everywhere. Alas, answers are scarce and we can tackle only a few special cases. We shall conclude this chapter with an application of the method of generic properties to give an alternative proof of the nilpotency of the Fitting subgroup of a stable group.

Model theory plays a more prominent rôle in the third chapter where we introduce the notions of "internal" and "foreign", due to Hrushovski, in order to distinguish what may be considered "large" and "small" definable sets; this is done first in general and then in the particular context of groups, and applied to the study of involutions and their centralizers. We proceed to define the Φ-connected component, which is designed to be a "large" subgroup with strong connectivity properties. The technical apparatus developed there enables us to circumvent regular types; its full thrust will become apparent in chapter 5. A first application of these methods in the fourth section of chapter 3 yields a structure theorem for a Φ-connected component without abelian normal subgroups: it decomposes into a finite direct sum of $\mathfrak{R}$-primary subgroups.

The last four substantive sections of chapter 3 introduce yet another technique for determining the size of a type with respect to a given family P of types, namely *localized P-rank*. Generalizing Lascar rank, this yields a unified approach both to considerations of monomial rank *à la* Berline-Lascar and to Hrushovski's p-weight machinery. In particular, we shall prove an Indecomposability Theorem, and analyse fields and bad groups.

Chapter 4 analyses the dependence relation given by forking. After introducing the notion of a pre-geometry in general, we study various additional conditions under which the notion behaves particularly nicely. The first one is *local modularity*; it distils the main properties from the abelian structures encountered in Example 0.3.1. Locally modular structures, unless their geometry is trivial (in a specified sense), behave essentially like modules, at least generically. To prove this necessitates various group existence theorems, the most important one being about the recovery of a group from geometric data, the so-called *group configuration*. We shall show that a locally modular group is abelian-by-finite, and explore the ring underlying its module structure (the ring of *quasi-endomorphisms*). In fact, this ring is also present in a more general context: in the fourth section we prove a dichotomy theorem between local finiteness and the existence of a connected-by-finite definable quotient.

The second condition is the weaker *CM-triviality*, introduced more recently by Hrushovski while analysing structures obtained from their finite substructures by an amalgamation process. Here we can give some general conditions under which CM-trivial groups must be nilpotent.

In the penultimate two sections, we shall analyse the forking relationship between different types. First we shall study *dimensional* theories, where every type is non-orthogonal to one of boundedly many regular types; these are related to theories categorical in big powers. The eighth section, finally, introduces binding groups, which describe the interconnection between two types. We shall use them to prove some general structure theorems for stable theories, in particular Hrushovski's Theorem that unidimensional theories are superstable.

In the fifth chapter we consider condition $\Re$. This property, derived from superstable groups but also common to small stable groups, provides the last glimpses of a rudimentary rank: generic types are ordered by their possibilities of algebraization. Nevertheless, many results using rank arguments can also be proved under the sole assumption of $\Re$, albeit under greater technical difficulties. $\Re$-fields are commutative, algebraically closed and possess only few automorphisms; $\Re$-groups have large abelian subgroups. Even the existence theorem for definable fields and the linearity theorem for certain group actions, cornerstones of the programme to solve the Cherlin-Zil'ber Conjecture, hold in this context.

The next two sections of chapter 5 deal with a soluble Φ-connected $\Re$-group G. Stretching definability to its limits by considering quotients of $\bigwedge$-definable groups by $\bigvee$-definable groups, we prove that G has a nilpotent derived subgroup and a definable hypercentre. Furthermore, if G has a nilpotent homomorphic image, then the kernel of this homomorphism has a nilpotent supplement. This is used to develop the theory of Carter subgroups. Using a structure theorem for the centre of the derived subgroup, we may also define some analogue of the Frattini subgroup, with the property that nilpotency modulo that subgroup implies nilpotency. Here the Carter subgroups find their first application when they replace the Sylow subgroups in structural investigations. The last section returns to Sylow 2-subgroups, which are shown to be normal-by-finite. This enables us to give a relatively short proof of the existence of infinitely many conjugacy classes in any infinite $\Re$-group.

0.1 Getting to Grips with Groups

A *semi-group* is a set G together with an associative binary function $*$: $G \times G \longrightarrow G$. Normally, we write gh instead of $g * h$. An element e_l in G is a *left unit* if $e_l g = g$ for all $g \in G$, and G has *left cancellation* if $gh = gk$ implies $h = k$ for all $g, h, k \in G$. Similar notions exist on the right; if both right and left units exist, they are unique and equal, and usually denoted by 1. The *inverse* of some element g is an element g^{-1} such that $gg^{-1} = 1$ and $g^{-1}g = 1$; if all elements have inverses, then one condition implies the other.

A *group* is a semi-group with unit and inverses; it is *abelian* or *commutative* if $gh = hg$ for all group elements g and h (in this case we often write the

group operation additively). A particular example of a group is the group of
permutations of a set X, denoted $\mathrm{Sym}(X)$, with composition as group law.
If X has n elements, we also denote $\mathrm{Sym}(X)$ by S_n, and the cardinality $|S_n|$
is $n!$. A *subgroup* H of a group G, denoted by $H \leq G$, is a subset of G closed
under multiplication and inversion; it is *normal* if for any $h \in H$ and $g \in G$
the *g-conjugate* $h^g := g^{-1}hg$ is again in H. The *order* $o(g)$ of an element
$g \in G$ is the least positive $n < \omega$ such that $g^n = 1$ (or ∞ if there is no such
n), and g generates a *cyclic subgroup* $\langle g \rangle$, which consists of all powers g^z with
$z \in \mathbb{Z}$ (putting $g^0 = 1$). This has size n and is also denoted by C_n (for $n = \infty$
it is countably infinite). More generally, if $g_1, \dots, g_n$ are elements of G, then
$\langle g_1, \dots, g_n \rangle$, the subgroup *generated* by $g_1, \dots, g_n$, is the smallest subgroup
of G containing all these elements, and consists of all finite products $\prod_j g_{i_j}^{z_j}$,
with $1 \leq i_j \leq n$ and $z_j \in \mathbb{Z}$.

If G is a group and p a prime, a *p-element* is an element whose order is
some power of p. The *exponent* $e(G)$ of G is the smallest $n < \omega$ such that
$g^n = 1$ for all elements g in G. If there is no such n, we put $e(G) = \infty$.
The group is *periodic* if every element has finite order; it is a *p-group* if every
element is a p-element. We may also consider the following subsets of G:

- the set $\mathrm{Tor}(G)$ of its elements of finite order, the *torsion* of G,

- the set $G[n]$ of its elements of order dividing n, for any $n < \omega$,

- the set G_p of its p-elements, the *p-part* of G, for any prime p, and

- the set $G_{p'}$ of its elements of finite order coprime to p, the *p'-part* of G,
  for any prime p.

If G is abelian, all of these are obviously subgroups.

A *homomorphism* of two groups G and H is a map $\varphi : G \longrightarrow H$ such that
$\varphi(g_1 g_2) = \varphi(g_1)\varphi(g_2)$ for all $g_1, g_2 \in G$. The *kernel* of φ is the normal subgroup $\{g \in G : \varphi(g) = 1\}$. Conversely, given any normal subgroup H, setting
$g_1 H * g_2 H = (g_1 g_2)H$ defines a group structure on the *coset space* G/H, and
the homomorphism $g \mapsto gH$ has kernel H. A homomorphism is injective iff its
kernel is reduced to the identity. A homomorphism from G to itself is called
an *endomorphism*; if it is bijective, it is an *automorphism*, and the collection
$Aut(G)$ of all automorphisms of G forms a group under composition. $Aut(G)$
has a normal subgroup $Inn(G)$ of *inner automorphisms*, i.e. those derived
from conjugation $\sigma_g : x \mapsto x^g$; the quotient $Aut(G)/Inn(G) =: Out(G)$ is
the group of *outer automorphisms* of G. A subgroup H of G is *characteristic*
in G if it is invariant under all automorphisms of G (whereas a normal subgroup need only be invariant under the inner ones); unlike normality, this is
a transitive relation: if N is characteristic in H and H is characteristic in G,
then N is characteristic in G. However, if N is characteristic in H and H is

normal in G, then N is at least normal in G. A quotient of a subgroup H of G by a normal subgroup N of H is called a *section* of G.

A *ring* is an abelian additive group R which is a semi-group under multiplication, such that the distributive laws $x(y + z) = xy + xz$ and $(x + y)z = xz + yz$ hold. It may have a unit ($x1 = 1x = x$ for all *non-zero* $x \in R$); R is called commutative if its multiplication is. A non-zero element r such that there is a non-zero $x \in R$ with $rx = 0$ is called a *left zero-divisor* (and x is a *right zero-divisor*); an element r such that there is $n < \omega$ with $r^n = 0$ is called *nilpotent*. R is *nil* if every element is nilpotent; it is *nil of finite nilexponent* n if $r^n = 0$ for all $r \in R$, and *nilpotent* of class n if $r_1 \cdots r_n = 0$ for all $r_1, \ldots, r_n \in R$.

A commutative ring with a unit and without zero-divisors is an *integral domain*. If every non-zero element of a ring has a multiplicative inverse, the ring is called a *skew field*, or just a *field* if it is commutative. Any integral domain R embeds uniquely into a smallest field F (up to isomorphism), its *field of fractions* or *quotient field*, and every element of F is of the form $r_1 r_2^{-1}$ for some $r_1, r_2 \in R$. Given any group G and a ring R we can form the *group ring* RG, which consists of all finite sums $\sum_i r_i g_i$, with $r_i \in R$ and $g_i \in G$, componentwise addition $\sum_i r_i g_i + \sum_i r'_i g_i = \sum_i (r_i + r'_i) g_i$ and multiplication $\sum_i r_i g_i \cdot \sum_j r_j g_j = \sum_g (\sum_{g_i g_j = g} r_i r_j) g$.

An additive subgroup I is a *left ideal* if $rI \leq I$ for every $r \in R$ (and similarly for right ideals); a (two-sided) ideal is both left and right ideal. For a two-sided ideal I we can form the *quotient* ring R/I by setting $(x + I) + (y + I) = (x + y) + I$ and $(x + I)(y + I) = xy + I$. A left ideal I is *left principal* if it is generated by a single element $r \in R$ (and similarly *right principal*), if R has a unit, then the left principal ideal (r) equals rR; it is *prime* if $xy \in I$ implies $x \in I$ or $y \in I$ for all $x, y \in R$, and *maximal* if there is no ideal J with $I < J < R$. Particular left ideals are *left annihilators* $\mathrm{ann}(r) := \{r' \in R : r'r = 0\}$; we say that r is *annihilated* (on the left) by r' (and similarly on the right).

If R' is another ring and φ an additive homomorphism $R \longrightarrow R'$, then φ is a *homomorphism* of rings if $\varphi(xy) = \varphi(x)\varphi(y)$ for all $x, y \in R$, and $\varphi(1) = 1$ (if the rings both have a unit). The *kernel* of φ is the two-sided ideal $\{r \in R : \varphi(r) = 0\}$; conversely, any two-sided ideal is the kernel of the homomorphism $R \longrightarrow R/I$ given by $r \mapsto r + I$. The quotient of a commutative ring by an ideal is an integral domain iff the ideal is prime, and a field iff the ideal is maximal. Clearly, a skew field F has no zero-divisors, and no ideals other than (0) and F itself.

If G is a group and X a set, then G *acts* on X if we have a homomorphism $\varphi : G \longrightarrow \mathrm{Sym}(X)$. In particular, $\varphi(g)$ is a permutation of X for every $g \in G$; we shall normally write x^g or gx instead of $\varphi(g)(x)$ (depending on whether we consider the symmetric group, and hence G, to act on the right or on the left). The *orbit* of some point $x \in X$ is the set $x^G := \{x^g : g \in G\}$. The action is *transitive* if $x^G = X$ for some (equivalently: for any) $x \in X$; it is *n-transitive*, if G maps any n-tuple of distinct points to any other n-tuple, and *sharply n-transitive* if for any two n-tuples there is exactly one group element mapping one tuple to the other (equivalently: if it is n-transitive and only the identity fixes an n-tuple). Clearly, the orbits partition X into disjoint subsets, and G acts transitively on each orbit. G acts *faithfully* if no $g \in G$ fixes every $x \in X$. The *centralizer* $C_G(x)$ of $x \in X$ in G is the set $\{g \in G : x^g = x\}$; this forms a subgroup of G. If G acts faithfully and transitively on a set X (on the left), then we can pick any element $x \in X$ and consider the action of G on $G/C_G(x)$ by left multiplication; this is canonically isomorphic to the original action via the bijection $gx \mapsto gC_G(x)$ between X and $G/C_G(x)$. Conversely, for any subgroup $H < G$ we have a transitive left G-action on the coset space G/H given by left multiplication.

In the case when X is already a group, we may in addition require that every φ_g is an endomorphism (or equivalently automorphism) of X. In this case, the identity of X is obviously fixed under the action, and hence discounted for transitivity or the number of fixed points. An important example for this is the action of a group on itself by conjugation, $\varphi_g : h \mapsto h^g$.

Proposition 0.1.1 *Let H be a subgroup of G of finite index n. Then H contains a subgroup N which is normal in G and has index at most $n!$.*

Proof: Consider the action of G on G/H. This induces a homomorphism φ of G into the group S_n of all permutations of G/H; let N be its kernel. Then G/N is isomorphic to the image of φ, of order at most $|S_n| = n!$, so N is a normal subgroup of index at most $n!$. Since $nH = H$ for all $n \in N$, it follows that N is contained in H. $\square$

If A is an abelian group, we may consider the set $\text{End}(A)$ of endomorphisms of A and put $(\varphi + \psi)(a) = \varphi(a) + \psi(a)$, for all $a \in A$. Then $\varphi + \psi$ is again an endomorphism of A, and $\text{End}(A)$ forms a ring under this addition and composition as multiplication. If $\varphi(a) = 0$ for some $a \in A$ and endomorphism φ, then a is *annihilated* by φ. In particular, if a group G acts on A, we may consider the subring of $\text{End}(A)$ generated by G, i.e. the collection of all endomorphisms of the form $\sum_i n_i \varphi_{g_i}$, for $n_i \in \mathbb{Z}$ and a finite summation. This is obviously a homomorphic image of the group ring $\mathbb{Z}G$ with kernel $\{\sum_i n_i g_i : \sum_i n_i \varphi_{g_i} = 0\}$.

Definition 0.1.1 *Let G be a group, H a subgroup, A a subset, and g and h elements of G.*

- The *commutator* of g and h is $[g, h] = g^{-1}h^{-1}gh$. If $g_1, g_2, \ldots$ are elements of G, we define inductively a (left normalized) *iterated com-*

mutator by $[g_1, \ldots, g_i, g_{i+1}] = [[g_1, \ldots, g_i], g_{i+1}]$; *repeated commutators* are defined via $[g,_0 h] = g$ and $[g,_{i+1} h] = [[g,_i h], h]$.

- We put $[g, A] := \{[g, a] : a \in A\}$ (a subset), and if B is another subset of G, we let $[A, B] = \bigcup_{a \in A}[a, B]$. However, if both A and B are subgroups, then $[A, B]$ will denote $\langle [a, b] : a \in A, b \in B \rangle$, the group generated by all commutators of elements of A with elements of B.

- The *g-conjugate* of A is $A^g = \{a^g : a \in A\}$. If $A^g = A$, we say that A is *g-invariant*; if A is *g-invariant* for all $g \in H$, we call it *H-invariant*. Note that H^g is again a subgroup. A *G-invariant* subset of G is usually just called *invariant*.

- The *A-closure* H^A of H is the group generated by $\{H^a : a \in \langle A \rangle\}$. It is normalized by $\langle A \rangle$, and $H^A = \langle H, [H, A] \rangle$.

- The *centralizer* of A in G is the subgroup $C_G(A) = \{g \in G : [g, A] = 1\} = \{g \in G : a^g = a$ for all $a \in A\}$.

- The *normalizer* of H in G is the subgroup $N_G(H) = \{g \in G : H^g = H\}$.

- The *centralizer of A in G modulo H* is $C_G(A/H) = \{g \in N_G(H) : [g, A] \subseteq H\}$.

- The *series of iterated centralizers* is defined inductively by $C_G^0(A/H) = H$, and $C_G^{i+1}(A/H) = C_G(A/C_G^i(A/H)) \cap \bigcap_{j \leq i} N_G(C_G^j(A/H))$.

- If N is a subgroup of G normalized by H, the *product* NH is the subgroup $\{nh : n \in N, h \in H\}$. The product is *central* if N and H commute; it is *direct* if N and H intersect trivially and commute. We write $N \times H$ for the direct product (or $N \oplus H$ in the additive case). The same definition holds for more than two factors (with every factor normalizing the previous partial product).

- If G is a group acting (on the right) on another group N as group of automorphisms, we can form the *semi-direct product* of N and G, denoted by $N \rtimes G$. Its underlying set is the Cartesian product $N \times G$, with multiplication given by $(n, g)(m, h) = (nm^{g^{-1}}, gh)$.

Definition 0.1.2 The *Prüfer rank* of a group G is the least $n \leq \omega$ such that every finitely generated subgroup has a generating set of cardinality at most n.

The *Prüfer* (or *quasi-cyclic*) *p-group* C_{p^∞} is the direct limit $\lim_n C_{p^n}$. Any periodic abelian group A decomposes as direct sum of its *p-parts* A_p; a countable abelian *p-group* is a direct sum of cyclic and quasi-cyclic groups, and its Prüfer rank equals the number of the summands. (Note, however, that $C_2 \oplus C_3 \cong C_6$ has Prüfer rank one.)

It is easily seen that

$$[x, yz] = [x, z][x, y]^z, \quad \text{and} \quad [xy, z] = [x, z]^y[y, z].$$

In particular, if A is a normal subgroup of G such that A is abelian or $[x, A]$ commutes with all elements of A, then $a \mapsto [x, a]$ is an endomorphism of A. Let x, y, and z be elements of a group G. Then the *Witt identity*

$$[x, y^{-1}, z]^y[y, z^{-1}, x]^z[z, x^{-1}, y]^x = 1$$

can be checked by direct computation, and has an important consequence, Philip Hall's *Three Subgroup Lemma*.

Lemma 0.1.2 *If K, L and M are three subgroups of a group G normalizing a subgroup N of G, then $[K, L, M] \leq N$ and $[L, M, K] \leq N$ together imply $[M, K, L] \leq N$.*

Proof: This follows easily from the Witt identity, since for any elements $k \in K$, $l \in L$ and $m \in M$ we have

$$[m, k, l] = \left([k^{-1}, l^{-1}, m]^l[l, m^{-1}, k^{-1}]^m\right)^{-k^{-1}} \in N. \quad \square$$

The study of relations between commutators is called *commutator calculus*.

Definition 0.1.3 Let G be a group.

- The *derived series* of G is defined inductively by $G^{(0)} = G$, and $G^{(i+1)} = [G^{(i)}, G^{(i)}]$. The first *derived group* $G^{(1)}$ is also called the *commutator subgroup* of G, denoted by G'; further elements of the series are *iterated derived* (or *commutator*) subgroups.

- The *descending central series* of G is defined inductively by $G^0 = G$, and $G^{i+1} = [G^i, G]$. (Note that $G^1 = G'$.) This is also called the *lower central series*.

- The *ascending central series* of G is defined inductively by $Z_0(G) = \{1\}$, and $Z_{i+1} = C_G(G/Z_i(G))$. Another name for it is *upper central series*. $Z_1(G)$ is simply called the *centre* of G and denoted by $Z(G)$; its elements are called *central*. The other elements of the series are the *iterated centres*. This series is often continued transfinitely, with unions at the limit stages. In particular, $Z_\omega(G) = \bigcup_{i<\omega} Z_i(G)$ is the ω-centre of G, and the hypercentre $Z_\infty(G)$ is the union of all iterated centres. Clearly, $Z_\infty(G) = \bigcup_{i<|G|^+} Z_i(G)$. Furthermore, $Z_\alpha(G) = C_G^\alpha(G)$.

- A group G is *soluble of derived length n* if $G^{(n)} = \{1\}$. It is *nilpotent of class n* if $G^n = \{1\}$. It is hypercentral of length α if $Z_\alpha(G) = G$.

Clearly, the elements of the derived series and the descending or ascending central series are characteristic subgroups; furthermore $G^{(i)}/G^{(i+1)}$ is abelian, G^i/G^{i+1} is central in G/G^{i+1}, and $Z_{i+1}(G)/Z_i(G)$ is central in $G/Z_i(G)$ for all i. In fact, it is easy to see that a group is soluble iff it has an *abelian series*, i.e. a series $G = G_0 \geq G_1 \geq \cdots \geq G_n = \{1\}$ such that every G_i is normal in G_{i-1}, with abelian quotients G_{i-1}/G_i; the derived length equals the minimal length n of such a series. Similarly, a group is nilpotent of class at most n iff it has a *central series* of length n (i.e. a series of normal subgroups as above, but with G_i/G_{i+1} central in G/G_{i+1} for all $i < n$), and hypercentral iff it has a *hypercentral series* (which is a central series of ordinal length, with $G_\lambda = \bigcap_{i<\lambda} G_i$ for limit ordinal λ). Finally, a group is nilpotent of class n iff every iterated commutator of length $n + 1$ is trivial, as these commutators generate G^n.

The following lemma is often useful in deriving properties for subgroups and quotients:

Lemma 0.1.3 *Let P be a property of groups, G a group, and M and N normal subgroups of G. Suppose P is preserved under subgroups and direct products. Then, whenever G/M and G/N both satisfy P, so does $G/(M \cap N)$.*

Proof: The map $g \mapsto (gM, gN)$ is a homomorphism from G to $G/M \times G/N$ with kernel $M \cap N$, so $G/(M \cap N)$ is isomorphic to a subgroup of $G/M \times G/N$. □

Examples for such a property are commutativity, solubility, nilpotency, and hypercentrality. In fact, the derived length (nilpotency class, hypercentral length) is equal to the maximal one for the single quotients.

Lemma 0.1.4 *Let M and N be two nilpotent normal subgroups of a group G, of class m and n, respectively. Then MN is nilpotent, of class at most $m + n$.*

Proof: Since any element of MN is a product of an element of M and an element of N, we can write any iterated commutator with entries in MN as a product of commutators of the same length or longer, and with entries in $M \cup N$. But the lower central series are characteristic in M and in N, so commutation with an element in $M \cup N$ maps M^i to M^i and N^i to N^i. Hence a commutator with entries in $M \cup N$ is trivial as long as it contains $m + 1$ entries in M or $n+1$ entries in N. It follows that every commutator of length $m + n + 1$ is trivial, and MN has nilpotency class $m + n$. □

It is easy to see that the same assertion holds with *soluble* instead of *nilpotent*.

Lemma 0.1.5 *Let P be a property of groups such that if M and N are normal subgroups satisfying P, then MN satisfies P. Then every finite group has a unique maximal normal subgroup satisfying P.*

Proof: As the group is finite, there is a maximal normal subgroup N satisfying P. If M is another one, then MN satisfies P, so $M = MN = N$ by maximality. $\square$

Definition 0.1.4 Let G be a group. The *soluble radical $R_s(G)$* is the group generated by all normal soluble subgroups. The *Fitting subgroup $F(G)$* is the group generated by all normal nilpotent subgroups.

It follows from Lemma 0.1.4 that the Fitting subgroup of a finite group is nilpotent; similarly the soluble radical of a finite group is soluble as well.

Definition 0.1.5 A subgroup H of G is *subnormal* if there is a sequence $G = G_0 > G_1 > \cdots > G_n = H$ such that G_{i+1} is normal in G_i for all $0 \le i < n$.

Proposition 0.1.6 *Let G be a finite group and H a subnormal soluble (resp. nilpotent) subgroup. Then H is contained in a normal soluble (resp. nilpotent) subgroup.*

Proof: We use induction on the length n of a subnormal chain. If it is one, then H is itself normal in G. Now H is normal in G_{n-1} and hence contained in the soluble radical R (the Fitting subgroup F) of G_{n-1}, which is characteristic in G_{n-1} and hence normal in G_{n-2}. By the inductive hypothesis, R (resp. F) is contained in a normal soluble (resp. nilpotent) subgroup of G, and so is H. $\square$

Lemma 0.1.7 *Let G be nilpotent of class n. Then $G^i \le Z_{n-i}(G)$, for $0 \le i \le n$.*

Proof: This is clearly true for $n = 0$ (so G is trivial) and $n = 1$ (as $G = Z(G)$ is abelian). So suppose that it holds for all nilpotent groups of class less than n. Then $[G^{n-1}, G] = G^n = \{1\}$, so $G^{n-1} \le Z(G)$, and $G/Z(G)$ is nilpotent of class $n - 1$. Hence by the inductive hypothesis,

$$G^i Z(G)/Z(G) = (G/Z(G))^i \le Z_{n-1-i}(G/Z(G)) = Z_{n-i}(G)/Z(G),$$

for all $0 \le i \le n - 1$. As $Z(G) \le Z_i(G)$ for $i \ge 1$, we get $G^i \le Z_{n-i}(G)$ for all $0 \le i \le n - 1$, and clearly $\{1\} = G^n = Z_0(G)$. $\square$

Corollary 0.1.8 *G is nilpotent of class n iff $Z_n(G) = G$.*

Proof: If G is nilpotent of class n, then $Z_n(G) \geq G^0 = G$. On the other hand, if $G = Z_n(G)$, then one sees inductively that $G^i \leq Z_{n-i}(G)$ for $0 \leq i \leq n$, whence $G^n = \{1\}$ and G is nilpotent of class n. $\square$

Lemma 0.1.9 *Suppose S is a normal soluble subgroup of G centralizing F, where F is the Fitting subgroup of G. Then $S \leq Z(F)$.*

Proof: Replacing S by some derived subgroup, we may assume that $S/Z(F)$ is abelian. But then S is nilpotent, since $Z(F) \leq F \leq C_G(S)$. Therefore $S \leq F$, whence $S \leq C_F(F) = Z(F)$. $\square$

Lemma 0.1.10 *Let G be nilpotent and N a non-trivial normal subgroup of G. Then N contains a non-trivial central element.*

Proof: Let i be minimal such that N contains a non-trivial element z in $Z_i(G)$. Then $[z, g] = z^{-1}z^g \in N \cap Z_{i-1}(G) = \{1\}$ for all $g \in G$, and z must be central. $\square$

We shall now use the Three Subgroup Lemma in order to establish the relationship between the descending central series and iterated commutators, which will be needed later.

Lemma 0.1.11 *Let H be a subgroup of G. Then for any $i < j$*

$$[H^i, C_G^j(H)] \leq C_G^{j-i-1}(H).$$

Proof: We use induction on $i + j$. For $i + j = 1$ we must have $i = 0$ and $j = 1$, so the claim is obviously true by definition of $C_G(H)$. So suppose the assertion holds for $i + j = n$, and consider i, j with $i + j = n + 1$. Then by the inductive hypothesis

$$[H^{i-1}, C_G^j(H), H] \leq [C_G^{j-i}(H), H] \leq C_G^{j-i-1}(H),$$

and also

$$[H, C_G^j(H), H^{i-1}] \leq [C_G^{j-1}(H), H^{i-1}] \leq C_G^{j-i-1}(H).$$

Since all groups in question normalize $C_G^{j-i-1}(H)$, we may apply the Three Subgroup Lemma 0.1.2 and get

$$[H^i, C_G^j(H)] = [H^{i-1}, H, C_G^j(H)] \leq C_G^{j-i-1}(H). \ \square$$

In particular, H^{i-1} and $C_G^i(H)$ commute.

Lemma 0.1.12 *Suppose $K \leq H \leq G$, and $C_G(K^i) = C_G(H^i)$ for all $i < j$. Then $C_G^j(K) = C_G^j(H)$.*

Proof: We use induction on j. For $j = 0$ and $j = 1$ this is obvious. So suppose it holds up to $j - 1$.

Claim. For all $i < j$ we have $[H^{j-i-1}, C_G^j(K)] \leq C_G^i(H)$.

Proof of Claim: We use induction on i. For $i = 0$ this amounts to showing that $C_G^j(K) \leq C_G(H^{j-1}) = C_G(K^{j-1})$ (using the assumption), which follows from Lemma 0.1.11. So we suppose that the claim holds up to $i - 1$. Then

$$[H^{j-i-1}, K, C_G^j(K)] \leq [H^{j-i}, C_G^j(K)] \leq C_G^{i-1}(H),$$

and by Lemma 0.1.11 and the inductive hypothesis,

$$[K, C_G^j(K), H^{j-i-1}] \leq [C_G^{j-1}(K), H^{j-i-1}] = [C_G^{j-1}(H), H^{j-i-1}] \leq C_G^{i-1}(H).$$

All groups in question normalize $C_G^{i-1}(K) = C_G^{i-1}(H)$, so the Three Subgroup Lemma yields

$$[H^{j-i-1}, C_G^j(K), K] \leq C_G^{i-1}(K).$$

But both H^{j-i-1} and $C_G^j(K)$ normalize $C_G^k(K) = C_G^k(H)$ for all $k < i$, whence

$$[H^{j-i-1}, C_G^j(K)] \leq C_G^i(K) = C_G^i(H). \ \square$$

In particular, $[H, C_G^j(K)] \leq C_G^{j-1}(H)$, and $C_G^j(K)$ normalizes $C_G^i(H) = C_G^i(K)$ for all $i < j$. Hence $C_G^j(K) \leq C_G^j(H)$. On the other hand, clearly $[K, C_G^j(H)] \leq C_G^{j-1}(H) = C_G^{j-1}(K)$, and $C_G^j(H)$ normalizes $C_G^i(K) = C_G^i(H)$ for all $i < j$. Therefore $C_G^j(H) \leq C_G^j(K)$. $\square$

We shall now look at nilpotency in the context of finite groups.

Proposition 0.1.13 *Let G be a group of order p^n for some prime p. Then G is nilpotent.*

Proof: Consider the action of G on itself by conjugation. This partitions G into orbits x^G for various $x \in G$, with $|x^G| = |G : C_G(x)| = |G|/|C_G(x)|$. Hence if x is not central in G, then p divides the order of x^G. But the sum of all the orbit sizes is the size of G and therefore divisible by p. It follows that the number of one-element orbits is divisible by p. As there is at least one such orbit, namely 1^G, there must be another one, and $Z(G) > \{1\}$. But now the order of $G/Z(G)$ is smaller and again a power of p, so the result follows by induction. $\square$

Although an infinite group of prime power exponent p^n need not be nilpotent except for very small values of p^n, a group of exponent 2 is even abelian: $[g, h] = g^{-1}h^{-1}gh = ghgh = (gh)^2 = 1$.

Definition 0.1.6 Let G be a finite group of order $p^n m$, where p does not divide m. A subgroup of order p^n is called a *p-Sylow subgroup* of G.

Theorem 0.1.14 SYLOW'S THEOREMS *Every finite group has p-Sylow sub-groups, for every prime p. Their number is congruent to one modulo p. If P is a p-Sylow subgroup and H is a subgroup of G of p-power order, then there is a conjugate P^g of P such that $H \leq P^g$. All p-Sylow subgroups are conjugate in G.*

In particular, there are $|G : N_G(P)|$ distinct p-Sylow subgroups.
Proof: See e.g. [60] Kapitel I, Paragraph 7. $\square$

Proposition 0.1.15 FRATTINI ARGUMENT *Let N be a normal subgroup of a finite group G and P a Sylow subgroup of N. Then $G = N_G(P)N$.*

Proof: For any $g \in G$, the conjugate P^g is again a Sylow subgroup of N by normality. Hence there is $n \in N$ with $P^g = P^n$, i.e. $gn^{-1} \in N_G(P)$. $\square$

Definition 0.1.7 A subgroup H of G is *self-normalizing* if $N_G(H) = H$.

Lemma 0.1.16 *Let P be a p-Sylow subgroup of a finite group G. Then $N_G(P)$ is self-normalizing.*

Proof: Consider $g \in G$ normalizing $N_G(P)$. Then P^g is again a normal p-Sylow subgroup of $N_G(P)$, and so PP^g is a p-group. By maximality, $PP^g = P$ and $P = P^g$, whence $g \in N_G(P)$. $\square$

Definition 0.1.8 A group G satisfies the *normalizer condition* if no proper subgroup $H < G$ is self-normalizing.

Theorem 0.1.17 *The following are equivalent for a finite group G:*

1. *G is nilpotent.*

2. *G satisfies the normalizer condition.*

3. *G is the direct product of its Sylow subgroups.*

4. *All Sylow subgroups of G are normal.*

Proof: If G satisfies the normalizer condition, then by Lemma 0.1.16 every Sylow subgroup is normal. But if P and Q are two different normal Sylow subgroups and $p \in P$, $q \in Q$, the commutator $[p, q] = q^{-p}q = p^{-1}p^q$ lies in $P \cap Q$, which is trivial. Hence if all Sylow subgroups are normal, they commute, and it is easy to see that G must be their direct product.

Now every Sylow subgroup is nilpotent by Proposition 0.1.13, and their direct product is then nilpotent of class the maximum of the classes of the

Sylow subgroups. Finally, let G be nilpotent and $H < G$. Then if i is maximal such that $Z_i(G) \le H$, we have $[Z_{i+1}(G), H] \le Z_i(G) \le H$. Hence $H \not\le Z_{i+1}(G) \le N_G(H)$ (and i does not equal the nilpotency class since H is proper in G). $\square$

Lemma 0.1.18 *Let G be a group, X a G-invariant subset, H a subgroup of G satisfying the normalizer condition, and K a subgroup of G intersecting H in I such that $H \cap X \not\subseteq K$. Then there is $h \in N_{H \cap X}(I \cap X) - I$.*

Proof: If the whole of $H \cap X$ normalizes $I \cap X$, we are done. Otherwise $N_H(I \cap X)$ is a proper subgroup of H, so $N_H(N_H(I \cap X)) > N_H(I \cap X)$ and there is $g \in H$ which moves $I \cap X$, but fixes $N_H(I \cap X)$. As g must also stabilize $N_{H \cap X}(I \cap X)$ by G-invariance of X, we get $N_{H \cap X}(I \cap X) \supset I \cap X$. $\square$

Theorem 0.1.19 BAER-SUZUKI THEOREM *Let G be a finite group and X a G-invariant subset such that every two elements of X generate a p-group. Then X generates a p-group.*

Proof: We may assume that X generates G and use induction on the cardinality of G. The assertion is obvious for $|G| \le 3$. So assume that every proper subgroup of G generated by a subset of X is a p-group. If G is not a p-group, then there are two maximal distinct p-groups H and K generated by subsets of X, such that $I := H \cap K \cap X$ is maximal. They are nilpotent and satisfy the normalizer condition; by Lemma 0.1.18 there are elements $h \in N_{H \cap X}(I) - I$ and $k \in N_{K \cap X}(I) - I$. So $N = \langle I \rangle$ is a p-group normalized by h and k; as $\langle h, k \rangle$ is a p-group, so is $\langle N, h, k \rangle$, and the latter must be contained in a maximal one which is generated by elements of X, say L. By maximality of I we get $H = L = K$, a contradiction. $\square$

A subgroup M of G is *maximal* if it is proper and there is no subgroup H with $M < H < G$.

Definition 0.1.9 The *Frattini subgroup* $\Phi(G)$ of a finite group G is the intersection of all maximal subgroups of G.

Lemma 0.1.20 *If G is a finite group, then $\Phi(G)$ is the set of all elements $g \in G$ which can be discarded from any generating set for G.*

Proof: Suppose $g \in \Phi(G)$ and let X be a subset of G with $\langle X, g \rangle = G$. Suppose $\langle X \rangle < G$, so this must be contained in a maximal subgroup M. But then $g \in M$ as well, contradicting the fact that $X \cup \{g\}$ generates G. Conversely, consider $g \notin \Phi(G)$, so there is a maximal subgroup M with $g \notin M$. Then $\langle M, g \rangle = G$ by maximality, and g cannot be discarded. $\square$

Theorem 0.1.21 *Let M and N be normal subgroups of the finite group G, and suppose $M \leq N$ and $M \leq \Phi(G)$. If N/M is nilpotent, so is N.*

Proof: Consider a p-Sylow subgroup P of N. Then PM/M is a p-Sylow subgroup of N/M (the index $|N/M : PM/M| = |N : PM|$ is coprime to p); since N/M is nilpotent, PM/M is characteristic in N/M, whence PM is normal in G. Now P is a Sylow in PM, so $G = N_G(P)PM = N_G(P)M$ by Proposition 0.1.15. Since $M \leq \Phi(G)$, we get $G = N_G(P)$; in particular P is normal in N. Theorem 0.1.17 now implies nilpotency of N. $\square$

Corollary 0.1.22 *Let G be a finite group. Then $\Phi(G)$ is nilpotent. If $G/\Phi(G)$ is nilpotent, so is G.* $\square$

Definition 0.1.10 Let G be a group and N a normal subgroup. A *supplement* of N in G is a subgroup H with $NH = G$; a *complement* of N in G is a supplement H such that in addition $H \cap N = \{1\}$.

Proposition 0.1.23 *Let P be a property of groups which holds for a finite group G iff it holds for $G/\Phi(G)$, and which is preserved under quotients. Then whenever N is a normal subgroup such that G/N satisfies P, there is a supplement H for N in G which satisfies P.*

Proof: Let H be a minimal supplement for N in G (note that there are supplements, notably G itself). Suppose $H \cap N \not\leq \Phi(H)$, so there is a maximal subgroup M of H with $H = (H \cap N)M$. Then M is a smaller supplement for N in G, contradicting minimality. Hence $H \cap N \leq \Phi(H)$. But $H/(H \cap N) \cong HN/N = G/N$ satisfies P. Therefore $H/\Phi(H)$ satisfies P, and so does H. $\square$

In particular, any normal subgroup N of a finite group G with nilpotent quotient G/N has a nilpotent supplement.

Definition 0.1.11 A subgroup C of a group G is a *Carter subgroup* if C is nilpotent and self-normalizing.

It follows from Theorem 0.1.17 that a Carter subgroup must be a maximal nilpotent subgroup.

Theorem 0.1.24 *Let G be a finite soluble group. Then G has a Carter subgroup, and all Carter subgroups of G are conjugate.*

Proof: We use induction on the order of G, the assertion being trivial for abelian groups. So consider the last non-trivial derived subgroup A of G. By the inductive hypothesis, G/A has a Carter subgroup H/A. Then A has a maximal nilpotent supplement C in H; the normalizer $N_H(C) = N_A(C)C$ is the product of two normal nilpotent subgroups and hence itself nilpotent; by maximality it equals C. But now any $g \in N_G(C)$ must normalize $CA/A = H/A$, and g must lie in H. Hence $g \in N_H(C) = C$, and C is self-normalizing.

If C_1 is another Carter subgroup of G, we consider $g \in N_G(C_1A)$. If $C_1A = G$, then C_1A/A is a Carter subgroup of G/A and hence conjugate to H/A. On the other hand, if $C_1A < G$ and $g \in N_G(C_1A)$, then both C_1 and $(C_1)^g$ are Carter subgroups of C_1A and there is $h \in C_1A$ with $(C_1)^h = (C_1)^g$, so $gh^{-1} \in N_G(C_1) = C_1$ and $g \in C_1A$. It follows that C_1A/A is a Carter subgroup of G/A; we may conjugate and assume $C_1A = H$. Furthermore, we may clearly assume that H is centreless, since any central elements of H must lie both in C and in C_1. But then if $C \cap A$ were non-trivial, it would contain an element $z \in Z(C)$ which also commutes with the whole of A, and therefore is central in $CA = H$, a contradiction. It follows that both C and C_1 are complements for A in H.

Now consider $1 \neq c \in Z(C)$. Then $A_1 := C_A(c)$ is invariant under C and A, and therefore normal in H. It is proper in A; if it is non-trivial, then we see as above that both CA_1/A_1 and C_1A_1/A_1 are Carter subgroups of H/A_1 and conjugate by the inductive hypothesis. So we may assume $CA_1 = C_1A_1$, but then C and C_1 are conjugate Carter subgroups of CA_1. We therefore may assume that c acts without fixed points on A.

If $[c,a] = [c,b]$ for some $a, b \in A$, then $c^a = c^b$ and $ab^{-1} \in C_A(c) = \{1\}$. Hence the endomorphism $a \mapsto [c,a]$ of A is injective, and therefore surjective. In particular, if $a \in A$ is such that $ca \in C_1$, there is $b \in A$ such that $a = [c,b]$, i.e. $ca = b^{-1}cb$. But then $c^b \in Z(H/A) \cap C_1$, so for any $c_1 \in C_1$ we get $[c^b, c_1] \in A \cap C_1 = \{1\}$, and actually $c^b \in Z(C^b) \cap Z(C_1)$. Since $C_H(c^b) = C_H(c)^b = C^b$ intersects A trivially, it must be a complement for A in H which equals both C^b and C_1. $\square$

Nilpotent and soluble groups clearly have a rather rich supply of normal subgroups (unless they have prime order, of course). We shall now look at the other extreme:

Definition 0.1.12 A group G is *simple* if it has no proper normal non-trivial subgroups. It is *quasi-simple* if $G = G'$ and $G/Z(G)$ is simple, and it is *semi-simple* if $G = G'$ and $G/Z(G)$ is a direct product of non-abelian simple groups.

Note that an abelian group is simple iff it has prime order (and hence is cyclic).

Theorem 0.1.25 *Let G be finite and semi-simple. Then G decomposes as a central product of quasi-simple groups. Any subnormal subgroup N of G is normal, and N' is semi-simple. In particular, a soluble normal subgroup is central in G.*

Proof: The Three Subgroup Lemma implies that if $[K, L, K] = 1$, then $[K', L] = 1$. So let $G/Z(G)$ be the direct product of non-abelian simple groups $G_i/Z(G)$. Put $H_i = G_i'$. Since $[G_i, G_j] \leq Z(G)$ for $i \neq j$, we get $[G_i, G_j, G_j] = 1$ and therefore $[G_j', G_i] = 1$. Hence the product $\prod_i H_i$ is central.

Now $G = G' = (\prod_i G_i)' = \prod_{i,j}[G_i, G_j] \leq Z(G) \prod_i H_i$, since $[G_i, G_j] \leq Z(G)$ for all $i \neq j$. Therefore $G = G' = (Z(G) \prod_i H_i)' = \prod_i H_i' \leq \prod_i H_i \leq G$, and equality must hold all the way through. Because $G_i/Z(G)$ is simple non-abelian, we get $H_i Z(G) = G_i' Z(G) = G_i$, and $H_i/(Z(G) \cap H_i)$ is isomorphic to $G_i/Z(G)$, i.e. simple and non-abelian. It follows that $H_i = H_i'$ is quasi-simple.

Now let N be normal in G, and let I be the set of indices i with $H_i \leq N$. Then for $i \notin I$ we have that $N \cap H_i \leq Z(H_i)$ by quasi-simplicity, so $[N, H_i, H_i] = 1$ and $[H_i', N] = 1$, whence N and $H_i = H_i'$ commute. It follows that $N' = \prod_{i \in I} H_i$ is semi-simple. Furthermore $C_G(H_i) = Z(G) \prod_{j \neq i} H_j$, so $N = C_N(G) \prod_{i \in I} H_i = C_N(G)N'$. If N is soluble, N' must be trivial and $N \leq Z(G)$; if N is quasi-simple, it must equal H_i for some $i \in I$, proving the uniqueness of the decomposition.

Finally, if M is normal in N, then semi-simplicity of N' implies that M' must be semi-simple as well, and equal to a product of some of the H_i, hence normal in G. Furthermore, M commutes with those H_i in N which are not contained in M, and N commutes with H_i for $i \notin I$. Therefore $[M, G] \leq M$ and M is normal in G. $\square$

Definition 0.1.13 The *socle* $S(G)$ of a group G is the group generated by all minimal normal subgroups. The *layer* $E(G)$ is the group generated by all normal semi-simple subgroups.

Note that minimal normal subgroups are disjoint and commute. If there are n non-abelian ones, $N_0, \ldots, N_{n-1}$, then for every i we have an element $h_i \in N_i$ such that $C_G(h_i)$ contains $\prod_{j \neq i} N_j$ but not N_i; by taking products we get for every $I \subset n$ an element h_I such that $C_G(h_I)$ contains N_i iff $i \notin I$.

Lemma 0.1.26 *If N is a minimal normal subgroup of a finite group G, then N is a direct product of simple groups.*

Proof: $S(N)$ is characteristic in N and hence normal in G, whence $S(N) = N$. Since minimal normal subgroups of N intersect trivially and commute, the product $\prod_i N_i$ of them must be direct once we leave out superfluous factors.

Similarly, for every N_i, we get $N_i = S(N_i)$ equals a direct product $\prod_j N_{ij}$ of minimal normal subgroups. But N_i commutes with N_j for $j \neq i$ and $N = \prod_j N_j$. Hence the action of N on the N_{ij} equals that of N_i, and N_{ij} is actually normal in N. It follows that N_i is simple. $\square$

Note that G acts by conjugation on the set of minimal normal subgroups of N as a group of permutations.

Lemma 0.1.27 *If M and N are normal and semi-simple, then so is MN.*

Proof: Clearly $MN \geq (MN)' = M'[M,N]N' \geq MN$, so equality holds. $C_{M \cap N}(M)$ is a normal abelian subgroup of N and thus central in N. As $[Z(M), N] \leq C_{M \cap N}(M)$, we get $[Z(M), N, N] = 1$ and hence $[N', Z(M)] = 1$. It follows that $Z(MN) \geq Z(M)Z(N)$.

Both $X := MZ(N)/Z(M)Z(N)$ and $Y := NZ(M)/Z(M)Z(N)$ are direct products of non-abelian simple groups. If H is one of these which lies in X and Y, then H commutes with all the others; if H lies only in X but not in $X \cap Y$, then H intersects Y trivially, and commutes with it by semi-simplicity. Thus $XY = MN/Z(M)Z(N)$ is the direct product of non-abelian simple subgroups, and MN is semi-simple. $\square$

Hence $E(G)$ is the unique maximal normal semi-simple subgroup of G, for any finite group G.

Definition 0.1.14 The *generalized Fitting subgroup* $F^*(G)$ is the product $F(G)E(G)$.

Theorem 0.1.28 *Let E be the layer and F the Fitting subgroup of a finite group G. Then $E \leq C_G(F)$, and in fact $EZ(F)/Z(F) = S(C_G(F)/Z(F))$. Furthermore, $C_G(F^*(G)) = Z(F)$, so $G/Z(F)$ embeds into $\mathrm{Aut}(F^*(G))$.*

Proof: Let N be normal and semi-simple in G. Then $[N, F]$ is nilpotent and normal in N, hence central. So $[F, N, N] = 1$, whence $1 = [N', F] = [N, F]$, and $N \leq C_G(F)$. Therefore $E \leq C_G(F)$.

Now $C_G(F)/Z(F)$ has no abelian normal subgroups by Lemma 0.1.9, and clearly $Z(F) = Z(C_G(F))$. So $Z(E) \leq Z(F)$ and $EZ(F)/Z(F)$ is a direct product of non-abelian simple groups. $C_G(F)$ permutes these groups by conjugation; since the decomposition is unique, the direct product of the groups in any given orbit is a minimal normal subgroup of $C_G(F)/Z(F)$. Therefore $EZ(F)/Z(F) \leq S(C_G(F)/Z(F))$. On the other hand, by Lemma 0.1.26 the socle $S(C_G(F)/Z(F))$ is a direct product of simple groups, none of which can be abelian. It follows that its pre-image S in $C_G(F)$ has a semi-simple commutator group S', and $S = S'Z(F)/Z(F)$. As $S' \leq E$, we get $S \leq EZ(F)/Z(F)$.

Both E and F centralize $Z(F)$, so clearly $Z(F) \leq C_G(F^*(G))$. If $C_G(F^*(G))$ does not equal $Z(F)$, there must be a minimal normal subgroup $N/Z(F)$ contained in $C_G(F^*(G))/Z(F)$. Then $N \leq S(C_G(F)/Z(F)) = EZ(F)/Z(F) \leq F^*(G)/Z(F)$. Therefore N is soluble, and hence contained in $Z(F)$, a contradiction.

Now G acts on $F^*(G)$ by conjugation, and the kernel of this action, i.e. of the homomorphism into $\mathrm{Aut}(F^*(G))$, is $C_G(F^*(G)) = Z(F)$. $\square$

0.2   Mastering Minor Model Theory

We shall start with a short introduction to first-order logic and elementary model theory, specifying the meaning of *formula* (and hence *definable*), *model*, and *elementary extension*, and expounding the basic logical tools we shall be using.

Following set-theoretic convention, we shall denote the set of natural numbers by ω; higher ordinal numbers will usually be denoted by α, β, etc. The sum $\alpha + \beta$ of two ordinals is the well-order type of the concatenation $\alpha^\frown\beta$, and the product $\alpha \cdot \beta$ is the well-order type of copies of α arranged in order-type β (so $2 \cdot \omega = \omega$, but $\omega \cdot 2 = \omega + \omega$). Ordinal exponentiation is defined inductively via $\alpha^0 = 1$, $\alpha^{\beta+1} = \alpha^\beta \cdot \alpha$ and $\alpha^\lambda = \bigcup_{\beta<\lambda} \alpha^\beta$ for a limit ordinal. (Hence $2^\omega = \lim_{n<\omega} 2^n = \omega$.)

We identify cardinal numbers with the smallest ordinal number of their cardinality and denote them by λ, κ, etc.; κ^+ will stand for the successor cardinal of κ (whereas $\alpha+1$ is the *ordinal* successor). The sum of two cardinals is the cardinality of their disjoint union, and their product is the cardinality of their Cartesian product. Cardinal exponentiation is denoted by λ^κ; this also stands for the set of all functions from κ to λ (which has cardinality λ^κ). So an expression like 2^ω is triply ambiguous – its meaning should however be clear from the context.

A *language* $\mathcal{L}$ is the (disjoint) union of sets $\mathcal{F}$ of function symbols, $\mathcal{R}$ of relation symbols, $\mathcal{C}$ of constant symbols, and $\mathcal{X}$ of variable symbols. To every function symbol $f \in \mathcal{F}$ and to every relation symbol $R \in \mathcal{R}$ we associate a natural number, the *arity*, which denotes the number of arguments the function or relation is intended to take. An $\mathcal{L}$-*structure* $\mathfrak{M}$ is given by a set M of elements, its domain, together with a function $f^{\mathfrak{M}} : M^j \longrightarrow M$ for every $f \in \mathcal{F}$ of arity j, a subset $R^{\mathfrak{M}} \subseteq M^j$ for every $R \in \mathcal{R}$ of arity j, and a constant $c^{\mathfrak{M}} \in M$ for every $c \in \mathcal{C}$. If A is a subset of M, we may expand the language by adding new constant symbols c_a for every $a \in A$, and obtain the language $\mathcal{L}(A)$. We can then turn $\mathfrak{M}$ into an $\mathcal{L}(A)$-structure by setting $c_a^{\mathfrak{M}} := a$. A particular relation symbol is equality; the corresponding subset of M^2 is usually just the diagonal $\{(m,m) \in M^2 : m \in M\}$.

An $\mathcal{L}$-*word* is a finite string of symbols in $\mathcal{L}$, together with the *(Boolean) connectives* $\neg$ (negation), $\wedge$ (and, conjunction), $\vee$ (or, disjunction), $\rightarrow$ (implication), $\leftrightarrow$ (equivalence), the *quantifiers* $\exists$ (there exists), $\forall$ (for all), and brackets $(,)$. The set of $\mathcal{L}$-*terms* is the smallest collection of words containing all the constant and variable symbols, and which is closed under applying any j-ary function symbol $f \in \mathcal{F}$ to terms $t_1, \dots, t_j$ already obtained: if $t_1, \dots, t_j$ are terms, so is $f(t_1, \dots, t_j)$. A term is *closed* if it contains no variable symbols; for an $\mathcal{L}$-structure $\mathfrak{M}$ we may then extend the *interpretation* $c^{\mathfrak{M}}$ of the constant symbols to all closed $\mathcal{L}$-terms t inductively via

$$f(t_1,\ldots,t_j)^{\mathfrak{M}} = f^{\mathfrak{M}}(t_1^{\mathfrak{M}},\ldots,t_j^{\mathfrak{M}}), \text{ for all } j\text{-ary } f \in \mathcal{F}.$$

In other words, terms are interpreted as elements of the domain of the $\mathcal{L}$-structure in question, in such a way that a function symbol f is interpreted as the function $f^{\mathfrak{M}}$. Thus, every element m of M which lies in the closure of the set of constants $\{c^{\mathfrak{M}} : c \in \mathcal{C}\}$ under all functions $\{f^{\mathfrak{M}} : f \in \mathcal{F}\}$ gets a (not necessarily unique) name, a term t with $t^{\mathfrak{M}} = m$.

The set of (first-order) $\mathcal{L}$-*formulæ* is the smallest collection of $\mathcal{L}$-words which contains all words of the form $R(t_1,\ldots,t_j)$, where $R \in \mathcal{R}$ is a j-ary relation symbol and $t_1,\ldots,t_j$ are terms (the *atomic formulæ*), and is closed under the connectives and quantification. In other words, whenever φ and ψ are formulæ and $x \in \mathcal{X}$ is a variable symbol, then $\neg\varphi$, $(\varphi \wedge \psi)$, $(\varphi \vee \psi)$, $(\varphi \to \psi)$, $(\varphi \leftrightarrow \psi)$, $\exists x\,\varphi$ and $\forall x\,\varphi$ are all formulæ as well (and the brackets ensure unique readability). The *degree* of a formula φ is the number of connectives and quantifiers; properties of formulæ are usually proved by induction on their degree. Any occurrence of x in a formula of the form $\forall x\,\varphi$ or $\exists x\,\varphi$ is *bound* by the quantifier $\forall$ or $\exists$, unless it is already bound by a quantifier in φ. A variable which is not bound is *free*; a *closed* formula (or *sentence*) is a formula without free variables. If a variable x occurs freely in a formula φ, we may indicate this by writing $\varphi(x)$ (and similarly for more than one variable). If a formula $\varphi(x_1,\ldots,x_n)$ is *quantifier-free* (i.e. contains no quantifiers), then $\forall x_1 \cdots \forall x_n\,\varphi$ is called a *universal* formula, and $\exists x_1 \cdots \exists x_n\,\varphi$ is called *existential*. If φ contains constants $c_{m_1},\ldots,c_{m_k}$ for elements $m_1,\ldots,m_k$ in some model, these elements are called the *parameters* occurring in φ. We usually abbreviate a tuple $m_1,\ldots,m_k$ as $\bar{m}$, and leave the length k unspecified (and similarly for tuples of variables).

If φ is a *closed* $\mathcal{L}$-formula, we define the *truth* of φ in $\mathfrak{M}$, denoted by $\mathfrak{M} \models \varphi$, inductively on the length of the formula via

1. $\mathfrak{M} \models R(t_1,\ldots,t_j)$ iff $(t_1^{\mathfrak{M}},\ldots,t_j^{\mathfrak{M}}) \in R^{\mathfrak{M}}$ for all j-ary $R \in \mathcal{R}$,

2. $\mathfrak{M} \models \neg\varphi$ iff $\mathfrak{M} \not\models \varphi$,

3. $\mathfrak{M} \models (\varphi \vee \psi)$ iff $\mathfrak{M} \models \varphi$ or $\mathfrak{M} \models \psi$ (and correspondingly for the other connectives), and finally

4. $\mathfrak{M} \models \forall x\,\varphi(x)$ iff for all $m \in M$ we have $\mathfrak{M} \models \varphi(c_m)$, where $\varphi(c_m)$ is the $\mathcal{L}(m)$-formula obtained by replacing all free occurrences of x in φ by c_m, and

5. $\mathfrak{M} \models \exists x\,\varphi(x)$ iff there is $m \in M$ such that $\mathfrak{M} \models \varphi(c_m)$.

If $\mathfrak{M} \models \varphi$, we say that $\mathfrak{M}$ *satisfies* φ, or φ is *true* in $\mathfrak{M}$. If $\bar{m}$ is an n-tuple from M and $\varphi(\bar{x})$ is a formula with n free variables $\bar{x}$, then $\bar{m}$ *satisfies* φ in $\mathfrak{M}$, written $\bar{m} \models_{\mathfrak{M}} \varphi$, if $\mathfrak{M} \models \varphi(c_{m_1}, \dots, c_{m_n})$. Finally, $\varphi(\mathfrak{M})$ (also denoted $\varphi^{\mathfrak{M}}$) is the set of $\bar{m} \in M^n$ satisfying φ. This said, we shall usually identify a model with its domain, and shall not distinguish between a constant, a relation, or a function, and its respective symbol. The distinction is nevertheless useful in that it allows us to change more easily between different $\mathcal{L}$-structures – once we are given the new interpretation, the same formulæ will make sense. It is often useful to have two additional formulæ, namely the true one $\top$ and the false one $\bot$ – they are true and false respectively in any model.

If Φ is a set of $\mathcal{L}$-sentences and $\mathfrak{M}$ is an $\mathcal{L}$-structure, then $\mathfrak{M}$ is a *model* for Φ if $\mathfrak{M} \models \varphi$ for all $\varphi \in \Phi$; we denote this by $\mathfrak{M} \models \Phi$. If Φ and Ψ are two sets of sentences, then Φ implies Ψ, written $\Phi \models \Psi$, if every model of Φ is also a model of Ψ. If both Φ implies Ψ and *vice versa*, then Φ and Ψ are equivalent; they are equivalent *modulo* Θ if $\Phi \cup \Theta$ and $\Psi \cup \Theta$ are equivalent.

We should note at this point that if we have a binary function $*$ and a constant 1, we may express by a formula that $(M, *, 1)$ is a group under $*$ with unit element 1, that it has exponent $e < \omega$, or that it is abelian (by the sentence $\forall x \forall y \; x * y = y * x$). We can express that it has at least n distinct elements by the formula $\exists x_1, \dots, x_n \bigwedge_{i \neq j} x_i \neq x_j$ (where $\bigwedge_{1 \leq i \leq n} \varphi_i$ is an abbreviation for $\varphi_1 \wedge \dots \wedge \varphi_n$, and similarly for $\bigvee$), and hence we may express in infinitely many sentences (the above for all $n < \omega$) that a structure is infinite. However, we cannot express that it is finite, or periodic. This is due to one of the fundamental theorems of model theory:

Theorem 0.2.1 COMPACTNESS THEOREM *If Φ is a set of sentences such that every finite subset has a model, then Φ has a model.*

Proof: Let I be the set of finite subsets of Φ, and for $i \in I$ let $\mathfrak{M}_i$ be a model of i. Consider the family $\mathcal{F}$ of those subsets of I which contain, for some $j \in I$, the set $I_j := \{i \in I : j \subseteq i\}$. Then $j \in I_j$ and $I_{j_1} \cap I_{j_2} = I_{j_1 \cup j_2}$. If we call a family of subsets of I a *filter* on I if it is closed under supersets and finite intersections and does not contain $\emptyset$, this means that $\mathcal{F}$ is a filter.

Now the set of filters on I containing $\mathcal{F}$ is partially ordered by inclusion, and every chain has a maximal element, namely the union of the chain. By Zorn's Lemma there must be a maximal filter $\mathcal{U}$ containing $\mathcal{F}$; it is easy to see from maximality that for any $J \subseteq I$ either J or its complement $I - J$ must belong to $\mathcal{U}$. (Such filters are called *ultrafilters*; an ultrafilter is *non-principal* if it is not of the form $\mathcal{U}_x := \{J \subseteq I : x \in J\}$ for some $x \in I$.) We shall now construct a model $\mathfrak{M}$ of Φ.

The domain M will be the product $\prod_{i \in I} M_i$ modulo the equivalence relation $(m_i)_I \equiv (m_i')_I$ iff $\{i \in I : m_i = m_i'\} \in \mathcal{U}$. The constant symbol c is interpreted by the equivalence class $[(c^{\mathfrak{M}_i})_I]$; in order to define the interpretation of a j-ary relation symbol R or function symbol f, we first define a relation R_0 on $(\prod_{i\in I} M_i)^j$ via

$$\left((m_i^1)_I,\dots,(m_i^j)_I\right) \in R_0 \quad\text{iff}\quad \{i \in I : (m_i^1,\dots,m_i^j) \in R^{\mathfrak{M}_i}\} \in \mathcal{U},$$

and a function f_0 from $(\prod_{i\in I} M_i)^j$ to M via

$$f_0\left((m_i^1)_I,\dots,(m_i^j)_I\right) = \left[\left(f^{\mathfrak{M}_i}(m_i^1,\dots,m_i^j)\right)_I\right].$$

The fact that $\mathcal{U}$ is a filter then implies that the equivalence $\equiv$ is a congruence for these relations and functions, so the definitions carry over to M, the set of equivalence classes. It can then be checked by induction on the length of a term that if $t(x_1,\dots,x_j)$ is a term and $(m_i^1)_I,\dots,(m_i^j)_I$ are sequences in $\prod_{i\in I} M_i$, then $t([(m_i^1)_I],\dots,[(m_i^j)_I])$ is interpreted as $[i \mapsto t(m_i^1,\dots,m_i^j)^{\mathfrak{M}_i}]$.
Claim. $\mathfrak{M}$ satisfies a sentence $\sigma([(m_i^1)_I],\dots,[(m_i^j)_I])$ iff $\{i \in I : \mathfrak{M}_i \models \sigma(m_i^1,\dots,m_i^j)\} \in \mathcal{U}$.
Proof of Claim: We use induction on the degree of the formula. For an atomic sentence the claim is obvious by construction of $\mathfrak{M}$. So suppose that the assertion holds for φ and ψ. Then

- $\mathfrak{M} \models \neg\varphi$ iff $\mathfrak{M} \not\models \varphi$ iff $\{i \in I : \mathfrak{M}_i \models \varphi\} \notin \mathcal{U}$ iff $\{i \in I : \mathfrak{M}_i \not\models \varphi\} \in \mathcal{U}$ iff $\{i \in I : \mathfrak{M}_i \models \neg\varphi\} \in \mathcal{U}$ (using the fact that $\mathcal{U}$ is an ultrafilter and contains a set iff it does not contain its complement),

- $\mathfrak{M} \models \varphi \wedge \psi$ iff $\mathfrak{M} \models \varphi$ and $\mathfrak{M} \models \psi$ iff $\{i \in I : \mathfrak{M}_i \models \varphi\} \in \mathcal{U}$ and $\{i \in I : \mathfrak{M}_i \models \psi\} \in \mathcal{U}$ iff $\{i \in I : \mathfrak{M}_i \models \varphi$ and $\mathfrak{M}_i \models \psi\} \in \mathcal{U}$ (using the fact that a filter contains the intersection of two sets iff it contains both of them) iff $\{i \in I : \mathfrak{M}_i \models \varphi \wedge \psi\} \in \mathcal{U}$, and lastly

- $\mathfrak{M} \models \exists x\, \varphi(x)$ iff there is $[(m_i)_I]$ in M with $\mathfrak{M} \models \varphi\left([(m_i)_I]\right)$ iff there are $m_i \in M_i$ for $i \in I$ with $\{i \in I : \mathfrak{M}_i \models \varphi(m_i)\} \in \mathcal{U}$ iff $\{i \in I : \mathfrak{M}_i \models \exists x\, \varphi(x)\} \in \mathcal{U}$.

The cases of the other connectives and the universal quantifier are similar (and they could also be expressed in terms of $\neg, \wedge, \exists$ alone). $\square$

In particular, since $\mathcal{F} \subseteq \mathcal{U}$, for any $\varphi \in \Phi$ the set $\{i \in I : \mathfrak{M}_i \models \varphi\} \supseteq I_{\{\varphi\}} \in \mathcal{U}$, and φ is true in $\mathfrak{M}$. $\square$

The model $\mathfrak{M}$ is called the *ultraproduct* $(\prod_I \mathfrak{M}_i)/\mathcal{U}$ of the family $(\mathfrak{M}_i)_I$. If $\mathcal{U} = \mathcal{U}_x$ is principal, then $(\prod_I \mathfrak{M}_i)/\mathcal{U}$ is isomorphic to $\mathfrak{M}_x$.

Remark 0.2.1 We can now see that properties like "G is a finite group" are not even expressible by infinitely many sentences. For if Φ were such a set of sentences expressing finiteness, then every finite subset of $\Phi \cup \{$"G has at least n elements" $: n < \omega\}$ would have a model (a finite group of sufficiently large size), and then there would be a model of the whole set, i.e. a finite group with infinitely many elements.

Remark 0.2.2 Due to the non-constructive nature of Zorn's Lemma, it is rather hard to give a concrete example of an ultraproduct. Cardinality considerations tell us that a non-principal ultraproduct $(\prod_I \mathfrak{M}_i)/\mathcal{U}$ usually has cardinality $\prod_I |\mathfrak{M}_i|$; in particular, if all $\mathfrak{M}_i$ have the same cardinality κ, the ultraproduct has cardinality $\kappa^{|I|}$. For example, the ultraproduct of countable algebraically closed fields of characteristic p, over all prime numbers p, with respect to any non-principal ultrafilter must be an algebraically closed field of characteristic zero and size continuum, and thus isomorphic to the field of complex numbers.

We shall call a set Φ of sentences *consistent* if it has a model. By the Compactness Theorem, consistency is a local property, so the union of a chain of consistent sets of formulæ is again consistent. By Zorn's Lemma every consistent set Φ of sentences is contained in a maximal consistent set $\tilde{\Phi}$; such a set is called a *(complete first-order) theory* and has the property that it contains every sentence or its negation. To any $\mathcal{L}$-structure $\mathfrak{M}$ we can associate a theory $\mathrm{Th}_{\mathcal{L}}(\mathfrak{M})$, namely the set of $\mathcal{L}$-sentences true in $\mathfrak{M}$.

Definition 0.2.1 Let $\mathfrak{M}$ and $\mathfrak{N}$ be $\mathcal{L}$-structures with $M \subseteq N$. We say that $\mathfrak{M}$ is an *elementary substructure* of $\mathfrak{N}$, denoted by $\mathfrak{M} \prec \mathfrak{N}$, if

1. all constants are interpreted in $\mathfrak{M}$ and in $\mathfrak{N}$ by the same elements,

2. all j-ary functions are interpreted in $\mathfrak{M}$ as the restrictions to M^j of the corresponding functions in $\mathfrak{N}$, for all $j < \omega$,

3. all j-ary relations are interpreted in $\mathfrak{M}$ as the intersection with M^j of the corresponding subsets of N^j, for all $j < \omega$, and

4. every $\mathcal{L}(M)$-sentence true in $\mathfrak{M}$ is true in $\mathfrak{N}$.

In particular, $\mathfrak{M}$ and $\mathfrak{N}$ have the same theory. We say that $\mathfrak{M}$ is *elementarily embedded* in $\mathfrak{N}$ if the image of $\mathfrak{M}$ under the embedding is an elementary substructure of $\mathfrak{N}$.

If only 1.–3. hold, then $\mathfrak{M}$ is a *substructure* of $\mathfrak{N}$.

Lemma 0.2.2 *If $\mathfrak{M}$ is a substructure of $\mathfrak{N}$, then every universal sentence with parameters in $\mathfrak{N}$ is true in $\mathfrak{M}$, and every existential sentence with parameters in $\mathfrak{M}$ is true in $\mathfrak{N}$.*

Proof: Note first that every quantifier-free sentence with parameters in $\mathfrak{M}$ is true in $\mathfrak{M}$ iff it is true in $\mathfrak{N}$, as is easily seen from the fact that $\mathfrak{M}$ inherits its structure from $\mathfrak{N}$. Now any witness for an existential sentence in $\mathfrak{M}$ remains a witness in $\mathfrak{N}$, and by the same vein any counter-example to a universal sentence in $\mathfrak{M}$ would be a counter-example in $\mathfrak{N}$. $\square$

A theory T has *quantifier elimination* if every formula is equivalent modulo T to a quantifier-free formula. Since preservation in sub- and superstructures follows from the syntactic form of a formula, quantifier elimination is useful when changing structures; furthermore formulæ of low quantifier complexity are often more easy to analyse.

Example 0.2.1 In the language with equality and addition, the group of even integers is a substructure of all integers (and even isomorphic to it), but not elementary, since e.g. the sentence $\exists x\,(x + x = 2)$ is true in the latter structure, but not in the former. However, $\mathbb{Z} \oplus \{0\}$ is elementary in $\mathbb{Z} \oplus \mathbb{Q}$, and any algebraically closed field sits elementarily in every algebraically closed superfield. Both these statements can be proven by a quantifier elimination result. We shall go into greater detail in section 0.3.

Remark 0.2.3 If $\mathfrak{M}$ is a model of a theory T and $\mathcal{U}$ is an ultrafilter on some set I, then $\prod_I \mathfrak{M}_i/\mathcal{U}$ (with $\mathfrak{M}_i = \mathfrak{M}$ for all $i \in I$) is called an *ultrapower* of $\mathfrak{M}$. It satisfies T, and $\mathfrak{M}$ elementarily embeds into it via the diagonal map.

Lemma 0.2.3 *If $(\mathfrak{M}_i)_{i \in I}$ is an elementary chain of models of T (i.e. for any two models in the chain, one is an elementary substructure of the other or vice versa), then $\mathfrak{M} := \bigcup_{i \in I} \mathfrak{M}_i$ is an elementary superstructure of every $\mathfrak{M}_i$ (for $i \in I$), with the obvious interpretation of $\mathfrak{L}$.*

Proof: We shall use induction on the degree of the sentence. Clearly every atomic sentence with parameters in M_i is true in $\mathfrak{M}_i$ iff it is true in $\mathfrak{M}$, as $\mathfrak{M}_i$ is a substructure of $\mathfrak{M}$. The case of Boolean connectives being equally obvious, we consider a sentence $\varphi := \forall x\,\psi(x)$ with parameters in M_i, and suppose that every formula of smaller degree than φ with parameters in some $\mathfrak{M}_i$ is true in $\mathfrak{M}_i$ iff it is true in $\mathfrak{M}$. If φ holds in $\mathfrak{M}$, then $\mathfrak{M} \models \psi(m)$ for all $m \in M$, and hence for all $m \in M_i$. By the inductive assumption, $\mathfrak{M}_i \models \psi(m)$ for all $m \in M_i$, so φ holds in $\mathfrak{M}_i$. If on the other hand φ does not hold in $\mathfrak{M}$, then there is some $m \in M$ such that $\mathfrak{M} \models \neg\psi(m)$. But then there is $j > i$ such that $m \in M_j$; by the inductive hypothesis $\mathfrak{M}_j \models \neg\psi(m)$, and therefore $\mathfrak{M}_j \models \neg\forall x\,\psi(x)$. As $\mathfrak{M}_i$ is elementary in $\mathfrak{M}_j$, this sentence $\neg\varphi$ is also true in $\mathfrak{M}_i$. The case of an existentially quantified formula is dealt with similarly. $\square$

Definition 0.2.2 Let $\mathfrak{M}$ be an $\mathfrak{L}$-structure, $\varphi(x_1, \ldots, x_k, \bar{y})$ a formula, and Φ a set of formulæ. A subset X of M^k is

1. *definable* over $\bar{c}$, or $\bar{c}$-*definable*, if there are a tuple $\bar{c} \subset M$ of parameters and an $\mathfrak{L}(\bar{c})$-formula $\psi(x_1, \ldots, x_k)$ such that $X = \psi^{\mathfrak{M}}$,

2. *type-definable*, if X is the intersection of an arbitrary family of definable subsets,

3. φ-*definable*, if X is definable using the formula $\varphi(\bar{x}, \bar{c})$ for some suitable parameters $\bar{c}$ (such a formula is called a φ-*formula*), and

4. Φ-*definable*, if X is φ-definable for some formula $\varphi \in \Phi$.

If all the parameters used for some (type-)definable set X are contained in a subset A of M, we say that X is (type-)definable *over A*, or A-(type-)definable. We shall often identify a formula with the set it defines (where the model is given implicitly).

Definition 0.2.3 Let $\mathfrak{M}$ be an $\mathfrak{L}$-structure and $A \subseteq M$.

1. An *n-type* over A is a maximal consistent extension of $\mathrm{Th}_{\mathfrak{L}(A)}(\mathfrak{M})$ in $\mathfrak{L}(A \cup \{\bar{x}\})$, where $\bar{x}$ is an n-tuple of variables, treated as new constant symbols. The set A is called the *domain* of the type. The collection of all n-types over A is denoted by $S_n(A)$.

2. If $\bar{m} \in M$, the *type of $\bar{m}$ over A in $\mathfrak{M}$*, denoted by $\mathrm{tp}(\bar{m}/A)$, is $\mathrm{Th}_{\mathfrak{L}(A \cup \bar{m})}(\mathfrak{M})$, where we replace $\bar{m}$ by variable symbols $\bar{x}$. (Note that $\mathfrak{M}$ is implicit in the notation $\mathrm{tp}(\bar{m}/A)$.)

Note that a type is just a minimal consistent type-definable set (i.e. one given by a maximal number of formulæ).

Remark 0.2.4 Clearly the type of $\bar{m}$ over A is a type p, and we say that p is *realized* in $\mathfrak{M}$ by $\bar{m}$, written $\bar{m} \models_{\mathfrak{M}} p$. Conversely, let $p \in S_n(A)$. Then any finite subset π of p must be consistent with $\mathrm{Th}_{\mathfrak{L}(A)}(\mathfrak{M})$, and therefore $\mathfrak{M} \models \exists \bar{x} \bigwedge \pi(\bar{x})$ (where $\bigwedge \pi$ is the conjunction of all the elements in π). Let $\bar{m}_\pi \in M$ satisfy π. Again, we consider the set I of finite subsets of p and an ultrafilter $\mathcal{U}$ on I. If we identify A with its diagonal image in the ultrapower $(\prod_I \mathfrak{M})/\mathcal{U}$, then $[(m_\pi)_{\pi \in I}]$ will satisfy p. So p is realized in an elementary extension of $\mathfrak{M}$.

$S_n(A)$ carries a natural topology, given by a basis of clopen sets $[\varphi] := \{p \in S_n(A) : \varphi \in p\}$. This topology is totally disconnected; the compactness theorem states exactly that the space $S_n(A)$ is compact in this topology. A condition will be *open* if it is expressible by a (possibly) infinite disjunction of formulæ, and *closed* if it expressible by an infinite conjunction; if it is both open and closed, then compactness states that it is expressible by a single formula.

A *type* is an n-type for some $n < \omega$, and the set of all types over A is denoted by $S(A)$; a *partial type* is just a subset of a type, and a φ-type is a maximal partial type which consists of φ-formulæ. We may also consider types of infinite tuples, called *-types: if $B \subseteq M$, we may enumerate B as $(b_i)_I$ and talk about

$$\text{tp}(B/A) = \text{tp}((b_i)_I/A) = \bigcup \{ \text{tp}((b_i)_{I_0}/A) : I_0 \subset I \text{ finite} \},$$

using variables $(x_i : i \in I)$. Note that the enumeration of B is implicit in the notation $\text{tp}(B/A)$.

Definition 0.2.4 A *partial (elementary) isomorphism* of two $\mathcal{L}$-structures $\mathfrak{M}$ and $\mathfrak{N}$ is a partial map σ from $M_0 \subseteq M$ to $N_0 \subseteq N$, such that $\text{tp}(M_0) = \text{tp}(N_0)$ (where the first type is taken in $\mathfrak{M}$ and the second one is taken in $\mathfrak{N}$).

Definition 0.2.5 A type $p(\bar{x})$ over A is *isolated* by a formula $\varphi(\bar{x}) \in p$ if $\varphi(\bar{x})$ proves $p(\bar{x})$, i.e. any realization of φ is also a realization of p. A subset A of a model $\mathfrak{M}$ is *atomic* if the type of every tuple $\bar{a} \in A$ over $\emptyset$ is isolated.

Remark 0.2.5 If $\text{tp}(\bar{m})$ is isolated by $\varphi(\bar{x})$ and $\text{tp}(\bar{n}/\bar{m})$ is isolated by $\psi(\bar{y}, \bar{m})$, then $\varphi(\bar{x}) \wedge \psi(\bar{y}, \bar{x})$ isolates $\text{tp}(\bar{m}\bar{n})$. Conversely, if $\varphi(\bar{x}, \bar{y})$ isolates $\text{tp}(\bar{m}, \bar{n})$, then $\varphi(\bar{x}, \bar{n})$ isolates $\text{tp}(\bar{m}/\bar{n})$ and $\exists \bar{x}\, \varphi(\bar{x}, \bar{y})$ isolates $\text{tp}(\bar{n})$.

Proposition 0.2.4 *Any two countable atomic models are isomorphic.*

Proof: Suppose $\mathfrak{M}$ and $\mathfrak{N}$ are countable and atomic, and let $(m_i : i < \omega)$ be an enumeration of M and $(n_i : i < \omega)$ be an enumeration of N. We claim that whenever M_0 is a finite subset of M and σ a partial isomorphism mapping M_0 to $N_0 \subset N$, then for any $m \in M$ there is $n \in N$ with $\sigma(\text{tp}(m/M_0)) = \text{tp}(n/N_0)$. We can then extend σ to $M_0 \cup \{m\}$ by setting $\sigma(m) = n$.

In order to prove the claim, consider $p := \text{tp}(m/M_0)$. Since $\mathfrak{M}$ is atomic, this type is isolated by a formula $\varphi(x, M_0)$ by Remark 0.2.5. But as $\text{tp}(M_0) = \text{tp}(N_0)$, the formula $\varphi(x, N_0)$ isolates the type $\sigma(p)$ over N_0, and $\mathfrak{N} \models \exists x\, \varphi(x, N_0)$. So we can choose any witness n for this existential sentence and obtain the result.

We can now finish by means of a *back-and-forth argument*: inductively we find a chain of partial isomorphisms σ_i from $\mathfrak{M}$ to $\mathfrak{N}$ such that $m_i \in \text{dom}(\sigma_{2i})$ and (by symmetry) $n_i \in \text{dom}(\sigma_{2i+1}^{-1})$. Then $\bigcup_{i<\omega} \sigma_i$ is the required isomorphism from $\mathfrak{M}$ to $\mathfrak{N}$. $\square$

Example 0.2.2 Suppose φ contains a formula φ which has only finitely many realizations. We may assume that the number of realizations is minimal among formulæ in p. Then it is easy to see that φ isolates p.

For a different example, consider the theory of dense linear orders with endpoints. Then any n-type over a finite set A (containing the endpoints) is determined by the formula which says between which successive elements of A each co-ordinate of the type falls. Hence every type over a finite set is isolated. It follows from Proposition 0.2.4 that all countable dense linear orders with endpoints are isomorphic.

Definition 0.2.6 A theory T is *small* if $S_n(\emptyset)$ is countable for all $n < \omega$ (and hence $S(\emptyset)$ is countable).

Lemma 0.2.5 *Smallness is preserved under adding finitely many parameters to the language.*

Proof: If A has size n, then distinct k-types $p(\bar{x}, A)$ over A give rise to distinct $(n + k)$-types $p(\bar{x}, \bar{y})$ over $\emptyset$. $\square$

Lemma 0.2.6 *Suppose T is small. Then over any finite set A of parameters any non-empty A-definable set $\varphi(x)$ lies in some isolated type $p \in S_1(A)$.*

Proof: Suppose not, and let $\varphi(x)$ be a counter-example (over some finite set A contained in a model $\mathfrak{M}$). Then no A-definable subset $\psi(x)$ isolates a type over A, i.e. we can always split $\psi(x)$ into two consistent A-definable subsets $\psi_1(x)$ and $\psi_2(x)$. Iterating this, we obtain an infinite 2-splitting tree of formulæ over A such that the (sets of formulæ along the) branches are consistent, and any two branches are contradictory. But there are continuum many branches and any branch can be completed to a type over A; this yields continuum many types over A, contradicting Lemma 0.2.5. $\square$

Proposition 0.2.7 *A small countable theory has an atomic model.*

Proof: Consider a model $\mathfrak{M}$. Take countably many new constant symbols $\{n_i : i < \omega\}$, and enumerate all $\mathfrak{L}(n_i : i < \omega)$-formulæ of the form $\exists x\, \varphi(x, \bar{n})$ in order type ω. Choose an element $m \in M$ with $\mathrm{tp}(m)$ isolated (which exists by Lemma 0.2.6), and interpret n_0 as m. Suppose we have already interpreted $n_0, \dots, n_k$. Take the first formula $\exists x\, \varphi(x, \bar{n})$ over $n_0, \dots, n_k$ in the enumeration such that $\varphi(x, \bar{n})$ has a witness in $M - \{n_0, \dots, n_k\}$. By Lemma 0.2.6 there is some isolated type $p \in S_1(n_0, \dots, n_k)$ containing $\varphi(x, \bar{n})$; this type must be realized in $\mathfrak{M}$ by some element m, and we interpret n_{k+1} by m. (If there is no such formula, we just set $n_{k+1} = n_k$.) Finally, we set $N := \{n_i : i < \omega\}$. Clearly, N is atomic by Remark 0.2.5.

Since $\mathfrak{M}$ satisfies the formulæ $\exists x\, c = x$ for any constant c and $\exists x\, f(\bar{n}) = x$ for any function f, the set N contains all the constants and is closed under all functions; restricting the functions and relations of $\mathfrak{M}$ to N makes N into a substructure $\mathfrak{N}$ of $\mathfrak{M}$. We claim that it is elementary. This is seen by induction on the degree of a formula, the case of an atomic formula or of a Boolean combination being obvious. Now, if $\mathfrak{M} \models \exists x\, \varphi(x, \bar{m})$ for some $\bar{m} \in N$, then there must be constant symbols $n_{i_0}, \dots, n_{i_k}$ which are interpreted as $\bar{m}$. But successively for all true existential formulæ over these parameters we have thrown witnesses into N; as we had ordered the formulæ in order-type ω, we must have reached φ at some point, and there is a witness for $\exists x\, \varphi(x, \bar{m})$

in N, whence $\mathfrak{N} \models \exists x\, \varphi(x, \bar{m})$. On the other hand, if $\mathfrak{M} \models \forall x\, \neg\varphi(x, \bar{m})$, then $\mathfrak{M} \models \neg\varphi(n_i, \bar{m})$ for all $i < \omega$, and hence $\mathfrak{N} \models \neg\varphi(n_i, \bar{m})$ for all $i < \omega$ by the inductive hypothesis. But this means that $\mathfrak{N} \models \forall x\, \neg(x, \bar{m})$. $\square$

While atomic models realize as few types as possible, the following notion captures the other extreme:

Definition 0.2.7 A model $\mathfrak{M}$ is κ-saturated for some cardinal κ if every type over a subset of cardinality less than κ is realized in $\mathfrak{M}$. A model $\mathfrak{M}$ is *saturated* if it is $|M|$-saturated.

By a union-of-chains argument, it is easy to see that every model has a κ-saturated elementary extension, for every cardinal κ. However, the existence of saturated models in general depends on set-theoretic assumptions.

Proposition 0.2.8 *If $\mathfrak{M}$ and $\mathfrak{N}$ are saturated of the same cardinality κ, and $A \subset M$ and $B \subset N$ have the same type over $\emptyset$ (in their respective models) and have cardinality less than κ, we can extend the partial isomorphism $\sigma : A \mapsto B$ to an isomorphism of $\mathfrak{M}$ and $\mathfrak{N}$.*

Proof: We shall use a back-and-forth argument again. Enumerate both $M - A$ and $N - B$ in order type κ (so that every initial segment has cardinality less than κ). We have to show that for every partial isomorphism $\sigma_0 : A_0 \to B_0$ extending σ with $|A_0| < \kappa$ and any $m \in M$ we can find $n \in N$ with $\sigma_0(\mathrm{tp}(m/A_0)) = \mathrm{tp}(n/B_0)$. But clearly such an n must exist by κ-saturation of $\mathfrak{N}$.

Taking unions at the limit stages, transfinite induction now yields an increasing sequence $(\sigma_i : i < \kappa)$ of partial isomorphisms from $\mathfrak{M}$ to $\mathfrak{N}$ extending σ, such that the domain of σ_i contains the i-th element of M and the range of σ_i contains the i-th element of N, and both have cardinality less than κ.

But now $\bigcup_{i<\kappa} \sigma_i$ is an isomorphism from $\mathfrak{M}$ to $\mathfrak{N}$ mapping A to B. $\square$

In particular, any two realizations of the same type over a set $A \subset M$ of cardinality less than $|M|$ in a saturated model $\mathfrak{M}$ are conjugate under an automorphism of $\mathfrak{M}$ fixing A – and any two subsets of M so conjugate must necessarily satisfy the same type. We may thus consider types over A as orbits under $Aut(\mathfrak{M}/A)$, the group of automorphisms of $\mathfrak{M}$ fixing A pointwise. The importance of κ-saturated models stems from the fact that type-definable sets will contain elements if they are type-defined by an intersection of cardinality less than κ.

Example 0.2.3 Consider again the group $\mathbb{Z}$ with definable subgroups $n\mathbb{Z}$ of index n, for $n < \omega$. The intersection $\bigcap_{n<\omega} n\mathbb{Z}$ is trivial in the model $\mathbb{Z}$, but in a general model the subgroup type-defined by $\{\exists y\, ny = x : n < \omega\}$ has index at most 2^ω, being the intersection of countably many subgroups of finite index. For instance, in the elementary superstructure $G := \mathbb{Z} \oplus \mathbb{Q}$ the intersection $\bigcap_{n<\omega} nG = \{0\} \oplus \mathbb{Q}$ is infinite, and has index ω in G.

Remark 0.2.4 really showed that a type over a model $\mathfrak{M}$ is realized in any non-principal ultrapower of $\mathfrak{M}$ over an index set of size $|\mathfrak{M}|$. So take a non-principal ultrafilter $\mathcal{U}$ on ω, and consider $\mathcal{Z} := (\prod_\omega \mathbb{Z})/\mathcal{U}$, a *non-standard* model of the integers. This behaves in many respects like the integers, except that for every set P of prime numbers there is a non-standard integer n_P which is divisible by a prime p iff $p \in P$ (this is easily expressed by a partial type, which must be realized in $\mathcal{Z}$). It follows that $\bigcap_{n<\omega} n\mathcal{Z}$ has index 2^ω in $\mathcal{Z}$. In fact, in any ω^+-saturated elementary extension of $\mathbb{Z}$ the intersection of the formulæ $\exists y \, ny = x$ for $n < \omega$ will be a subgroup of index exactly 2^ω.

Next, we shall embark upon model-theoretic stability.

Definition 0.2.8 A theory T is λ-*stable* if for every set A of parameters of cardinality λ (inside any model $\mathfrak{M}$) there are at most λ types over A. A theory is *stable* if it is λ-stable for some cardinal λ; it is *superstable* if it is λ-stable for all sufficiently big λ.

Clearly an ω-stable theory is small.

Remark 0.2.6 By counting first the number of possibilities for the type of a_0 over A, then for the type of a_1 over Aa_0, etc., it is easy to see that a theory is λ-stable iff there are at most λ different 1-types over any set of parameters of cardinality λ.

Definition 0.2.9 Let T be a theory and $\varphi(\bar{x}, \bar{y})$ a formula.

- φ has the *order property* if there are a model $\mathfrak{M}$ of T and tuples $\bar{m}_i, \bar{n}_i$ in M for $i < \omega$ such that $\mathfrak{M} \models \varphi(\bar{m}_i, \bar{n}_j)$ iff $i < j$.

- φ has the *strict order property* if φ partially orders a definable subset of tuples.

- φ has the *independence property* if there are a model $\mathfrak{M}$, tuples $\bar{m}_i \in \mathfrak{M}$ for $i < \omega$, and tuples $\bar{n}_I \in \mathfrak{M}$ for $I \subset \omega$, such that $\mathfrak{M} \models \varphi(\bar{m}_i, \bar{n}_I)$ iff $i \in I$.

It is clear that a formula with the strict order property or the independence property has the order property. Conversely, Shelah has shown that if there is a formula with the order property, then there is also a formula with the strict order property or the independence property. We should note that if φ orders $(\bar{m}_i, \bar{n}_i : i < \omega)$ and we set $\bar{a}_i := \bar{m}_i \bar{n}_i$, then the formula $\psi(\bar{x}\bar{y}, \bar{u}\bar{v}) := \varphi(\bar{x}, \bar{v})$ holds of $\bar{a}_i \bar{a}_j$ iff $i < j$. The following combinatorial theorem is often very useful:

Theorem 0.2.9 RAMSEY'S THEOREM *Suppose X is an infinite set and the set of unordered n-tuples of X is painted in k different colours. Then there is an infinite monochromatic subset Y, i.e. a subset Y whose n-tuples all have the same colour.*

Proof: We shall use induction on n. If $n = 1$, then since X is coloured in finitely many colours, there must be an infinite monochromatic subset. So suppose the assertion holds for all colourings of n-tuples, and consider a colouring of the $(n + 1)$-tuples of X.

Fix $x_0 \in X =: X_0$. This induces a colouring of the n-tuples of $X_0 - \{x_0\}$: an n-tuple $\bar{x}$ is ascribed the colour of $\{x_0, \bar{x}\}$, and by the inductive hypothesis there is an infinite monochromatic subset X_1. Pick any $x_1 \in X_1$, consider the colouring of n-tuples of $X_1 - \{x_1\}$ induced by x_1, and find an infinite monochromatic subset X_2; and repeat this ω many times to obtain a sequence $X_0 \supset X_1 \supset X_2 \supset \cdots$ of infinite subsets of X and elements $x_i \in X_i - X_{i+1}$ such that all subsets $\{x_i, \bar{x}\}$ with $\bar{x} \in X_{i+1}$ have the same colour c_i. As there are only finitely many colours, one colour c must have been used infinitely often, say whenever $i \in I$ for some infinite index set $I \subseteq \omega$.

We claim that $Y := \{x_i : i \in I\}$ is our monochromatic subset, of colour c. Indeed, if $\{x_{i_0}, \dots, x_{i_n}\}$ is an $(n + 1)$-tuple from Y with $i_0 < i_1 < \cdots < i_n$, then $x_{i_j} \in X_{i_0}$ for $1 \le j \le n$, and the tuple has colour c. $\square$

Remark 0.2.7 We should note that the compactness theorem now implies that for every triple (k, n, m) of natural numbers there is some natural number $R(k, n, m)$ such that whenever n-tuples of a set of size at least $R(k, n, m)$ are painted in k colours, then there is a monochromatic subset of size m. For suppose not. We use a language with constants $\{c_i : i < \omega\}$ and k different n-ary relation symbols r_i to express that

- for every n-tuple exactly one of the relations holds, and

- no m distinct constants form a monochromatic set (this is a sentence for every choice of the m constants).

By assumption, this is finitely satisfiable, and it has a model by compactness. But the infinitely many constants in this model do not have a monochromatic subset of size m, let alone an infinite one, contradicting Theorem 0.2.9. There are, however, constructive proofs of the finite version of Ramsey's Theorem which actually give bounds on the function $R(k, n, m)$.

Lemma 0.2.10 *Suppose φ and ψ do not have the order property. Then neither does $\neg\varphi$, or $\varphi \vee \psi$. In particular, no Boolean combination of formulæ without the order property can have the order property.*

Proof: If $\neg\varphi(\bar{x},\bar{y})$ has the order property, then by compactness there are a model $\mathfrak{M}$ and tuples $\bar{m}_i,\bar{n}_i$ in $\mathfrak{M}$ for $i \in \mathbb{Z}$, such that $\mathfrak{M} \models \neg\varphi(\bar{m}_i,\bar{n}_j)$ iff $i < j$. But then $\mathfrak{M} \models \varphi(\bar{m}_{-i},\bar{n}_{-j+1})$ iff $i < j$, and φ has the order property.

Now suppose $\varphi \vee \psi$ orders $(\bar{m}_i,\bar{n}_i : i < \omega)$ in some model $\mathfrak{M}$. Colour pairs $(i,j) < \omega^2$ red if $\mathfrak{M} \models \varphi(\bar{m}_i,\bar{n}_j)$ and green if $\mathfrak{M} \models \psi(\bar{m}_i,\bar{n}_j)$. By assumption, every pair (i,j) with $i < j$ is either red or green; by Ramsey's Theorem there is an infinite subset $I \subset \omega$ such that the pairs $(i,j) \in I^2$ with $i < j$ are either all red or all green. As $\mathfrak{M} \models \neg\varphi(\bar{m}_i,\bar{n}_j) \wedge \neg\psi(\bar{m}_i,\bar{n}_j)$ for $i \geq j$, the tuples $(\bar{m}_i,\bar{n}_i : i \in I)$ are ordered by φ in the first case, and by ψ in the second. So φ or ψ has the order property. $\square$

Definition 0.2.10 Let A and B be subsets of a model $\mathfrak{M}$, and $\varphi(\bar{x},\bar{y})$ a formula. A type $p(\bar{x})$ over A is φ-*definable over* B if there is a formula $d_p\bar{x}\,\varphi(\bar{x},\bar{y})$ over B such that for any $\bar{a} \in A$ we have $\mathfrak{M} \models d_p\bar{x}\,(\bar{x},\bar{a})$ iff $\varphi(\bar{x},\bar{a}) \in p$. (Note that $d_p\bar{x}\,\varphi(\bar{x},\bar{y})$ is just a formula with free variables $\bar{y}$. The reason for the notation is that $d_p\bar{x}$ acts like some kind of quantifier and can be read as "for all $\bar{x}$ satisfying p, ...", as clearly $\bar{a} \models d_p\bar{x}\,\varphi(\bar{x},\bar{y})$ iff $\bar{a} \models \varphi(\bar{m},\bar{y})$ for any realization $\bar{m}$ of p.) We call $d_px\varphi$ a φ-*definition* for p. Finally, p is *definable* if it is φ-definable over its domain, for all formulæ φ.

Theorem 0.2.11 *If α is an ordinal, $\varphi(\bar{x},\bar{y})$ a formula, and π a partial type, put*

$$\Gamma_\varphi^\alpha(\pi) := \left\{ \varphi^{\mu^{(n)}}(\bar{x}_\mu,\bar{y}_{\mu\lceil n}) : \mu \in 2^\alpha, n \in \alpha \right\} \cup \left\{ \pi(\bar{x}_\mu) : \mu \in 2^\mu \right\}.$$

Then the following are equivalent:

1. *T is stable,*

2. *every type (over any set of parameters) is definable,*

3. *no formula has the order property, and*

4. *for every formula $\varphi(\bar{x},\bar{y})$ the set $\Gamma_\varphi^\alpha(\pi)$ is inconsistent for $\alpha = \omega$ and $\pi = \top$, where $\varphi^0 = \varphi$ and $\varphi^1 = \neg\varphi$.*

Note that by compactness $\Gamma_\varphi^\alpha(\pi)$ is inconsistent for every infinite α iff $\Gamma_\varphi^\omega(\pi)$ is inconsistent iff $\Gamma_\varphi^k(\pi_0)$ is inconsistent for some $k < \omega$ and a finite subset π_0 of π.

Proof: (2. $\Rightarrow$ 1.) If every type is definable, then for any model of cardinality λ and any formula φ there are only λ many choices for the definition $d_p\bar{x}\,\varphi$. As two types agree iff they have the same definition for every formula, there are only $\lambda^{|T|}$ many different types over $\mathfrak{M}$, and T is stable for any λ with $\lambda = \lambda^{|T|}$. (There are cardinal numbers with this property, namely $\kappa^{|T|}$ for any cardinal κ.)

(1. $\Rightarrow$ 3.) Suppose that φ has the order property. Then by compactness for every cardinal λ there are a model $\mathfrak{M}$ of cardinality λ and elements $\bar{m}_i, \bar{n}_i \in M$ for some linearly ordered index set I of cardinality λ with 2^λ cuts, such that $\mathfrak{M} \models \varphi(\bar{m}_i, \bar{n}_j)$ iff $i < j$. So for any initial segment J of I the set

$$\{\varphi(\bar{m}_i, \bar{y}) \wedge \neg\varphi(\bar{x}, \bar{n}_i) : i \in J\} \cup \{\neg\varphi(\bar{m}_i, \bar{y}) \wedge \varphi(\bar{x}, \bar{n}_i) : i \in I - J\}$$

is consistent and can be completed to a type $p_J(\bar{x}\bar{y})$ over M. These types are clearly different, and $\mathfrak{M}$ is not λ-stable.

(3. $\Rightarrow$ 4.) $\Gamma^\omega_\varphi(\pi)$ is consistent if there are a model $\mathfrak{M}$ and a rooted tree T of height ω with nodes $\bar{m}_\eta$, for $\eta \in 2^{<\omega}$, and elements $\bar{n}_\mu \models \pi$ in M for $\mu \in 2^\omega$ as branches, such that $\mathfrak{M} \models \varphi(\bar{n}_\mu, \bar{m}_{\mu\restriction n})$ iff $\mu(n) = 0$. We shall prove inductively on t that there is a sequence

$$\bar{n}_0, \bar{m}_0, \bar{n}_1, \bar{m}_1, \dots, \bar{n}_{s-1}, \bar{m}_{s-1}, T_t, \bar{n}_s, \bar{m}_s, \dots, \bar{n}_{t-1}, \bar{m}_{t-1}$$

of tuples $\bar{n}_i, \bar{m}_i \in M$ and 2-splitting subtrees T_t of T of height ω, such that

1. $\mathfrak{M} \models \varphi(\bar{n}_i, \bar{m}_j)$ iff $i \leq j$,

2. if $\bar{m}$ is a node of T_t, then $\mathfrak{M} \models \varphi(\bar{n}_i, \bar{m})$ iff $0 \leq i < s$, and

3. if $\bar{n}$ is a branch in T_t, then $\mathfrak{M} \models \varphi(\bar{m}_i, \bar{n})$ iff $s \leq i < t$.

For $t = 0$ this holds because of the existence of T. Suppose now that we have found such a sequence of length t. Pick a branch $\bar{n}$, and consider $W = \{\bar{m}' \in T_t : \mathfrak{M} \models \varphi(\bar{n}, \bar{m}')\}$ and $B = \{\bar{m}' \in T_t : \mathfrak{M} \models \neg\varphi(\bar{n}, \bar{m}')\}$. This 2-colours T, and it is easy to see that either W or B contains an infinite 2-splitting subtree T', with root $\bar{m}$, say. If T' is in W, just insert $\bar{n}, \bar{m}, T_{t+1}$ instead of T_t to get a sequence of length $t + 1$, where T_{t+1} is the (infinite 2-splitting) subtree of T' whose branches $\bar{n}'$ satisfy $\neg\varphi(\bar{n}', \bar{m})$. If T' is in B, insert $T_{t+1}, \bar{n}, \bar{m}$, where T_{t+1} is the (infinite 2-splitting) subtree of T' whose branches $\bar{n}'$ satisfy $\mathfrak{M} \models \varphi(\bar{n}', \bar{m})$. It follows that φ has the order property.

(4. $\Rightarrow$ 2.) Let p be a type over A. If Γ^ω_φ is inconsistent, there is a maximal $k < \omega$ such that $\Gamma^k_\varphi(p)$ is consistent, and there is a finite set $\pi \subset p$ such that $\Gamma^{k+1}_\varphi(\pi)$ is inconsistent. If $\varphi(\bar{x}, \bar{a}) \in p$, then $\Gamma^k_\varphi(\pi \cup \{\varphi(\bar{x}, \bar{a})\})$ is consistent; otherwise $\neg\varphi(\bar{x}, \bar{a}) \in p$ and $\Gamma^k_\varphi(\pi \cup \{\neg\varphi(\bar{x}, \bar{a})\})$ is consistent and by inconsistency of $\Gamma^{k+1}_\varphi(\pi)$ it cannot be that $\Gamma^k_\varphi(\pi \cup \{\varphi(\bar{x}, \bar{a})\})$ is consistent as well. So $\varphi(\bar{x}, \bar{a}) \in p$ iff $\Gamma^k_\varphi(\pi \cup \{\varphi(\bar{x}, \bar{a})\})$ is consistent, but the latter is expressible by a formula $d_p\bar{x}\varphi(\bar{x}, \bar{a})$. $\square$

Corollary 0.2.12 *A stable theory T is λ-stable for all λ such that $\lambda^{|T|} = \lambda$.*

Proof: This follows from the proof of (2. $\Rightarrow$ 1.) in Theorem 0.2.11. $\square$

Remark 0.2.8 In fact, an ω-stable theory is λ-stable for all $\lambda \geq |T|$. This can be seen by constructing a 2-splitting tree of height ω with consistent branches (and hence continuum many types over the parameters used for these formulæ) from a set A of cardinality λ with $|S(A)| > \lambda$. A theory is *superstable* if it is λ-stable for all $\lambda \geq 2^{|T|}$.

Example 0.2.4 Consider the theory of dense linear orders without endpoints. Clearly, over a countable model there are continuum many cuts, and hence continuum many types, which must all differ by an inequality. So there cannot be an $(x < y)$-definition for most of them. On the other hand, over the real numbers $\mathbb{R}$ every type *is* definable, and of one of five kinds:

1. isolated by $x = r$ for some real r,

2. induced by $\{x > q : q < r\} \cup \{x < r\}$ for some real r,

3. induced by $\{x > r\} \cup \{x < q : q > r\}$ for some real r,

4. induced by $\{x < r : r \in \mathbb{R}\}$, or

5. induced by $\{x > r : r \in \mathbb{R}\}$.

Types in 1. are realized and nothing new; types in 4. and 5. are the infinitely large negative and positive elements, respectively. The type in 2. says that x is infinitely close *below* r, whereas in 3. it is infinitely close *above* r. Nevertheless, this is not sufficient to make the theory stable (and it is obvious that it does have the order property).

Example 0.2.5 A theory is *strongly minimal* if for every formula $\varphi(x, \bar{y})$ there is some $n_\varphi < \omega$ such that for any set $\bar{a}$ of parameters in any model, the set defined by $\varphi(x, \bar{a})$ has at most n_φ elements, or excludes at most n_φ elements. (In fact, it is sufficient to make this requirement for all parameters $\bar{a}$ coming from any *fixed* model.) Then over any set A of parameters there are two kinds of types:

1. those which specify that x lies in a finite definable set with parameters from A (the *algebraic* types), and

2. those which specify that x lies in no finite definable set with parameters from A (the *transcendental* type).

There clearly is only one type of the latter kind, and any type p under 1. is isolated by the smallest A-definable set $\psi_p(x)$ contained in p. So there are at most $|A \cup \mathfrak{L}|$ types over A, and the theory of a strongly minimal set is ω-stable.

An algebraic 1-type p has $\varphi(x,\bar{y})$-definition $\forall x\, \psi_p(x) \to \varphi(x,\bar{y})$; the transcendental type has φ-definition

$$\exists x_0 \cdots x_{n_\varphi} \bigwedge_i \varphi(x_i,\bar{y}) \wedge \bigwedge_{i\neq j} x_i \neq x_j.$$

An example for a strongly minimal set is a pure set without structure, an elementary abelian p-group, a torsion-free divisible abelian group, or an algebraically closed field of any characteristic.

Proposition 0.2.13 *A λ-stable theory T has a saturated model of cardinality λ, for any regular cardinal $\lambda \geq |T|$.*

In fact, the proposition holds even without the regularity assumption.
Proof: Let $\mathfrak{M}$ be a model of T of cardinality λ. Then there are at most λ many types over M, and they are all realized in an elementary extension $\mathfrak{N}$ of cardinality λ. It follows that we can construct an elementary chain $(\mathfrak{M}_i : i < \lambda)$ of models of cardinality λ, such that $\mathfrak{M}_{i+1}$ realizes all types over $\mathfrak{M}_i$ (for $i < \lambda$), and at limit ordinals the model is the union of its predecessors. Then the union of these is again a model $\mathfrak{M}$ of cardinality λ. If A is a subset of M of cardinality less than λ, then by regularity there must be some $i < \lambda$ such that $A \subseteq M_i$, so all types over A are already realized in $\mathfrak{M}_{i+1}$, hence in $\mathfrak{M}$. $\square$

We shall now expand our set-up, in order to be able to deal with elements which are virtually present in a model, albeit not as real elements. For instance, consider the quotient of a group by a definable subgroup. Unlike the elements of a subgroup, the elements of a quotient group are not elements of our original model, but classes modulo some definable equivalence relation. The following construction will allow us to treat them as if they were real elements.

Definition 0.2.11 Let $\mathfrak{M}$ be a model of a theory T. For every $\emptyset$-definable equivalence relation $E(\bar{x},\bar{y})$ we add a new unary predicate $P_E(x)$ to our language, together with a new function π_E. This will form the language $\mathfrak{L}^{eq}$; the model $\mathfrak{M}^{eq}$ is the disjoint union of domains M_E consisting of all the equivalence classes of tuples in M modulo E, for all $\emptyset$-definable E. We shall identify the original model $\mathfrak{M}$ with the set of classes modulo equality, and call it the *home sort*; the original constants, functions and relations will live there. All other elements will be *imaginary sorts*. Finally, we shall interpret P_E as M_E, and π_E as the projection from a tuple in the home sort to its equivalence class modulo E. The theory T^{eq} is then the $\mathfrak{L}^{eq}$-theory of $\mathfrak{M}^{eq}$.

It can be checked that T^{eq} does not depend on the choice of our original model $\mathfrak{M}$. In general, a model of T^{eq} will have elements satisfying no predicate P_E;

these elements form the *superfluous sort* and carry no structure at all. In particular, the structure obtained by simply omitting superfluous elements will be an elementary substructure. Finally, any model of T can be uniquely expanded to a model of T^{eq} without superfluous elements. It follows that T is λ-stable iff T^{eq} is λ-stable.

Particular sorts of T^{eq} will be

- *permutation sorts*, i.e. classes of M^j modulo E, where $(m_1, \ldots, n_j)$ and $(n_1, \ldots, n_j)$ are equivalent modulo E iff one tuple is a permutation of the other, and

- *tuple sorts*, i.e. classes modulo E, where $(m_1, \ldots, m_j)E(n_1, \ldots, n_j)$ holds iff the tuples are the same.

For the rest of this section (and whenever necessary), we shall assume that we work in T^{eq}. As finite tuples now are nothing else but (imaginary) elements, we shall henceforth notate them without the bar.

If $E_m(x, y)$ is an m-definable equivalence relation, we may consider the $\emptyset$-definable equivalence relation $(x, s)E(y, t)$ iff $s = t$, and either E_s is an equivalence relation and $E_s(x, y)$, or else E_s is not an equivalence relation and $x = y$. Then the classes of E_m are just the same as the classes of E which come from pairs with second co-ordinate m. In particular, if T is a theory of groups and H is a subgroup defined by some formula $\varphi(x, m)$, the (right) cosets of H are imaginary elements, namely classes of the equivalence relation $E(x, s; y, t)$ given by

$$s = t \wedge [x = y \vee [\varphi(xy^{-1}, s) \wedge \forall u \forall v ((\varphi(u, s) \wedge \varphi(v, s)) \to \varphi(uv^{-1}, s))]]$$

with second co-ordinate m; left cosets are dealt with similarly.

It will be very convenient to work in a universal domain for our theory T. Let $\mathfrak{C}$ be a saturated model of very big cardinality with domain $\mathcal{C}$, and consider only elementary substructures of $\mathfrak{C}$ of smaller cardinality. Since every model of smaller cardinality can be elementarily embedded into $\mathfrak{C}$, this does not restrict our choice of models, but for any set A of parameters (a subset of $\mathcal{C}$ of smaller cardinality) and any element a (in $\mathcal{C}$) the type $\operatorname{tp}(a/A)$ is unambiguously defined as the type of a over A in $\mathfrak{C}$. We shall simply write $\models$ instead of $\mathfrak{C} \models$. This $\mathfrak{C}$ is generally called the *monster model*. For example, the complex numbers may be taken as monster model for the theory of algebraically closed fields of characteristic zero, if we restrict ourselves to considerations of countable subsets.

Definition 0.2.12 Let A be a set of parameters, and a an element.

- a is *algebraic* over A if there is an A-formula $\varphi(x)$ such that $\varphi^{\mathfrak{C}}$ is finite and contains a.

- a is *definable* over A if there is an A-formula $\varphi(x)$ such that $\varphi^{\mathfrak{C}} = \{a\}$.

- A set X is *A-invariant* if it is stabilized setwise under $Aut(\mathfrak{C}/A)$.

- Two sets X and Y are *A-conjugate* if there is an automorphism of $\mathfrak{C}$ fixing A which maps X to Y.

The *algebraic closure* of A, denoted by $acl(A)$, is the set of all elements algebraic over A; the *definable closure* $dcl(A)$ of A is the set of all elements definable over A.

Lemma 0.2.14 *Let A and a be as above. Then a is algebraic over A iff its orbit under $Aut(\mathfrak{C}/A)$ is finite; it is definable over A iff the orbit is reduced to a itself iff $\{a\}$ is A-invariant.*

Proof: $\Rightarrow$ is obvious in both cases, as $\varphi^{\mathfrak{C}}$ is invariant under $Aut(\mathfrak{C}/A)$. For the other direction, consider $p(x) = tp(a/A)$. By saturation of $\mathfrak{C}$, it must be inconsistent to say $p(x_0) \cup \cdots \cup p(x_n) \cup \{x_i \neq x_j : i \neq j\}$ for n equal to the size of the orbit of a. But this means that a finite part π is inconsistent, and the conjunction of the formulæ in π has the required properties. $\square$

Hence a is algebraic over A iff the stabilizer S_a of a in $Aut(\mathfrak{C}/A)$ has finite index; and definable over A iff S_a equals $Aut(\mathfrak{C}/A)$. It is now easy to see that acl and dcl are idempotent, i.e. true closure operators.

If a subset X of $\mathcal{C}$ is defined by a formula $\varphi(x, m)$, we consider the equivalence relation $E_\varphi(x, y)$ given by $\forall z\, \varphi(z, x) \leftrightarrow \varphi(z, y)$, and we define the *canonical base* $Cb(X)$ of X to be the definable closure of the class of m modulo E_φ. This is in fact independent of the choice of formula defining X; an automorphism of $\mathfrak{C}$ stabilizes X setwise iff it fixes $Cb(X)$ pointwise.

Definition 0.2.13 *Let p be a type over A and $\varphi(x, y)$ a formula. Then $R_\varphi(p)$ is the maximal k such that there are a model $\mathfrak{M}$ containing A and an extension q of p over $\mathfrak{M}$ with $\Gamma_\varphi^k(q)$ consistent.*

If q is an extension of p, we say that q does not fork over A if q has an extension q' to a model such that q' is definable over $acl(A)$.

By Theorem 0.2.11, in a stable theory $R_\varphi(p)$ is finite for every type p and every formula $\varphi(x, y)$. Clearly, if q extends p, then $R_\varphi(q) \leq R_\varphi(p)$.

Lemma 0.2.15 *Let $\mathfrak{M}$ be an elementary substructure of $\mathfrak{N}$, and $p \in S(\mathfrak{M})$ definable. Then p has a unique non-forking extension q to $\mathfrak{N}$, and $d_p x \varphi$ is equivalent to $d_q x \varphi$ for all formulæ φ. Furthermore, $R_\varphi(q) = R_\varphi(p)$.*

Proof: If a non-forking extension q exists, then $d_p x \varphi$ and $d_q x \varphi$ are definable over $\mathfrak{M}$ (a model is algebraically closed) and agree on $\mathfrak{M}$, so they are equivalent. It follows that the non-forking extension must be unique. But if there were no non-forking extension, there would be formulæ $\varphi_1, \ldots, \varphi_k$ and elements $n_1, \ldots, n_k$ in N such that $\models d_p x\, \varphi_i(x, n_i)$ for $1 \leq i \leq k$ and $\bigwedge_{i \leq k} \varphi_i(x, n_i)$ is inconsistent. However,

$$\mathfrak{M} \models \forall y_1, \ldots, y_k \, (\bigwedge_{i \leq k} d_p x\, \varphi_i(x, y_i) \to \exists x \bigwedge_{i \leq k} \varphi_i(x, y_i)),$$

and so does $\mathfrak{N}$, a contradiction.

Suppose there is a formula φ such that $k = R_\varphi(q) < R_\varphi(p)$. Then there are a formula $\psi(x, n) \in q$ and a finite part $\pi(\bar{x}, n)$ of $\Gamma_{k+1}(\varphi, \psi(x, n))$ such that $\pi(\bar{x}, n)$ is inconsistent. Hence

$$\mathfrak{N} \models \exists y \, [d_p x \psi(x, y) \wedge \forall \bar{x}\, \neg \pi(\bar{x}, y)].$$

As $\mathfrak{M}$ is elementary in $\mathfrak{N}$, it must also satisfy this sentence, so there is an element $m \in M$ witnessing the existential quantifier. But this means that $\psi(x, m) \in p$ and $\Gamma_{k+1}(\varphi, \psi(x, m))$ is inconsistent, contradicting $R_\varphi(p) > k$. $\square$

Lemma 0.2.16 *Let $\mathfrak{M}$ be a saturated model of a stable theory, $p(x)$ and $q(y)$ types over $\mathfrak{M}$, and $\varphi(x, y)$ a formula. Then $d_p x\varphi(x, y) \in q$ iff $d_q y\varphi(x, y) \in p$.*

Proof: Suppose, say, that $d_p x\varphi \in q$ and $d_q y\varphi \notin p$, and let $d_p x\varphi$ and $d_q y\varphi$ be defined over a finite subset A of M. By saturation we can find sequences a_i and b_i (for $i < \omega$) such that $a_n \models p{\restriction}A \cup \{b_i : i < n\}$ and $b_n \models q{\restriction}A \cup \{a_i : i \leq n\}$ for all $n < \omega$. Then for all i we must have $\mathfrak{M} \models d_p x\, \varphi(x, b_i)$ and $\mathfrak{M} \models \neg d_q y\, \varphi(a_i, y)$. Hence for $i > j$ we have $\mathfrak{M} \models \varphi(a_i, b_j)$, and for $i \leq j$ we get $\mathfrak{M} \models \neg\varphi(a_i, b_j)$, so φ has the order property, contradiction. $\square$

Lemma 0.2.17 *Suppose $A \subseteq N$ for some stable model $\mathfrak{N}$, and $q \in S(\mathfrak{N})$ extends $p \in S(A)$ with $R_\varphi(q) = R_\varphi(p) < \infty$. Then q is φ-definable over $\operatorname{acl}(A)$. In particular, for every formula $\varphi(x, y)$ there is an extension q of p to $\mathfrak{N}$ which is φ-definable over $\operatorname{acl}(A)$.*

Proof: Let $\mathfrak{M}$ be a model containing A. By Lemma 0.2.15 we may assume that $\mathfrak{N}$ is a big, $(2^{|M|})^+$-saturated model containing $\mathfrak{M}$: just replace q by its equi-definable extension. If $d_q x\varphi$ were not algebraic over A, then by saturation there would be more than $2^{|M|}$ different A-conjugates of q with distinct φ-types; as the number of types over $\mathfrak{M}$ is bounded by $2^{|M|}$, there is an extension $p_0 \in S(\mathfrak{M})$ of p which has two extensions q_0 and q_1 over $\mathfrak{N}$ with distinct φ-types, such that $R_\varphi(p) = R_\varphi(p_0) = R_\varphi(q_0) = R_\varphi(q_1)$. But q_0 and q_1 differ in a formula $\varphi(x, n)$ for some $n \in N$; this implies

$$R_\varphi(p) \geq R_\varphi(p_0) \geq R_\varphi(q_0) + 1 = R_\varphi(q) + 1 = R_\varphi(p) + 1,$$

a contradiction. Hence $d_{q_\varphi} x \varphi$ is definable over $\mathrm{acl}(A)$.

For the second assertion, let $\mathfrak{M}$ be a model containing A and q' an extension of p to $\mathfrak{M}$ of the same R_φ-rank. Then by Lemma 0.2.15 again, q' has an equi-definable extension to a model containing both $\mathfrak{M}$ and $\mathfrak{N}$ of the same R_φ-rank, which restricts to the required type over $\mathfrak{N}$. $\square$

Theorem 0.2.18 *Suppose* $\mathrm{Th}(\mathfrak{C})$ *is stable. Let* $p \in S(A)$, *and* $A \subseteq B$. *Then* p *has a non-forking extension to* B. *Any type over an algebraically closed set has a unique non-forking extension to any superset* B *of* A.

Proof: Let $\mathfrak{M}$ be a model containing B. By Lemma 0.2.17, for every formula $\varphi(x, y)$ there is an extension q_φ of p over $\mathfrak{M}$ which is φ-definable over $\mathrm{acl}(A)$. Clearly, we may assume that A is algebraically closed, as a non-forking extension of *any* extension of p to $\mathrm{acl}(A)$ will be a non-forking extension of p.

Now let $q' \in S(\mathfrak{M})$ be another extension of p with $d_{q'} x \varphi$ definable over A. Pick any $m \in M$. By the above argument (for $\mathrm{tp}(m/A)$ instead of p), there is an extension r of $\mathrm{tp}(m/A)$ to $\mathfrak{M}$ such that $d_r y \, \varphi(x, y)$ is definable over A. Then $\varphi(x, m) \in q_\varphi$ iff $\mathfrak{M} \models d_{q_\varphi} x \, \varphi(x, m)$ iff $d_{q_\varphi} x \, \varphi(x, y) \in \mathrm{tp}(m/A)$ iff $d_{q_\varphi} x \, \varphi \in r$ iff $d_r y \, \varphi \in q_\varphi$ iff $d_r y \, \varphi \in p$. Similarly $\varphi(x, m) \in q'$ iff $d_r y \, \varphi \in p$, so $d_{q_\varphi} x \varphi$ and $d_{q'} x \varphi$ are equivalent. This means that there is exactly one φ-definition which works for all extensions of p to $\mathfrak{M}$ which are φ-definable over A. As it clearly works for p as well, we denote it by $d_p x \varphi$. In particular, a non-forking extension of p to $\mathfrak{M}$ is unique; and thus non-forking extensions of p are unique.

Consider the set $q_0 = \{\varphi(x, m) : \varphi \in \mathfrak{L}, \mathfrak{M} \models d_p x \varphi(x, m)\}$. For any formulæ $\varphi_1(x, y_1), \ldots, \varphi_k(x, y_k)$ consider their conjunction $\varphi(x, y_1, \ldots, y_k) = \bigwedge_{i \le k} \varphi_i(x, y_i)$. Now $d_p x \, \varphi = d_{q_\varphi} x \, \varphi$ is clearly equivalent to $\bigwedge_{i \le k} d_{q_\varphi} x \, \varphi_i(x, y_i)$, which means that $d_{q_\varphi} x \varphi_i$ is invariant under A-automorphisms and hence definable over A. So $d_{q_\varphi} x \varphi_i$ is equivalent to $d_p x \varphi_i$ for all $1 \le i \le k$. It follows that q_0 is consistent and complete, and hence a non-forking extension. $\square$

Definition 0.2.14 Let A, B, and C be sets of elements. We say that A and B are *independent* over C, denoted by $A \underset{C}{\downarrow} B$, if for all tuples $a \in A$ $\mathrm{tp}(a/BC)$ does not fork over C.

Theorem 0.2.19 PROPERTIES OF INDEPENDENCE *Let* $\mathrm{Th}(\mathfrak{C})$ *be stable.*

1. *For any* A, B, C *there is a* C-conjugate A' *of* A *with* $A' \underset{C}{\downarrow} B$ *(existence).*

2. *If* $a \underset{C}{\not\downarrow} B$, *there is some* $b \in B$ *with* $a \underset{C}{\not\downarrow} b$ *(finiteness).*

3. $A \downarrow_C B$ iff $B \downarrow_C A$ *(symmetry)*.

4. $A \downarrow_C B$ and $A \downarrow_{BC} D$ iff $A \downarrow_C BD$ *(transitivity)*.

5. For any a and A there is $A_0 \subseteq A$ of cardinality at most $|T|$ such that $a \downarrow_{A_0} A$ *(small basis)*.

6. For any $p \in S(A)$ and $B \supseteq A$ there are at most $2^{|T|}$ non-forking extensions of p to B *(boundedness)*.

Proof: Clearly, we may assume that C is algebraically closed.

1. Let $\mathfrak{M}$ be a model containing BC. For any (finite tuple) $a \in A$ there is a unique non-forking extension of $\operatorname{tp}(a/C)$ to $\mathfrak{M}$, so these must fit together to give a unique non-forking extension of $\operatorname{tp}(A/C)$ to $\mathfrak{M}$, and the restriction to CB does not fork over C.

2. Suppose $a \downarrow_C b$ for all $b \in B$. Then $\models \varphi(a,b)$ iff $\models d_{\operatorname{tp}(a/C)} x\, \varphi(x,b)$ by uniqueness of the non-forking extension (and hence the defining scheme for non-forking extensions), so this is also a defining scheme for $\operatorname{tp}(a/B)$, and $a \downarrow_C B$.

3. Let $a \in A$ and $b \in B$, and suppose $A \downarrow_C B$. Then clearly $a \downarrow_C b$; we consider a saturated model $\mathfrak{M}$ containing a, b, C. Let $p(x)$ be the non-forking extension of $\operatorname{tp}(a/C)$ to $\mathfrak{M}$ and $q(y)$ the non-forking extension of $\operatorname{tp}(b/C)$ to $\mathfrak{M}$. For any formula φ the uniqueness of the non-forking extension and the independence $a \downarrow_C b$ imply that $\mathfrak{M} \models \varphi(a,b)$ iff $\mathfrak{M} \models d_p x\, \varphi(x,b)$ iff $d_p x\, \varphi(x,y) \in q(y)$ iff $d_q y\, \varphi(x,y) \in p(x)$ (by Lemma 0.2.16) iff $\varphi(a,y)$ is in the non-forking extension of $\operatorname{tp}(b/C)$ to aC. Hence $b \downarrow_C a$ for all $a \in A$, whence $b \downarrow_C A$, and finally $B \downarrow_C A$.

4. Clearly, if $\operatorname{tp}(A/BCD)$ is definable over C, then it is definable over BC and $\operatorname{tp}(A/BC)$ is definable over C. On the other hand, for any $a \in A$ and any formula $\varphi(x,y)$ there is a φ-definition $dx\, \varphi(x,y)$ over C which works for non-forking extensions of $\operatorname{tp}(a/C)$ to any model containing C. But since $a \downarrow_C B$, this definition $dx\, \varphi$ is also a φ-definition of $\operatorname{tp}(a/BC)$ over BC; as it yields a consistent type over any model containing BC, it must be a φ-definition for non-forking extensions of $\operatorname{tp}(a/BC)$. As $a \downarrow_{BC} D$, it is a φ-definition of $\operatorname{tp}(a/BCD)$, whence $a \downarrow_C BD$.

5. For every formula φ there is an $a_\varphi \in A$ such that the φ-definitions for non-forking extensions of $\operatorname{tp}(a/A)$ are over $\operatorname{acl}(a_\varphi)$, and we may choose A_0 to be the set of all these a_φ, for $\varphi \in \mathfrak{L}$.

6. Fix AB and choose a subset A_0 of A of cardinality at most $|T|$ such that $a \underset{A_0}{\downarrow} A$. There are at most $2^{|T|}$ types over $\mathrm{acl}(A_0)$, but any non-forking extension of $\mathrm{tp}(a/A)$ to AB must be the unique non-forking extension of its restriction to $\mathrm{acl}(A_0)$. $\square$

Theorem 0.2.20 *Let $p \in S(A)$. Then for any formula $\varphi(x,y)$, there are only finitely many inequivalent φ-definitions for non-forking extensions of p to any model $\mathfrak{M} \supseteq A$.*

Proof: As any two models are contained in a common third one and types over models have unique non-forking extensions, it is enough to consider a single saturated model $\mathfrak{M}$. Consider two non-forking extensions q and q' of p to $\mathfrak{M}$ and realize $p := q\lceil_{\mathrm{acl}(A)}$ by a and $p' := q'\lceil_{\mathrm{acl}(A)}$ by a'. As $\mathrm{tp}(a/A) = \mathrm{tp}(a'/A) = p$, there is an A-automorphism of $\mathfrak{M}$ taking a to a'. This automorphism must map $\mathrm{acl}(A)$ to $\mathrm{acl}(A)$, and p to p'. Hence it must map the unique φ-definition for all non-forking extensions of p to the unique φ-definition for all non-forking extensions of p', so $d_q x \varphi$ and $d_{q'} x \varphi$ are conjugate. As they are definable over $\mathrm{acl}(A)$, they have only finitely many conjugates under A-automorphisms. $\square$

Note that this yields another proof for the bound of the number of non-forking extensions of p.

Corollary 0.2.21 OPEN MAPPING THEOREM *Let $A \subseteq B$, and $S_n^{nf}(B)$ be the set of n-types over B which do not fork over A. Then the restriction map from $S_n^{nf}(B)$ to $S_n(A)$ is open.*

Note that the map is obviously continuous.
Proof: Let $[\varphi(x,b)]$ be a basic open set, and $d_1 y\,\varphi(x,y), \ldots, d_n y\,\varphi(x,y)$ the finitely many inequivalent φ-definitions for non-forking extensions of $\mathrm{tp}(b/A)$ to a model. Clearly, $\psi(x) := \bigvee_i d_i z\,\varphi(x,y)$ is A-invariant and hence A-definable, and $\models \psi(a)$ iff for some $b'B' \models \mathrm{tp}(bB/A)$ independently of a we have $\models \varphi(a,b')$. Now there is an A-automorphism σ with $\sigma(b'B') = bB$. Then $\mathrm{tp}(\sigma(a)/B)$ lies in $[\varphi(x,b)] \cap S_n^{nf}(B)$, and clearly $\mathrm{tp}(\sigma(a)/A) = \mathrm{tp}(a/A)$. Hence the image of $[\varphi(x,b)]$ is $[\psi(x)]$. $\square$

Definition 0.2.15 A type $p \in S(A)$ is *stationary* if it has exactly one non-forking extension to every superset B of A. In this case, we denote the non-forking extension by $p|B$.

In particular, types over models and algebraically closed sets are stationary. Note that even for a stationary type $p \in S(A)$ a φ-definition need not be shared by the non-forking extensions: if X is an A-definable set disjoint from A, we may add it to or subtract it from the φ-definition of any type over A. However, for a stationary type $d_p x \varphi$ will always denote the φ-definition which works for the non-forking extensions; this is again definable over A.

Definition 0.2.16 Suppose $p \in S(A)$ is stationary, and $q \in S(A)$ arbitrary. The *product* $p \otimes q$ is the type $\mathrm{tp}(a, b/A)$ for any independent realizations a of p and b of q.

Note that this is well-defined: If a', b' are also independent realizations of p and q, then firstly there is an A-automorphism mapping b to b'. But this must map the unique non-forking extension of p over A, b to the unique non-forking extension of p over A, b'; the former is realized by a and the latter by a'.

Definition 0.2.17 The *strong type* of a over A, denoted by $\mathrm{stp}(a/A)$, is the type of a over $\mathrm{acl}(A)$.

A stationary type extends to a unique strong type; in general a type extends to at most $2^{|T|}$ different strong types.

Corollary 0.2.22 FINITE EQUIVALENCE RELATION THEOREM *Suppose* $\mathrm{tp}(a/A) = \mathrm{tp}(a'/A)$. *Then a and a' have the same strong type over A iff they agree on all A-definable equivalence relations with a finite number of classes.*

Proof: Suppose there are $b \in \mathrm{acl}(A)$ and some formula φ with $\models \varphi(a, b) \wedge \neg\varphi(a', b)$, where b lies in the finite A-definable set B. Consider the relation E defined by xEy iff $\forall z \in B \, (\varphi(x, z) \leftrightarrow \varphi(y, z))$. This clearly is an A-definable equivalence relation with a finite number of classes, and $\neg aEa'$. $\square$

Definition 0.2.18 The *canonical base* of a strong type $p \in S(A)$ is the definable closure of $\{\mathrm{Cb}(d_p x \, \varphi) : \varphi \in \mathfrak{L}\}$.

This is well-defined and always contained in $\mathrm{acl}(A)$; clearly p does not fork over $\mathrm{Cb}(p)$. We shall write $\mathrm{Cb}(a/A)$ for $\mathrm{Cb}(\mathrm{stp}(a/A))$.

Lemma 0.2.23 *Suppose $a \in \mathrm{dcl}(bc)$ and $c' = \mathrm{Cb}(c/abd)$. Then $a \in \mathrm{dcl}(bc')$.*

Proof: Consider some a' with $\mathrm{tp}(a'/bc') = \mathrm{tp}(a/bc')$, and let c'' realize the unique non-forking extension of $\mathrm{tp}(c/c')$ to $c'baa'$. Then

$$\mathrm{tp}(a'c''bc') = \mathrm{tp}(ac''bc') = \mathrm{tp}(acbc'),$$

so as $a \in \mathrm{dcl}(bc)$ we get $a = a' \in \mathrm{dcl}(bc'')$. $\square$

Definition 0.2.19 Suppose $A \subseteq B$, and $p \in S(B)$.

- p is *finitely satisfiable* in A if every formula $\varphi(x, b) \in p$ has a realization $m \in A$.

- p is a *coheir* of $p\!\restriction_A$ if it is finitely satisfiable in every model containing A.

- A formula $\varphi(x,y)$ is represented in p if there is $b \in B$ with $\varphi(x,b) \in p$.

- p is an *heir* of $p{\restriction}_A$ if every formula represented in p is represented in every extension of $p{\restriction}_A$ to a model.

- The *class* of a type p over a model $\mathfrak{M}$ is the set $\{\varphi(x,y) \in \mathfrak{L} : \varphi(x,m) \in p$ for some $m \in \mathfrak{M}\}$. The *fundamental order* is the set of all classes of types over models, ordered by reverse inclusion.

- The *bound* $\beta(p)$ of a type $p \in S(A)$ is the maximal element in the fundamental order among the classes of types over models extending p.

The fundamental order is closed under unions and intersections (by compactness in the theory with an additional predicate for an elementary substructure).

Theorem 0.2.24 *Let* $\mathrm{Th}(\mathfrak{C})$ *be stable. Then there is a unique bound for every type. Furthermore, the following are equivalent for a type* $p \in S(A)$ *with an extension* $q \in S(B)$:

1. q does not fork over A.

2. q is a coheir of p.

3. q is an heir of p.

4. $\beta(q) = \beta(p)$.

5. $R_\varphi(q) = R_\varphi(p)$ for all formulæ φ.

Proof: The existence of a unique bound follows once we prove that non-forking extensions are heirs, as every type over A has a non-forking extension to any model containing A.

($1. \Rightarrow 2.$) Consider any formula $\varphi(x,b) \in q$ and a model $\mathfrak{M}$ containing A, and let q be realized by a, with $a \underset{B}{\downarrow} \mathfrak{M}$. Then $a \underset{A}{\downarrow} B\mathfrak{M}$, whence $a \underset{\mathfrak{M}}{\downarrow} b$ and $r := \mathrm{tp}(b/\mathfrak{M}a)$ is definable over $\mathfrak{M}$. Therefore $d_r y\,\varphi(x,y)$ is consistent and satisfied by some element $m \in \mathfrak{M}$. Hence $\models \varphi(m,b)$, and q is a coheir of p. ·

($3. \Rightarrow 4.$) Consider any formula $\varphi(x,y) \in \beta(q)$. If $\mathfrak{M}$ is a model containing B, then $\mathrm{Th}_{\mathfrak{L}(M)}(\mathfrak{M}) \cup q \cup \{\neg\varphi(x,m) : m \in M\}$ is inconsistent, hence a finite part π is inconsistent, i.e. there are a sentence $\vartheta(b,m_1,\dots,m_k) \in \mathrm{Th}_{\mathfrak{L}(M)}(\mathfrak{M})$, and a formula $\psi(x,b) \in q$, such that $\vartheta(b,\bar{m}) \wedge \psi(x,b) \wedge \bigwedge_{i \le k} \neg\varphi(x,m_i)$ is inconsistent. But this means that the formula

$$[\exists \bar{y}\,\vartheta(z,\bar{y}) \wedge \psi(x,z)] \wedge \left(\forall \bar{y}\,\forall x\,[\vartheta(z,\bar{y}) \wedge \psi(x,z)] \to \bigvee_{i \le k} \varphi(x,y_i)\right)$$

is represented in q (by b), and must be in $\beta(p)$. Therefore $\varphi \in \beta(p)$ as well.

(4. $\Rightarrow$ 5.) Consider a formula $\varphi(x, y)$ and let q' be an extension of p to some model $\mathfrak{M}$ with $R_\varphi(q') = R_\varphi(p)$. If $k = R_\varphi(q) < R_\varphi(p)$ for some formula φ, then this is due to some formula $\psi(x, m) \in q$. However, since the class of q' contains the class of q, there must be $m' \in \mathfrak{M}$ such that $R_\varphi(\psi(x, m')) < k+1$ (this is a first-order condition) and $\psi(x, m') \in q'$, contradicting $R_\varphi(q') > k$.

(5. $\Rightarrow$ 1.) This is Lemma 0.2.17.

We can now close the circle by noting that $\operatorname{tp}(a/Ab)$ is a coheir of $\operatorname{tp}(a/A)$ iff for every formula φ with parameters in A such that $\models \varphi(a, b)$ and any model $\mathfrak{M}$ there is $m \in M$ with $\models \varphi(m, b)$ iff $\operatorname{tp}(b/Aa)$ is an heir of $\operatorname{tp}(b/A)$ in the language $\mathcal{L}(A)$, and using forking symmetry. In particular $a \underset{A}{\downarrow} B$ implies $B \underset{A}{\downarrow} a$, so $\operatorname{tp}(B/Aa)$ is a coheir of $\operatorname{tp}(B/A)$ and $\operatorname{tp}(a/B)$ is an heir (in $\mathcal{L}(A)$ and *a fortiori* in $\mathcal{L}$) of $\operatorname{tp}(a/A)$: we get (1. $\Rightarrow$ 3.) and the existence of bounds. $\square$

Note that the fundamental order may *a priori* be affected by naming parameters, as we shall then have to consider more formulæ. However, Theorem 0.2.24 shows that this does not happen in a stable theory.

Corollary 0.2.25 *Let T be a stable theory. Any chain of forking extensions of types in T has cardinality at most $2^{|T|}$. Any well-ordered or anti-well-ordered chain has cardinality at most $|T|$.*

Proof: A chain of forking extensions yields a chain in the fundamental order. But there are at most $|T|$ formulæ; this yields the bound on the length of a chain. If the chain is well-ordered, we associate with the α-th class c_α a formula in $c_\alpha - c_{\alpha+1}$ (and analogously for an anti-well-ordered chain). $\square$

Corollary 0.2.26 *Let T be stable and $A \subseteq B$. The subset of types in $S_n(B)$ which do not fork over A is closed.*

Proof: Let Φ be the set of formulae $\varphi(\bar{x}, b)$ which are not finitely satisfiable in every model containing A. Then $p \in S(B)$ does not fork over A iff it lies in $[\neg\varphi : \varphi \in \Phi]$. $\square$

Definition 0.2.20 Let A be a set of parameters. A sequence $(a_i : i \in I)$ indexed by some ordered set I is *order indiscernible* over A if for all $k < \omega$ and $i_1 < i_2 < \cdots < i_k$ the type $\operatorname{tp}(a_{i_1} \ldots a_{i_k}/A)$ does not depend on the choice of indices. The sequence is *indiscernible* over A if this type does not even depend on the order of the indices in I (as long as they are distinct).

Proposition 0.2.27 *Suppose $\mathfrak{M}$ is $(|A|^+ + \kappa)$-saturated. Then there is an indiscernible sequence of length κ over A.*

Proof: Consider the type Σ in variables $(x_i : i < \kappa)$ saying that all the x_i are distinct, together with

$$\{\varphi(\bar{x}) \leftrightarrow \varphi(\bar{y}) : \varphi \in \mathcal{L}; \bar{x}, \bar{y} \subset (x_i : i < \kappa) \text{ with increasing indices}\}.$$

Claim. This partial type is consistent.

Proof of Claim: Consider any finite subset Σ_0. This finite subset mentions only k distinct formulæ for some $k < \omega$, which we may suppose all have n free variables (adding dummy variables if necessary). So we can colour the n-tuples of $\mathfrak{M}$ in 2^k different colours according to which of the k formulæ they satisfy. By Ramsey's Theorem there is an infinite monochromatic subset of $\mathfrak{M}$, which must satisfy Σ_0. So Σ is consistent by compactness. $\square$

But now by saturation we can realize any completion of Σ to a complete type over A one by one in $\mathfrak{M}$. $\square$

Proposition 0.2.28 *A theory T is stable iff every infinite order indiscernible sequence is indiscernible.*

Proof: Suppose $(a_i : i \in I)$ is order indiscernible over A, but not indiscernible. By compactness, we may assume that $I = \omega \cdot 3$. Consider a formula $\varphi(\bar{x})$ over A such that $\varphi(a_1, \ldots, a_{i-1}, a_i, a_{i+1}, a_{i+2}, \ldots, a_k)$ holds, but not $\varphi(a_1, \ldots, a_{i-1}, a_{i+1}, a_i, a_{i+2}, \ldots, a_k)$. We may choose $\bar{a} = a_1, \ldots, a_{i-1}$ in the first copy of ω and $\bar{a}' = a_{i+2}, \ldots, a_k$ in the last copy; then $\varphi(\bar{a}, x, y, \bar{a}')$ orders the set $\{a_j : j \text{ in the middle copy of } \omega\}$. (Note that $\varphi(\bar{a}, a_i, a_i, \bar{a}')$ holds either for all or for no i in the middle copy of ω.) So T is unstable.

Conversely, if T is unstable, there is a formula $\varphi(x, y)$ which orders an infinite set $\{a_i : i \in I\}$. Consider the partial type Σ from the proof of Proposition 0.2.27, and add the condition

$$\{\varphi(x_i, x_j) : i < j < \kappa\} \cup \{\neg\varphi(x_j, x_i) : i \le j < \kappa\}.$$

Again any finite subset can be realized in $\{a_i : i \in I\}$ by Ramsey's Theorem, so by compactness there is a realization $(b_i : i \in I)$ of the whole type. But this sequence is order indiscernible (over the parameters used in φ) and ordered by φ, and hence cannot be fully indiscernible. $\square$

Lemma 0.2.29 *Suppose the sequence $(a_i : i < \omega)$ is indiscernible, and for every n there is b_n such that $\varphi(a_i, b_n)$ holds for at least n indices i and does not hold for at least n other indices. Then the theory has the independence property.*

Proof: By indiscernibility, for every two finite disjoint subsets I and J of ω we find b_I^J such that $\varphi(a_i, b)$ is true if $i \in I$, and false if $i \in J$. But then by compactness we find such a b_I for every subset $I \subset \omega$ and $J := \omega - I$. $\square$

Definition 0.2.21 A *Morley sequence* of a strong type p over A is an infinite independent sequence of realizations of p.

Theorem 0.2.30 *An infinite sequence $(a_i)_I$ over A in a stable theory is a Morley sequence over A iff it is indiscernible and for all $i \in I$ we have $a_i \underset{A}{\cancel{\smile}} \{a_j : j < i\}$.*

Proof: Clearly a Morley sequence is indiscernible (there is just one non-forking extension of a strong type) and independent. Conversely, let us show first that $a_i \underset{A}{\smile} \{a_j : j < i\}$ for all $i \in I$ implies independence of $(a_i)_I$. Suppose it were not independent, so there would be $i \in I$ such that $a_i \underset{A}{\cancel{\smile}} \{a_j : j \neq i\}$. Then there is a minimal finite I_0 such that $\{a_i : i \in I_0\}$ is not independent. Let k be the maximal index in I_0, and $I_1 = I_0 - \{k\}$. Then $\{a_i : i \in I_1\}$ is independent by the minimality of I_0, and $a_k \underset{A}{\smile} \{a_i : i \in I_1\}$ by hypothesis. Hence $\{a_i : i \in I_0\}$ is independent by symmetry and transitivity of independence.

It remains to show that all a_i have the same strong type over A. Let $a \in \text{acl}(A)$, and $\bar{a}$ be the tuple of its A-conjugates. Let $\varphi(x, y)$ be any formula and define an equivalence relation: xEz iff $\forall y \in \bar{a}\ \varphi(x, y) \leftrightarrow \varphi(z, y)$. This is A-definable and has finitely many classes. By indiscernibility, either all elements of the sequence lie in the same class, or all lie in different classes. But clearly the latter is impossible. $\square$

Definition 0.2.22 If a and b are elements and A is a set of parameters, we say that a *forks* with b over A *via* φ if $\varphi(x, y)$ is in the bound of neither $p(x) = \text{tp}(a/A)$ nor $q(y) = \text{tp}(b/A)$, but $\models \varphi(a, b)$.

Theorem 0.2.31 *Let $(a_i)_I$ be a Morley sequence over A in a stable theory. Then for any formula $\varphi(x, y)$ there is some $n < \omega$ such that for any b there are at most n distinct $i \in I$ such that a_i forks with b over A via φ.*

Proof: Suppose otherwise. Then by compactness we get a Morley sequence $(a_i)_I$ of arbitrary length such that b forks with every element of the sequence via φ. But then $a_i \underset{A}{\smile} \{a_j : j < i\}$ and $a_i \underset{A}{\cancel{\smile}} b$ together yield $b \underset{A\{a_j : j < i\}}{\cancel{\smile}} a_i$ and we get arbitrarily long forking chains, contradicting Corollary 0.2.25. $\square$

Corollary 0.2.32 *Suppose $(a_i : i < \omega)$ is a Morley sequence for p in a stable theory. Then $\text{Cb}(p) \subset \text{dcl}(a_i : i < \omega)$.*

Proof: By Lemma 0.2.29 for every formula $\varphi(x, y)$ there is some $n_\varphi < \omega$ such that for any b the sentence $\varphi(a_i, b)$ holds for at most or all but at most n_φ indices i. Then

$$\bigvee \left\{ \bigwedge_{j=0}^{n_\varphi} \varphi(a_{i_j}, y) : 0 \leq i_0, \ldots, i_{n_\varphi} \leq 2n_\varphi + 1 \right\}$$

is a φ-definition for p by Theorem 0.2.31. $\square$

We shall finish this section with a short discussion of ranks. If On is the class of all ordinals, let On^+ be the class On together with a new symbol ∞, with the convention that ∞ is greater than every ordinal (and $\alpha + \infty = \infty + \alpha = \infty$).

Definition 0.2.23 The *Lascar rank* is the smallest function U from the class of all types (over any subset of $\mathfrak{C}$) to On^+ satisfying

$U(p) \geq \alpha + 1$ iff p has a forking extension q with $U(q) \geq \alpha$.

The *Shelah rank* is the smallest function RC from the class of all types to On^+ satisfying

$RC(p) \geq \alpha + 1$ iff every formula in p contains arbitrarily many distinct types of Shelah rank α (over some suitable parameter set).

The *Morley rank* is the smallest function RM from the class of all types to On^+ satisfying

$RM(p) \geq \alpha + 1$ iff p has an extension p' such that every formula $\varphi(x) \in p'$ contains infinitely many pairwise contradictory types p_i with $RM(p_i) \geq \alpha$.

The *Morley degree* of a type p is the minimum over all formulæ $\varphi \in p$ of the maximal number of pairwise contradictory types of Morley rank $RM(p)$ in $[\varphi]$.

Remark 0.2.9 It turns out that a countable theory is ω-stable iff every type has ordinal Morley rank, and a theory is superstable iff every type has ordinal Lascar rank iff every type has ordinal Shelah rank. An extension q of a type p is non-forking iff $U(q) = U(p)$ iff $RC(p) = RC(q)$ iff $RM(q) = RM(p)$, provided the rank in question has ordinal value. Furthermore, ranks are preserved under definable finite-to-one maps. $RM(p) \geq RC(p) \geq U(p)$ for any type p, Morley degree is well-defined and finite, and for every formula $\varphi(x)$ the Shelah and Morley ranks of φ, namely $\sup\{RC(p) : p \in [\varphi]\}$ and $\sup\{RM(p) : p \in [\varphi]\}$, are attained by a type in $[\varphi]$ (over the parameters of φ). In particular, a superstable theory has n-types of maximal Shelah rank (although types of maximal Lascar rank need not exist) and an ω-stable theory has n-types of maximal Morley rank for every $n < \omega$. The transcendental type of a strongly minimal set has Morley rank one and Morley degree one; conversely a set of Morley rank and degree one is strongly minimal.

A set (or type) of Shelah rank one is called *weakly minimal*.

Definition 0.2.24 A theory is λ-*categorical* if all its models of cardinality λ are isomorphic.

Morley's Theorem states that a theory is λ-categorical for some $\lambda > |T|$ iff it is λ-categorical for all such λ. (Actually, he only proved it for countable theories; the full result is due to Rowbottom, Ressayre and Shelah.) An uncountably categorical countable theory has finite Morley rank. For groups, Zil'ber [130] has shown a partial converse: a simple group of finite Morley rank is uncountably categorical; we shall look at this in Section 4.7.

0.3 Examples

In general, it is very difficult to prove stability for a theory, unless one has particular information about the definable subsets, such as quantifier elimination.

Example 0.3.1 Let R be a ring, and $\mathcal{L}$ consist of addition $+$, a constant 0, and scalar multiplication λ_r for every $r \in R$. Then any left R-*module* $\mathfrak{M}$ becomes naturally an $\mathcal{L}$-structure (and similarly for right R-modules). A *positive primitive* formula is one of the form

$$\exists \bar{y} \bigwedge_k \left(\sum_i \lambda_k^i x_i + \sum_j \lambda_k^j y_j = a_k \right);$$

it is easy to see that it defines a coset of an $\emptyset$-definable additive subgroup of the appropriate power of M (and if all $a_k = 0$, then this coset is a subgroup). By results of Szmielev [108], Monk [74], Baur [13], and Garavaglia every formula is equivalent to a Boolean combination of positive primitive formulæ.

More generally, we define an *abelian structure* to be an abelian group A with some predicates for subgroups of powers of A. Any module is an abelian structure, as a scalar λ may be identified with the subgroup $\{(a, \lambda a) : a \in A\}$ of A^2. We shall prove (Theorem 4.2.8) that in an abelian structure, every definable set is equal to a Boolean combination of cosets of $\mathrm{acl}(\emptyset)$-definable subgroups. Now consider a formula of the form $(\bar{x}\bar{y}) \in S$, where S is a coset of the subgroup H in the appropriate power of the group, and suppose that it has the order property. Then there are tuples $\bar{m}_i, \bar{n}_i$ in some model G such that $(\bar{m}_i, \bar{n}_j) \in S$ iff $i < j$. But now $(\bar{m}_0, \bar{n}_1) \in S$, $(\bar{m}_0, \bar{n}_2) \in S$ and $(\bar{m}_1, \bar{n}_2) \in S$, whence $(0, \bar{n}_1 - \bar{n}_2) \in H$ and $(\bar{m}_1 - \bar{m}_0, 0) \in H$. Hence $(\bar{m}_1, \bar{n}_1) \in S$, contradicting our choice of $\bar{m}_1, \bar{n}_1$. Therefore this formula cannot have the order property.

By Lemma 0.2.10 no formula has the order property. Hence an abelian structure is stable; it is superstable iff there is no infinite descending sequence of definable subgroups each of infinite index in its predecessor, and it is ω-stable iff it satisfies the minimal condition on definable subgroups.

The model theory of modules has been an area of intense study; Prest [98] is a good introduction.

Example 0.3.2 Any *algebraically closed field* K (of given characteristic) has full elimination of quantifiers (Tarski [109] and Chevalley), in the natural language of fields. Hence every definable subset of K (in one free variable!) is finite or co-finite, and F has Morley rank and Morley degree one. Conversely, any field with elimination of quantifiers is algebraically closed, as is any superstable (skew) field.

Example 0.3.3 A *differential field* K is a field with an additional function δ, which is interpreted as a *derivation* on K and satisfies $\delta(x + y) = \delta(x) + \delta(y)$ and $\delta(xy) = \delta(x)y + x\delta(y)$. It is *differentially closed* if every system consisting of a differential equation in x of degree n and a differential inequality of order less than n, both with coefficients in K, has a solution in K. A differentially closed field of characteristic zero has quantifier elimination and Morley rank ω, a result due to Seidenberg [101] and Shelah [104]. Furthermore, both K and the subfield $C = \{k \in K : \delta(k) = 0\}$ of constants are algebraically closed, and C carries no structure other than the field structure. It is an open question whether any superstable differential field must be differentially closed.

Example 0.3.4 A field K is *separably closed* if it has no proper separable field extension. In characteristic zero a separably closed field is algebraically closed, but in characteristic $p \neq 0$ we get a chain of subfields $K \geq K^p \geq K^{p^2} \geq \cdots$. The transcendence degree of $[K : K^p]$ is called the *Ershov invariant*. Separably closed fields have quantifier elimination in a suitably enriched language, due to Ershov [40], and are stable (see Wood [127]). Since they have an infinite descending chain of subgroups of infinite index (unless they are algebraically closed), they cannot be superstable.

Another possibility for obtaining stable structures is a direct construction from finite substructures, via an amalgamation process initially due to Ehrenfeucht, Fraïssé and Jónsson (see [47]). It was adapted by Hrushovski [57] to give stable relational structures, and finally by Baudisch for nilpotent groups of exponent p and class 2. In particular, in [11] Baudisch has constructed a totally categorical non-(abelian-by-finite) group which does not interpret a field, and in [10] he has extracted a stable, non-superstable subgroup of it. It is noteworthy in this context that Hrushovski has also managed to adapt his amalgamation method to amalgamate two strongly minimal sets (with an additional technical condition, definability of multiplicity) which need not be locally finite; in particular he has constructed a strongly minimal set on which two different fields live completely independently.

In certain theories it is also possible to establish stability by counting types. Baudisch [8] proved that the free nilpotent group (on infinitely many generators) of exponent p^n and class $c < p$ is ω-stable; Chapuis has constructed centreless soluble ω-stable groups of derived length n for every $n < \omega$ in [31].

Another method is given by a construction of Mekler [72], who associates to every theory a nilpotent group of class 2 and prime exponent which shares many of the model-theoretic properties of the original structure. In particular, λ-stability is preserved for all λ, although finiteness of rank (Lascar or Morley) is not. If we allow extra structure on the group, such an interpretation is actually quite easy:

Example 0.3.5 Let $\mathfrak{M}$ be any structure in a language $\mathfrak{L}$, and consider the underlying domain M to be a basis for an abelian group A of exponent p, or a transcendence basis for an algebraically closed field K of arbitrary characteristic p. Let P be a new unary predicate, put $\mathfrak{L}_A := \mathfrak{L} \cup \{P, 0, +\}$ and $\mathfrak{L}_K := \mathfrak{L} \cup \{P, 0, 1, +, \cdot\}$, and make A into an $\mathfrak{L}_A$-structure by interpreting P as M and the functions and relations of $\mathfrak{L}$ on $P^A = M$ as they were interpreted on $\mathfrak{M}$; similarly K becomes an $\mathfrak{L}_K$-structure. Then for any elementary superstructure $\mathcal{A}$ (or $\mathcal{K}$) the subset $P^A =: \mathfrak{N}$ (or $P^K =: \mathfrak{N}$) is an elementary superstructure of $\mathfrak{M}$ and N is a linearly (or algebraically) independent subset. Hence any automorphism of $\mathfrak{N}$ can be extended to an automorphism of $\mathcal{A}$ or $\mathcal{K}$. There are two kinds of elements: those algebraic over N, which are determined by finitely many elements in N, and a unique *generic* type which is transcendental over N. It follows that $S(\mathfrak{N})$, $S(\mathcal{A})$ and $S(\mathcal{K})$ all have the same cardinality, and $\mathfrak{M}$ is λ-stable iff A is λ-stable iff K is. However, $U(A) = U(K) = U(\mathfrak{M}) \cdot \omega$ (which may of course be ∞).

Finally, the most important method by far is the interpretation of a group in a stable structure.

Definition 0.3.1 An $\mathfrak{L}$-structure $\mathfrak{M}$ is *interpretable* in an $\mathfrak{L}'$-structure $\mathfrak{N}$ if there is an $\mathfrak{L}'$-definable subset M' of N^{eq}, for every constant of $\mathfrak{L}$ an element of M', for every j-ary function of $\mathfrak{L}$ an $\mathfrak{L}'$-definable function $(M')^j \to M'$, and for every j-ary relation of $\mathfrak{L}$ an $\mathfrak{L}'$-definable subset of $(M')^j$ such that all this induces on M' a structure isomorphic to $\mathfrak{M}$.

If T has a model which is interpretable in a model of T', we say that T is interpretable in T'.

So $\mathfrak{M}$ is interpretable in $\mathfrak{N}$ iff (an isomorphic copy of) $\mathfrak{M}$ is definable in $\mathfrak{N}^{eq}$. If T is interpretable in T' then every model $\mathfrak{M}$ of T will have an elementary superstructure which is interpretable in a model of T', although in general $\mathfrak{M}$ itself need not be interpretable in a model of T'. However, by counting types, it still follows that a theory interpretable in a λ-stable theory is again λ-stable, and we can even get a bound for the ranks.

Example 0.3.6   1. A group interpretable in an algebraically closed field
K has finite Morley rank. (In fact, by Weil, van den Dries, and
Hrushovski [125, 110], see [95], these are exactly the algebraic groups
over K.)

  2. A group interpretable in a differentially closed field K is ω-stable. (In
  fact, by Pillay [86], these are exactly the differentially algebraic groups
  over K.)

  3. A group interpretable in a separably closed field K is stable. (In fact,
  by Messmer [73], these are exactly the algebraic groups over K.)

  4. Let K be an algebraically closed field with a unary predicate U which
  holds exactly for the roots of unity. Zil'ber [135] has proven that (K, U)
  has Morley rank ω, so any group interpretable in this structure is ω-
  stable. The same conclusion holds if we choose U to be some finitely
  generated subgroup of $\mathbb{C}$, by Grünenwald and Haug [43].

However, not only do we get interpretable groups in these classical contexts,
but more surprisingly, deep results of Zil'ber [131] and Hrushovski [51] show
that structural properties of a stable theory may imply the interpretability of
certain groups of automorphisms of the structure.

0.4   Historical and Bibliographical Remarks

Apart from the references already cited in section 0, the big encyclopaedia on
stability is of course Shelah's *Classification Theory* [105]; Baldwin [3] aims at
the same comprehensiveness, but greater comprehensibility. The introduction
to stability theory owes much to the lecture notes by Ziegler [129], with the
argument (3. ⇒ 4. in Theorem 0.2.11) due to Hodges. Other introductions to
stability theory are by Pillay [83] and Lascar [67]; for a general introduction
to model theory the reader may consult the classic book by Chang and Keisler
[30], or the recent book by Hodges [50].

Chapter 1

Groups & Generality

In section 0, we shall introduce the various chain conditions which are implied by stability: the ascending and the descending chain conditions for intersections of uniformly definable families of subgroups (these intersections will then also be uniformly definable), and the $|T|^+$-chain condition for arbitrary $\bigwedge$-definable subgroups. As we shall want these conditions to hold in saturated models as well, they are necessarily uniform. We shall also mention chain conditions which hold under stronger stability assumptions: the descending chain condition for definable subgroups in ω-stable groups, the descending chain condition for connected $\bigwedge$-definable subgroups in the superstable case, and the ascending chain condition for connected definable subgroups in theories of finite rank.

However, chain conditions are important in their own right, not only in connection with stability; once they are established, no further use of stability will be made in that chapter (with the exception of Theorem 1.1.13, which has been included in this section because of its proximity to the nilpotency results there). In fact, a more algebraic setting might consist of a group together with a family of sections comprising all centralizers and closed under taking normalizers, quotients, and – in the case of an abelian group – the subgroups of n-divisible elements and elements of order n for all $n < \omega$; we would then require the uniform chain condition on intersections of uniform subfamilies. For ease of terminology we shall not take this path but rather assume that the group in question is substable; an examination of the proof will yield the exact chain conditions needed for each theorem.

A first consequence (section 1) of the chain conditions is the existence of an $\bigwedge$-definable connected component as bounded intersection of definable subgroups of finite index. In general, however, this is not a definable subgroup, and we shall also study various approximations, namely components which are connected with respect to certain formulæ. After that, we shall discuss definability of certain basic subgroups (centres, normalizers, etc.) and properties (solubility, nilpotency), as well as some results on groups generated by uniformly soluble (or nilpotent) normal subgroups and the definable layer.

Groups with chain conditions on centralizers form a particularly important class, denoted by $\mathfrak{M}_c$. They are studied in section 2, where we derive results about their nilpotent and soluble subgroups. In particular, an $\mathfrak{M}_c$-group has a unique maximal normal nilpotent subgroup. A further analysis of nilpotent periodic $\mathfrak{M}_c$-groups leads to a study of substable nilpotent groups in section 3.

Having already encountered commutator conditions in the preceding two sections, we deal with the special case of Engel conditions in section 4. We prove that a substable group generated by Engel elements is nilpotent, and the set of bounded left Engel elements equals the Fitting subgroup. Bounded right Engel elements are more difficult to handle, but partial results about their hypercentrality are available.

Section 5 deals with Sylow subgroups. For $p = 2$ the results are particularly nice: the 2-Sylow subgroups are locally nilpotent, nilpotent-by-finite, and conjugate whenever the group is periodic or ABD. For odd p we shall need the stronger assumption of bi-finiteness. This will then lead to the existence of supplements: if a quotient G/N is a p-group, then it is the image modulo N of any p-Sylow subgroup of G. Finally we shall generalize the Baer-Suzuki theorem: an invariant subset with the property that any two of its elements generate a finite p-group must generate a locally nilpotent p-group, which is necessarily normal.

1.0   Chain Conditions

A *stable group* is a stable $\mathcal{L}$-structure $\mathfrak{M}$ such that there is a definable binary function $*$ which makes $(M, *)$ into a group. Note that there may be additional structure in $\mathfrak{M}$. More generally, the domain of a group G may be a definable, type-definable, or merely interpretable subset of $\mathfrak{M}$, in which case we shall speak of a definable, type-definable, or interpretable stable group (and the same goes for semi-groups). Note that a group definable or interpretable in a stable structure is stable in its own right.

Definition 1.0.1   • A group G is $\bigwedge$-*definable* in a structure $\mathfrak{M}$ if it is the intersection of definable *groups* in $\mathfrak{M}$.

• A group G is *substable* if there is a stable group $\mathcal{G}$ such that G is isomorphic to a subgroup of $\mathcal{G}$.

So $\bigwedge$-definable subgroups are type-definable, using just formulæ of a particular kind (those defining groups). We should note that the definition of substability only requires the existence of a stable supergroup, but does not specify it any further. In particular, since this supergroup need not be unique, a substable G may be isomorphic to subgroups of two completely different stable groups! If we do want to stress the ambient stable group, we may say that G is substable *in* $\mathcal{G}$ (and identify G with its isomorphic copy inside $\mathcal{G}$).

Definition 1.0.2 Let G be substable in $\mathcal{G}$, and H a subgroup of G. Then H is *($\bigwedge$-)definable relative to G,* or *relatively ($\bigwedge$-)definable,* if it is the intersection of G with a $(\bigwedge)$-definable subgroup of $\mathcal{G}$.

Note that this definition requires the formula(e) which relatively define H not only to intersect G in a subgroup, but actually to define a group on the whole of $\mathcal{G}$. It will be shown in chapter 2 that if the group H is the intersection with G of a definable *set*, then H is the intersection with G of a definable *group*. *A priori* the collection of relatively definable subsets of a substable group G may vary with the ambient group $\mathcal{G}$. In practice, however, we shall usually only make use of relatively definable subsets which are relatively definable in any ambient stable structure (such as centralizers).

We shall want properties of $\bigwedge$- and type-definable groups to be preserved under elementary extensions, and hence in general shall need a certain amount of saturation of G when dealing with them. This presents no problem when dealing with first-order conditions, as they are preserved when moving to an elementary sub- or superstructure (say, a saturated, or somewhat saturated one). Substability, on the other hand, is a property of a particular group G, not a theory, and requires a particular model $\mathcal{G}$ of a stable supergroup – another model of $\mathrm{Th}(\mathcal{G})$ might not contain a copy of G, although of course any elementary superstructure will. Even worse: another model of $\mathrm{Th}(G)$ might not even be substable! Obviously a subgroup of a substable group is again substable, but the quotient of two substable groups need not be substable as well. (Since all free groups are substable, preservation under quotients would imply that all groups are substable.) However, as we shall see in Lemma 1.1.6, the quotient of a substable group by a relatively definable subgroup is again substable. (The analogue for definable stable groups holds trivially: definable subgroups and quotients of definable subgroups by definable subgroups are all interpretable, hence stable as well.)

The following easy lemma will be quite helpful later on:

Lemma 1.0.1 *A stable type-definable semi-group M with left and right cancellation, or with left cancellation and right unit, is a group.*

Note that the hypotheses of the lemma are supposed to hold in all elementary superstructures.

Proof: By compactness, there is a formula $\varphi(x)$ containing M on which multiplication is defined and associative, and such that if a, b, c satisfy φ, and $ab = ac$, then $b = c$. (However, φ need not be closed under multiplication.) Now let a be an arbitrary element of the semi-group. The sequence $a, a^2, a^3, \ldots$ cannot be ordered by the formula $\exists x \, (\varphi(x) \wedge x_1 x = x_2)$. Therefore there are some b satisfying φ, and $n > m$ with $a^n b = a^m$. Put $e = a^{n-m} b$, so $a^m e = a^m$.

If φ' is a formula contained in φ and containing M, then for the same reason there are $n' > m'$ and b' such that $a^{n'} b' = a^{m'}$. Put $e' = a^{n'-m'} b'$. Then $a^{m+m'} e = a^{m'} a^n b = a^{m'+m} = a^m a^{n'} b' = a^{m+m'} e'$, whence $e = e'$ by left cancellation. It follows that e lies in M.

But now consider any element c of the semi-group. Then $a^m c = a^m e c$, whence $c = ec$ and e is a left unit. By symmetry (or merely by assumption) there is also a right unit, which must equal e. Then $a^{n-m-1} b$ is the inverse of a. $\square$

Thus any stable associative ring without left or right zero-divisors is a division ring. In particular, a type-definable subring of a stable field is a subfield.

Lemma 1.0.2 *Let σ be an automorphism of the stable structure $\mathfrak{M}$. Then for any definable subset A of M the image σA cannot be properly contained in A. The same conclusion holds if σ is only a permutation of M, but lies in a definable group of permutations.*

Proof: Suppose A is defined by a formula $\varphi(x, a)$. If $\sigma A \subset A$, then the sequence $a, \sigma(a), \sigma^2(a), \ldots$ would be ordered by the formula $\forall x \, [\varphi(x, x_1) \rightarrow \varphi(x, x_2)]$.

If σ lies in a definable group of permutations, then the images $\sigma^i(A)$ need no longer be defined by the formula $\varphi(x, \sigma^i(a))$. However, they are defined uniformly by the formula $\psi(x, \sigma^i) := \exists y \, [\sigma^i(y) = x \wedge \varphi(x, a)]$, so $\forall x \, [\psi(x, x_1) \rightarrow \psi(x, x_2)]$ orders the powers of σ. $\square$

As a first application we show the following.

Theorem 1.0.3 *An infinite group G with only one non-trivial conjugacy class is unstable.*

Proof: If any group element has order 2, the group has exponent 2, is abelian, and has infinitely many conjugacy classes. Hence there is some non-trivial element g in G with $g \neq g^{-1}$. But both g and g^{-1} lie in the same conjugacy class, so there is h in G with $g^h = g^{-1}$. Now h^2 centralizes g but h does not, whence $C_G(h^2)$ properly contains $C_G(h)$. As h is conjugate to g and cannot have order 2 either, there must be some k in G with $h^k = h^2$ and we get $C_G(h) < C_G(h^k) < C_G(h^{k^2}) < \cdots$, contradicting stability. $\square$

We shall now look at chain conditions on families of subgroups of a stable group.

Definition 1.0.3   • The groups in a family $\{H_i : i \in I\}$ of subgroups of G are *uniformly definable* if there is a single formula φ such that all H_i in the family are φ-definable.

• The *ucc* is the uniform chain condition on families of uniformly definable subgroups: for any formula φ there is some $m_\varphi < \omega$ such that any chain of φ-definable subgroups of G has length at most m_φ.

• The *icc* is the uniform chain condition on intersections of uniformly definable subgroups: for any formula φ there is some $n_\varphi < \omega$ such that any chain of intersections of φ-definable subgroups of G has length at most n_φ.

Note that by uniformity, both chain conditions are ascending as well as descending.

Proposition 1.0.4 *Let G be a group without the strict order property. Then G satisfies the ucc.*

Proof: Suppose there were a formula $\varphi(x, \bar{y})$ and parameters $\bar{a}_i$ for $i < \omega$, such that $\varphi(x, \bar{a}_i)$ defined a subgroup H_i of G, with H_{i+1} properly contained in H_i for all $i < \omega$. Then the formula $\forall x\,[\varphi(x, \bar{x}_1) \to \varphi(x, \bar{x}_2)]$ defines an order on the set $\{\bar{a}_i : i < \omega\}$. $\square$

We can also say something about intersections of uniformly definable groups.

Theorem 1.0.5 *Let G be a group without the independence property. Then for any formula $\varphi(x, \bar{y})$ there is some number n such that any intersection of a finite family $\{H_1, \dots, H_m\}$ of φ-definable subgroups of G is the intersection of a subfamily of cardinality at most n.*

Proof: Suppose otherwise. Then there are a formula φ and for all $m < \omega$ a family $\{H_i = \varphi(G, \bar{a}_i) : 1 \leq i \leq m\}$ of subgroups of G, such that the intersection of the family does not equal a proper subintersection. So for any $1 \leq i \leq m$ there is an element $b_i \in \bigcap_{j \neq i} H_j - H_i$, and for any subset $I \subseteq \{1, \dots, m\}$ the element $b_I = \prod_{i \in I} b_i$ lies in $\bigcap_{j \notin I} H_j - \bigcup_{j \in I} H_j$ (where $b_\emptyset = 1$). So $\varphi(b_I, \bar{a}_i)$ holds iff $i \notin I$; by compactness φ has the independence property. $\square$

Corollary 1.0.6 *Let G be a group without the independence property. Then there is a finite bound n on the number of pairwise commuting non-abelian normal subgroups of G. In particular, there are only finitely many minimal, or $\bigwedge$-definable minimal, normal non-abelian subgroups.*

Proof: Let $\{N_i : i \in I\}$ be a family of pairwise commuting non-abelian subgroups, and let $n_i \in N_i - Z(N_i)$ for $i \in I$. Then $\bigcap_{i \in I} C_G(n_i)$ does not equal a proper subintersection, as for any $i \in I$ the element n_i centralizes $\bigcap_{j \in I} C_G(n_j)$, but not $\bigcap_{j \neq i} C_G(n_j) \geq N_i$. By Theorem 1.0.5 the cardinality of I must be bounded by some finite n.

Now if N and M are two distinct non-abelian ($\bigwedge$-definable) normal subgroups, then $[N, M] \leq M \cap N$. But by minimality $M \cap N$ must be a proper subgroup of both M and N, and hence trivial. $\square$

Corollary 1.0.7 *A stable group has the icc.*

Proof: If $\{H_i : i \in I\}$ is a family of uniformly definable subgroups of G, then by Theorem 1.0.5 there is $n < \omega$ such that the intersection of any finite subfamily equals the subintersection of at most n of its members. But intersections of n members of $\{H_i : i \in I\}$ form a family of uniformly definable subgroups of G, which has a minimal element $H = \bigcap_{j=1}^{n} H_{i_j}$, by the ucc. So for any H_i (with $i \in I$) the intersection $H \cap H_i$ is again an intersection of n subgroups in $\{H_{i_1}, \ldots, H_{i_n}, H_i\}$, and must contain H by minimality. It follows that H_i contains H, and $H = \bigcap_{i \in I} H_i$. $\square$

Definition 1.0.4 A group has the κcc if any chain of $\bigwedge$-definable subgroups has cardinality less than κ.

Corollary 1.0.8 *In a stable group G any $\bigwedge$-definable group $H = \bigcap_{i \in I} H_i$ equals a subintersection of size at most $|T|$. Furthermore, G satisfies the $|T|^+ cc$.*

Proof: For any formula $\varphi(x, \bar{y})$ we consider the intersection H_φ of all those H_i which are φ-definable. By the icc, this is a finite intersection, and clearly $H = \bigcap_{\varphi \in \mathcal{L}} H_\varphi$.

Now consider a proper chain H^i of $\bigwedge$-definable subgroups of G. If it had cardinality $|T|^+$, then for at least one formula φ there would be a subchain of the H_φ^i of cardinality $|T|^+$, contradicting the icc. $\square$

For groups, we get a $\bigwedge$-definable analogue of Lemma 1.0.2:

Lemma 1.0.9 *Let σ be a model-theoretic automorphism of the stable group G, and let H be a $\bigwedge$-definable subgroup of G. Then H cannot properly contain σH. The same conclusion holds if σ is only a group-theoretic automorphism, but lies in a definable group of automorphisms of G.*

Proof: Suppose $H > \sigma H$, and consider a formula $\varphi(x, \bar{a})$ defining a group which contains σH but not H. Then the intersection H_φ of all φ-definable groups containing H is mapped by σ to the intersection of all φ-definable groups containing σH, which is a proper subset, contradicting Lemma 1.0.2.

If σ lies in a definable group Σ of automorphisms of G, then we may replace H_φ by the intersection $\hat{H}$ of all conjugates of φ-definable groups under Σ which contain H. Defining $\widehat{\sigma(H)}$ similarly, we see that σ maps $\hat{H}$ to $\sigma(\hat{H}) = \widehat{\sigma(H)} < \hat{H}$; this contradicts Lemma 1.0.2. $\square$

Under stronger stability assumptions we get stronger chain conditions (see e.g. [95]):

Fact 1.0.10 • *A totally transcendental group satisfies the ωdcc: any properly descending chain of definable subgroups has finite length.*

• *A superstable group satisfies the ωdcc⁰: any descending chain of definable subgroups, each of infinite index in its predecessor, has finite length.*

• *A group of finite U-rank satisfies the ωacc⁰: any ascending chain of definable groups, each of infinite index over its predecessor, has finite length.*

Obviously, the ωdcc implies that any $\bigwedge$-definable group is definable, and the ωdcc⁰ implies that for any $\bigwedge$-definable group H there is a definable supergroup H_0 such that H is the intersection of subgroups of finite index in H_0.

We should remark that the chain conditions are clearly inherited by definable subgroups and quotients by definable subgroups. Less clear is the following proposition, which shows that the icc is also preserved under taking traces of uniformly definable groups on any substable set.

Proposition 1.0.11 *If G is a group with icc and S an arbitrary subset of G, then chains of intersections of uniformly definable subgroups with S will also satisfy the uniform chain condition.*

Proof: Suppose $H_i \cap S$ is a strictly ascending chain for $i \in I$, where the H_i are uniformly definable. Then the chain $(\bigcap_{j \geq i} H_j)_{i \in I}$ is strict as well, and uniformly definable by the icc. By the ucc, I must be finite. $\square$

Finally we shall remark that unstable groups may also satisfy various chain conditions.

Example 1.0.1 A structure $\mathfrak{M}$ is *o-minimal* if the domain M carries a linear order, and every definable subset is a finite union of open intervals with endpoints in $M \cup \{\pm\infty\}$, and points. Examples of such structures are dense linear orders, real closed fields (Tarski [109]), real closed fields with exponentiation (Wilkie [126]), and restricted analytic fields with exponentiation (van den Dries, Macintyre and Marker [111]). O-minimality implies a cell-decomposition theorem for higher-dimensional definable subsets, and in particular the existence of a well-behaved dimension theory. It follows that

groups interpretable in an o-minimal theory must satisfy the ωdcc and the
ωacc^0. (In fact, Pillay [85] has shown that any group interpretable in a real
closed field is a Lie group over that field.)

Finally, we should mention that there is a wider class of theories allowing a
notion of independence than the stable ones, the *simple* theories of Shelah
[106], also studied by Kim and Pillay [64, 65]. Among them are pseudo-
finite fields, or more generally bounded pseudo-algebraically-closed fields [55],
algebraically closed fields with a generic automorphism [32], and smoothly
approximable structures [36]. They do not have the strict order property and
hence satisfy the ucc; however, they may have the independence property and
need not satisfy the icc.

1.1   Connected Components and Definability

In this section we shall define the connected component of a substable group.
Furthermore, we shall consider definable versions of basic group-theoretic
properties, and generalize results of section 0.1 to the present context.

Definition 1.1.1 Let G be a substable group.

- The φ-*connected component* G^0_φ is the intersection of all relatively φ-
definable subgroups H of finite index.

- The *connected component* G^0 is the intersection of all G^0_φ, for $\varphi \in \mathfrak{L}$.

- G is φ-*connected* if it is equal to its φ-connected component; it is *connected* if it is equal to its connected component.

Warning 1.1.1 If the group is type-definable in a model which is not sufficiently saturated, it may happen that a relatively definable subgroup has
finite index only accidentally, but not in elementary superstructures. For example, we may again consider the group type-defined in $\mathbb{Z}$ via the formulæ
$\{\exists y \, ny = x : n < \omega\}$; in this particular model it reduces to the trivial group
and *every* subgroup has finite index. In order to remedy this problem for a
group H defined by a partial type π, we must either

- require the ambient model to be $|T|^+$-saturated, or

- consider only the intersection with H of a definable group such that
finitely many cosets cover some *finite* subtype π_0.

Lemma 1.1.1 *Suppose H is a definable group which intersects the substable
group G in a subgroup of index n in G. Then there is a definable supergroup
$\bar{H}$ of G such that H intersects $\bar{H}$ in a subgroup of index n in $\bar{H}$, and all the
cosets in $\bar{H}/H$ have representatives in G.*

Proof: Let H_0 be the intersection of all G-conjugates of H. By the icc this is a finite intersection of subgroups intersecting G in a subgroup of finite index, so H_0 intersects G in a subgroup of finite index, and it is normalized by G. Therefore $H_0 G =: \bar{H}$ is a definable group. But then a system of representatives of the cosets of $G \cap H_0$ in G is also a system of representatives of the cosets of H_0 in $\bar{H}$. As H contains H_0, the result follows. $\square$

It now follows that the two approaches for the connected component of a group G type-defined by some partial type π are equivalent. Suppose first that the ambient group is $|T|^+$-saturated. If H is a definable group which intersects G in a subgroup of finite index, then $\bar{H}$ contains G and hence a finite subtype π_0 by saturation, so finitely many translates of H_0 cover π_0. As H_0 is a subgroup of H, finitely many cosets of H cover π_0. Conversely, it is clear that if finitely many cosets of H cover a finite subtype π_0 of π, then H will intersect G in a subgroup of finite index in any model. We should note at this point that a definable property holds on the connected component of a type-definable group (in a saturated model) iff it holds on a subgroup of finite index.

Remark 1.1.1 By the icc every φ-connected component is a finite intersection and relatively definable, hence has finite index in G. Therefore the connected component is relatively $\bigwedge$-definable and has index at most $2^{|T|}$, so for a definable group we may go to a big model and obtain a big connected component. However, if G is merely substable or a $\bigwedge$-definable group in a model which is not sufficiently saturated, it may happen that the set of realizations (in that particular model) of the connected component is too small – it may even be trivial.

We should note that only parameters are needed which are also necessary to define G (i.e. the whole of G for a substable group, or the parameters needed for the defining formula(e) of G). To see this, we go to a $|T|^+$-saturated extension and consider an automorphism fixing those parameters. This automorphism maps a subgroup $\varphi(G, \bar{a})$ of finite index in G to a subgroup of the same form and of the same finite index. Hence G^0_φ is invariant under all those automorphisms; as it is definable relative to G, it is definable over the given parameters, and so is G^0. The terminology is borrowed from algebraic geometry: the connected component of an algebraic group is the smallest closed subgroup of finite index.

As a conjugate of a relatively definable subgroup of finite index is again relatively definable of finite index, the connected component of G is normal in G. This need not be true for a φ-connected component; however, if the conjugate of a φ-definable subgroup is again φ-definable (e.g. if φ is of the form $\varphi(y^{-1}xy, \bar{y})$), then the φ-connected component is normal as well. The connected component of a stable group is again connected. This is also valid for φ-connectivity, but the meanings of φ in G and in G_φ^0 (i.e. the relativization of φ to G_φ^0) will in general be different, unless φ is quantifier-free. Of particular interest is the formula $xy = yx$ (centralizers); the corresponding φ-connected component is called the *centralizer-connected component* G^{cc} of G. Another important formula is $x^y \in G$ (conjugation) for some (relatively) definable G, which gives rise to the *conjugacy-connected component* G^{co}. And if we choose φ to be a formula which (relatively) defines G, then G_φ^0 is the *locally connected component* G^{lo}. (This component G^{lo} may in general depend on the formula which defines G; if φ and ψ both define G, then there is no reason why G_φ^0 should equal G_ψ^0. The choice of defining formula is implicit in the use of the local component.) Note that if G is defined in terms of the group law alone (e.g. as a centralizer or normalizer), then G^{lo} is conjugacy-connected: any conjugate of G^{lo} is defined in the same way as G^{lo} itself, just using the conjugates of the parameters.

Finally, any quotient of a connected group is again connected; as for φ-connectivity, we encounter the usual problems about the relativization of φ. Conversely, for a relatively $\bigwedge$-definable normal subgroup N, connectivity of G/N and of N implies connectivity of G. And if N and H are substable groups such that H normalizes N, then it is easy to see that $(NH)^0 = N^0H^0$.

In addition to divisible groups (which do not have any proper subgroups of finite index, definable or not), the following lemma yields further examples of connected groups:

Lemma 1.1.2 *An infinite $\bigwedge$-definable group G without relatively definable proper non-trivial normal subgroups (this will be called* definably simple*), and the additive group of an infinite division ring D with icc, are connected.*

Proof: Let H be a relatively definable subgroup of finite index n in G. By Proposition 0.1.1, the intersection N of all G-conjugates of H is a subgroup of index at most $n!$, hence relatively definable as a finite subintersection. So it cannot be proper, and $n = 1$. (Note that the intersection of the conjugates of H under *all* – model-theoretic – automorphisms of G need not have finite index in general. However, these conjugates still form a family of uniformly relatively definable subgroups. Therefore, if G satisfies Theorem 1.0.5, then this intersection equals a finite subintersection, which must have finite index in G.)

Now let H be a relatively definable subgroup of finite index in D^+. The intersection I of all dH, for d in D, is a finite intersection, hence a left ideal of finite (additive) index. So it must equal D. $\square$

(In the next chapter we shall also prove the multiplicative connectivity of a substable division ring. This, however, goes beyond chain conditions; for instance the multiplicative group of the non-zero real numbers has a subgroup of index 2, namely the positive reals.)

Proposition 1.1.3 *A finite G-invariant subset A of a centralizer-connected group G is central.*

Proof: Any element a in A has only finitely many conjugates. Therefore the index of its centralizer $C_G(a)$ in G is finite; by centralizer-connectivity $C_G(a) = G$ and a is central. $\square$

Corollary 1.1.4 *A centralizer-connected group with finitely many commutators is abelian.*

Proof: For g in G the commutator set $g^G g^{-1}$ is finite, hence g^G is finite as well. As it is also G-invariant, g must be central. $\square$

We shall now consider the relative definability of various subgroups of a substable group.

Theorem 1.1.5 *Let G be an $(\bigwedge\text{-})$definable subgroup of a stable group, H a $(\bigwedge\text{-})$definable subgroup of G, and A a subset of G. Then the following groups are $(\bigwedge\text{-})$definable as well:*

1. *for any $g \in G$ the conjugate H^g,*

2. *the centralizer of A in G,*

3. *the normalizer of H in G,*

4. *the elements $Z_i(G)$ of the ascending central series of G for $i < \omega$, and*

5. *the iterated centralizers $C_G^i(A/H)$ of A in G modulo H, for $i < \omega$.*

Furthermore, if G is merely substable and H is relatively $(\bigwedge\text{-})$definable, then the groups in 1.–5. are relatively $(\bigwedge\text{-})$definable.

Proof: 1. is obvious: We replace x by gxg^{-1} in the relevant formulæ.

3. is obvious for definable H. So suppose $G = \bigcap G_i$ and $H = \bigcap H_i$, and let $\bar{H}_i$ be the intersection of all conjugates of H_i which contain H. Then $\bar{H}_i$ is a definable supergroup of H whose normalizer $N_{G_i}(\bar{H}_i)$ contains the normalizer $N_{G_i}(H)$, for all i. Furthermore H is also the intersection of all $\bar{H}_i$, whence $N_G(H) = \bigcap_{i,j} N_{G_j}(\bar{H}_i)$.

2. and 4. are special cases of 5., and 5. follows inductively from the first step, together with definability of the normalizers. Firstly, consider definable groups G and H. Then $C_G(A/H) = \bigcap_{a \in A} C_G(a/H)$ is definable by the icc, as all centralizers $C_G(a/H)$ are uniformly defined by the formula $x \in N_G(H) \wedge x^{-1}a^{-1}xa \in H$. Now for $G = \bigcap G_i$ and $H = \bigcap H_j$ we get $C_G(a/H) = \bigcap C_{G_i}(a/H_j)$.

If G is merely substable, then 1. is again obvious, as for $g \in G$ clearly $h \in G$ iff $h^g \in G$. For 5. we note that if $x, a \in G$, then necessarily $[x, a] \in G$, so we just have to intersect the centralizer formula with G. As for the normalizer, if $H = G \cap K$ for some definable group K, we consider $\bar{K} := \bigcap_{g \in N_G(H)} K^g$. Then $H = G \cap \bar{K}$ and the normalizer of $\bar{K}$ intersects G in $N_G(H)$. The case of relatively $\bigwedge$-definable H follows. $\square$

Note that we do not need any new parameters for the definitions apart from those needed for G, H, A, or g. We should also remark that the elements of the derived series and the descending central series, and the transfinitely iterated ascending central subgroups and centralizers, will in general not be relatively definable.

We can now prove the promised lemma on the substability of the quotient of a substable group by a relatively definable normal subgroup.

Lemma 1.1.6 *Let N be a relatively definable normal subgroup of a substable group G. Then G/N is substable.*

Proof: Let $\bar{N}$ be a definable group intersecting G in N, and consider $\bar{H} := \bigcap_{g \in G} \bar{N}^g$. Then $\bar{H}$ is definable by the icc, and intersects G in N. Furthermore, the normalizer K of $\bar{H}$ is definable and contains G. Therefore $G/N \cong G\bar{H}/\bar{H} \leq K/\bar{H}$. $\square$

Definition 1.1.2 Let G be a substable group. The *definable hull* $\mathrm{dc}(G)$ is the intersection of all definable supergroups of G (in the ambient stable group).

It follows immediately from the icc that the normalizer of $\mathrm{dc}(G)$ contains the normalizer of G, for any substable group G. Furthermore, Lemma 1.1.1 implies that $\mathrm{dc}(G)^0$ is the intersection of all definable subgroups which intersect G in a subgroup of finite index. It intersects G in G^0, but whereas G^0 can be too small and even trivial, $\mathrm{dc}(G)^0$ will have the right definable properties.

Lemma 1.1.7 *Let G and H be substable in $\mathcal{G}$. If K is a relatively definable subgroup of finite index in $[G, H]$, then there is a relatively definable subgroup L of finite index in G with $[L, H] \leq K$. In particular, $\mathrm{dc}[G, H]^0$ contains $[\mathrm{dc}(G)^0, \mathrm{dc}(H)]$, and if G is connected, then $\mathrm{dc}[G, H]$ is connected.*

Proof: Let K be a relatively definable subgroup of $[G, H]$ of finite index. By the icc the intersection of all GH-conjugates of K is a finite subintersection, and still has finite index in $[G, H]$; we may thus assume that K is GH-invariant. Then for all elements h in H the set $[h, G]K = h^{-1}h^G K$ is finite, and so is $h^G K$. Hence $C_G(h/K)$ has finite index in G, and so does $L := C_G(H/K)$ by the icc. Clearly $[L, H] \leq K$.

It now follows that $\mathrm{dc}(G)^0$ centralizes H modulo every definable supergroup of $\mathrm{dc}[G, H]^0$, and H is contained in $C_{\mathcal{G}}(\mathrm{dc}(G)^0/\mathrm{dc}[G, H]^0)$, and so is $\mathrm{dc}(H)$. The result follows. $\square$

Corollary 1.1.8 *If $(G_i : i < n)$ is an abelian (central) series for G, then $(G_i^0 : i < n)$ is an abelian (central) series for G^0.*

Proof: Immediate from Lemma 1.1.7. $\square$

Lemma 1.1.9 *Let G be a substable nilpotent group and H a relatively $\bigwedge$-definable subgroup of infinite index. Then H has infinite index in its normalizer in G.*

Proof: As H has infinite index in G, there is a minimal k such that the index $|Z_k(G) : H \cap Z_k(G)|$ is infinite. Then $|Z_{k-1}(G) : H \cap Z_{k-1}(G)|$ is finite. But a relatively $\bigwedge$-definable subgroup of finite index in a relatively definable subgroup is relatively definable; since $Z_{k-1}(G)$ is G-invariant, the intersection $N := \bigcap_{g \in G} (H \cap Z_{k-1}(G))^g$ is a relatively definable G-invariant subgroup of finite index in $Z_{k-1}(G)$.

Since G/N is substable, by Lemma 1.1.7 there is a relatively definable subgroup K of finite index in $Z_k(G)$ with $[K, G] \leq N \leq H$; this yields $[K, H] \leq H$ and $K \leq N_G(H)$. As H has infinite index in Z_k, it has infinite index in $N_G(H)$. $\square$

Remark 1.1.2 If G is an abelian group, then

- $G[q]$ is definable,

- G_p is definable iff there is a bound on the exponent of p-elements,

- $G_{p'}$ is definable iff there is a bound on the exponent of p'-elements (i.e. elements of finite order coprime to p).

Definition 1.1.3 A group G is

- *definably soluble* of derived length n, if there is a sequence $G = G_0 \geq G_1 \geq \cdots \geq G_n = \{1\}$ of relatively definable normal subgroups with abelian quotients G_i/G_{i+1} for all $i < n$,

- *definably nilpotent* of class n, if there is a sequence $G = G_0 \geq G_1 \geq \cdots \geq G_n = \{1\}$ of relatively definable normal subgroups, such that G_i/G_{i+1} is central in G/G_{i+1} for all $i < n$,

- *definably hypercentral* of length α, if there is an ascending sequence $(G_i : i \leq \alpha)$ of relatively definable normal subgroups of G, with $\{1\} = G_0$ and $G = G_\alpha$, whose quotients G_{i+1}/G_i are central in G/G_i for all $i < \alpha$, and such that $G_\lambda \leq \bigcup_{i < \lambda} G_i$ for any limit ordinal $\lambda \leq \alpha$.

A subgroup H of G is *definably characteristic* in G, if it is invariant under all definable automorphisms of G.

Theorem 1.1.10   *1. Let S be a soluble substable group. Then there is a definable, definably soluble supergroup S_0 of S of the same derived length, and the defining formula only depends on the derived length.*

2. Let N be a nilpotent substable group. Then there is a definable, definably nilpotent supergroup N_0 of N of the same class, and the defining formula only depends on the nilpotency class.

Proof:

1. We call the ambient stable group $\mathcal{G}$, and use induction on the derived length of S. If it is zero, then $S = \{1\}$ and the result is trivial. So suppose not, and consider the last non-trivial iterated commutator subgroup A of S. Then $B := Z(C_{\mathcal{G}}(A))$ is definable (by a formula independent of A), abelian, contains A, and is $N_{\mathcal{G}}(A)$-invariant. In particular, the normalizer $N_{\mathcal{G}}(B)$ contains S, so $N_{\mathcal{G}}(B)/B$ is a definable group containing SB/B, which has smaller derived length than S. By the inductive hypothesis there is a (uniformly in the derived length) definable supergroup C/B of SB/B in $N_{\mathcal{G}}(B)/B$, which is definably soluble of the same derived length as SB/B. The pre-image C is the required group.

2. The case of a nilpotent group is similar; alternatively we could use the relative definability of the ascending central series. $\square$

Corollary 1.1.11 *A soluble (nilpotent) substable group G is definably soluble (nilpotent).* $\square$

Theorem 1.1.12 *Let G be a substable group and $\mathfrak{N} = \{N_i : i \in I\}$ a family of normal nilpotent subgroups of class k. Then $\mathfrak{N}$ generates a nilpotent (normal) subgroup. The result also holds with soluble instead of nilpotent.*

Proof: By Theorem 1.1.10 the N_i are contained in uniformly definable subgroups $\bar{N}_i$; after replacing every $\bar{N}_i$ by the intersection of its G-conjugates, they are all normalized by G. But then the intersection $\bar{G}$ of all normalizers of the $\bar{N}_i$ (for $i \in I$) is definable and contains G. In particular we may replace G by $\bar{G}$ and N_i by $\bar{N}_i \cap \bar{G}$ and assume that G is stable and that the N_i are uniformly definable normal subgroups.

We first treat the nilpotent case. We put $N_i^j = Z_{k-j+1}(N_i)$, and $N_i^0 = G$. Then N_i^j is normal and uniformly definable for all $j \leq k + 1$ and $i \in I$, and nilpotent of class at most $(k - j + 1)$ for $j > 0$. Let n be a bound for the length of a descending chain of intersections of the N_i^j, and consider a sequence $(a_s)_{s<\omega}$ with $a_s \in N_{i(s)}$. Put $b_0 = a_0$ and $b_{s+1} = [b_s, a_{s+1}]$.

If t is such that $b_s \in N^t_{i(s+1)} - N^{t+1}_{i(s+1)}$, then $b_{s+1} \in N^{t+1}_{i(s+1)}$, and because of normality b_{s+1} lies in all N^j_i which contain b_s. Hence $\bigcap \{N^j_i : b_s \in N^j_i\}$ properly contains $\bigcap \{N^j_i : b_{s+1} \in N^j_i\}$. But the sequence of these intersections can descend at most n times, whence $b_n \in N^{k+1}_{i(n+1)}$, which is trivial. As any commutator in elements of $\langle \mathfrak{N} \rangle$ can be written as a product of (conjugates of) commutators whose entries come from the N_i and whose length is not smaller, it follows that $\langle \mathfrak{N} \rangle$ is nilpotent of class n.

In the soluble case, we use induction on the derived length, the case of abelian groups having just been dealt with. So we may assume that the commutator subgroups $\{N'_i : i \in I\}$ generate a soluble normal subgroup N, which we may assume to be definable. But then $\{N_i N / N : i \in I\}$ is a family of normal abelian subgroups of G/N and generates a nilpotent group; the result follows. $\square$

In particular, a stable group G has a normal definable soluble subgroup $R_k(G)$ containing all normal soluble subgroups of derived length at most k. Similarly there is a definable normal nilpotent subgroup $F_k(G)$ containing all normal nilpotent subgroups of class at most k. Therefore the soluble radical is the union of an ascending sequence of characteristic soluble subgroups and the Fitting subgroup is the union of an ascending sequence of characteristic nilpotent subgroups. In the next section we shall prove that the Fitting subgroup $F(G)$ is nilpotent, so the latter sequence $(F_k)_{k<\omega}$ eventually becomes stationary at $F(G)$. As the proof will not make use of Theorem 1.1.12, we shall obtain a second, and in fact more general, proof of that theorem. The analogous question for the soluble radical, however, remains open.

Surprisingly, we can still prove nilpotency of the group generated by a family of normal subgroups, even if we do not assume that the normal subgroups are nilpotent, provided their intersection is trivial. We should note that this is the only theorem in this section which uses stability in a stronger form than the icc; it will not be used in the sequel and may safely be skipped at a first reading.

Theorem 1.1.13 *Let $H(a)$ be an a-definable normal subgroup of a stable group, and suppose that for any infinite independent family $\{a_i \models \mathrm{tp}(a) : i < \omega\}$ the intersection $\bigcap_{i<\omega} H(a_i)$ is trivial. Then $H(a)$ is nilpotent.*

Proof: Suppose otherwise. We clearly may assume that $\mathrm{tp}(a)$ is stationary, as the hypothesis holds for any stationarization of it, and the nilpotency class of $H(a)$ then forms part of $\mathrm{tp}(a)$. Furthermore, we may work in a saturated model G of our ambient stable group. Assume for a contradiction that $H(a)$ is not nilpotent.

Let $H = \langle H(a') : a' \models \mathrm{tp}(a) \rangle$ be the group generated by all $\emptyset$-conjugates of $H(a)$, and let $Z := Z_\omega(H)$ be its ω-centre. Note that H is in general not at all definable, but its iterated centres are relatively definable as iterated centralizers, as in any substable group. By the icc there is a minimal intersection N of conjugates of $H(a)$ not contained in Z (where the parameters for the conjugates in the intersection are not necessarily independent). If $N \leq H(a')$ for some a' realizing p, then a' must fork with the parameters needed for the definition of N (otherwise independent conjugates of $H(a')$ over N would be independent over $\emptyset$, with non-trivial intersection). Hence $[N, H(a')] \leq Z$ for generic a' by minimality of N, and by compactness this commutator group is already contained in some iterated centralizer $C := C^i_G(H)$ of H.

But now consider two series $(a_i)_{i<\omega}$ and $(b_i)_{i<\omega}$ of independent realizations of $\mathrm{tp}(a)$. Let $(c_i)_{i<\omega}$ be a third such series which is independent of $(a_i)_\omega \cup (b_i)_\omega$. Then both $(a_i)_\omega {}^\frown (c_i)_\omega$ and $(b_i)_\omega {}^\frown (c_i)_\omega$ are series of independent realizations of $\mathrm{tp}(a)$; in particular they are indiscernible series of the same type. But there is n such that the centralizer modulo C of such a series is the centralizer of any n elements, which we may take in $(a_i)_\omega$, or $(c_i)_\omega$, or $(b_i)_\omega$. Therefore $(a_i)_\omega$ and $(b_i)_\omega$ have the same centralizer modulo C. It follows that the centralizer modulo C of any $H(a'')$ with $a'' \models \mathrm{tp}(a)$ contains the centralizer modulo C of the series $(a_i)_\omega$, say. Hence the centralizer modulo C of H, which is the centralizer modulo C of $\{H(a'') : a'' \models \mathrm{tp}(a)\}$, equals the centralizer modulo C of a subfamily independent of the parameters needed for the definition of N. Therefore it must contain N, contradicting the choice of N. $\square$

Note that by Theorem 1.1.12 the group H generated by all $H(a')$, for $a' \models p$, will also be nilpotent.

If G were superstable and φ a formula such that for any subformula $\psi \subseteq \varphi$ of the same Shelah rank the intersection $\bigcap\{H(a) : a \models \psi\}$ were trivial, then any completion of φ to a type of the same rank would satisfy the hypotheses of the theorem, so groups $H(a')$ with a' of maximal rank are nilpotent. But now by compactness there is a bound n on the nilpotency class (to have maximal rank is a closed condition on $\mathrm{tp}(a)$, while "$H(a)$ is nilpotent" is open), and there exists a formula ψ of lower rank such that $H(a')$ is nilpotent of class n for any $a' \models \varphi \wedge \neg\psi$.

Definition 1.1.4 A subgroup H of G is *definably subnormal* if there is a sequence $G = G_0 \rhd G_1 \rhd \cdots \rhd G_n = H$ of relatively $\bigwedge$-definable subgroups.

Lemma 1.1.14 *Let G be substable and H a relatively $\bigwedge$-definable subnormal subgroup of G. Then H is definably subnormal in G.*

Proof: Suppose $G = G_0 \rhd G_1 \rhd \cdots \rhd G_n = H$ is a series connecting G and H, and put $\bar{G}_i = \mathrm{dc}(G_i)$. Then $\bar{G}_0 = \mathrm{dc}(G)$ and $\bar{G}_n \cap G = H$. Furthermore, G_i is contained in the normalizer of $\bar{G}_{i+1}$. But the latter is $\bigwedge$-definable, whence $\bar{G}_i$ normalizes $\bar{G}_{i+1}$. Now the intersections $\bar{G}_i \cap G$ witness the definable subnormality of H. $\square$

Lemma 1.1.15 *Let N and H be substable groups, with H normalizing N. Then $\mathrm{dc}(NH) = \mathrm{dc}(N)\mathrm{dc}(H)$.*

Proof: First we note that $\mathrm{dc}(N)$ is normalized by H; as its normalizer is $\bigwedge$-definable, $\mathrm{dc}(H)$ must normalize $\mathrm{dc}(N)$. Clearly, $\mathrm{dc}(N)$ and $\mathrm{dc}(H)$ are both contained in $\mathrm{dc}(NH)$, whence $\mathrm{dc}(N)\mathrm{dc}(H) \leq \mathrm{dc}(NH)$. On the other hand, $NH \leq \mathrm{dc}(N)\mathrm{dc}(H)$ and the latter is an $\bigwedge$-definable group, so $\mathrm{dc}(NH) \leq \mathrm{dc}(N)\mathrm{dc}(H)$. $\square$

Definition 1.1.5 A group G is

- *definably simple* if G has no proper non-trivial relatively definable normal subgroups;

- *definably quasi-simple* if $G \leq \mathrm{dc}(G')$ and $G/Z(G)$ is definably simple; and

- *definably semi-simple* if $G \leq \mathrm{dc}(G')$ and $G/Z(G)$ decomposes as a direct product of finitely many non-abelian definably simple groups.

We shall normally consider $\bigwedge$-definable groups G, in which case the inequalities become equalities: $G = \mathrm{dc}(G')$.

Lemma 1.1.16 *Let G be definably semi-simple and $\bigwedge$-definable. Then G can be uniquely decomposed as a central product of definably quasi-simple $\bigwedge$-definable groups G_i. If N is a subnormal $\bigwedge$-definable subgroup of G, then there is a definably semi-simple group H (a finite product of some of the G_i) such that G is the central product of H and N. In particular N is normal in G. Furthermore, $\mathrm{dc}(N')$ is definably semi-simple, and G is also a central product of H and $\mathrm{dc}(N')$. Any soluble normal subgroup S of G is central.*

The proof of this is very similar to the proof of the corresponding fact (Theorem 0.1.25) for finite groups, but we have to check that the groups in question are $\bigwedge$-definable. Remember that Lemma 0.1.2 gives that $[K, L, K] = \{1\}$ implies $[K', L] = \{1\}$.

Proof: Consider a decomposition of $G/Z(G)$ into definably simple groups $\bar{G}_i/Z(G)$, and consider $G_i = \mathrm{dc}(\bar{G}_i')$. For $i \neq j$ the commutator subgroup $[\bar{G}_i, \bar{G}_j]$ is contained in $Z(G)$, so $[\bar{G}_i, \bar{G}_j, \bar{G}_j] = 1$. Hence $[\bar{G}_j', \bar{G}_i] = 1$ and $\bar{G}_j'$ is centralized by $\bar{G}_i$. Therefore G_j is contained in $C_G(\bar{G}_i)$; as $G_i \leq \bar{G}_i$, the product $\prod_i G_i$ is central.

The $\bar{G}_i$ commute modulo $Z(G)$, so

$$G = \mathrm{dc}(G') = \mathrm{dc}((\prod_i \bar{G}_i)') = \mathrm{dc}(\prod_{i,j}[\bar{G}_i, \bar{G}_j])$$
$$= \prod_{i,j} \mathrm{dc}[\bar{G}_i, \bar{G}_j] \leq Z(G)\prod_i \mathrm{dc}(\bar{G}_i').$$

Since the G_i commute, it follows that

$$G = \mathrm{dc}(G') \leq \mathrm{dc}((Z(G)\textstyle\prod_i G_i)') = \mathrm{dc}(\textstyle\prod_{i,j}[G_i, G_j])$$
$$= \mathrm{dc}(\textstyle\prod_i G_i') = \textstyle\prod_i \mathrm{dc}(G_i') \leq \textstyle\prod_i G_i \leq G,$$

and clearly equality must hold.

Now $\bar{G}_i/Z(G)$ is non-abelian and definably simple, and therefore $\bar{G}_i = Z(G)\mathrm{dc}(\bar{G}_i') = Z(G)G_i$. Hence $G_i = \mathrm{dc}(\bar{G}_i') = \mathrm{dc}((Z(G)G_i)') = \mathrm{dc}(G_i')$. Furthermore $\bar{G}_i/Z(G)$ is isomorphic to $G_i/C_{G_i}(G)$. But $C_{G_i}(G) \leq Z(G_i)$, and definable simplicity of $\bar{G}_i/Z(G)$ implies $C_{G_i}(G) = Z(G_i)$. Hence $G_i/Z(G_i)$ is non-abelian and definably simple, and G_i is definably quasi-simple.

Let now N be an $\bigwedge$-definable normal subgroup of G. Let I be the set of indices i such that $N \geq G_i$. Then $N \cap G_i \leq Z(G_i)$ for any $i \notin I$ by definable quasi-simplicity of G_i, so $\mathrm{dc}[N, G_i] \leq N \cap G_i \leq Z(G_i)$, whence $[N, G_i, G_i] = 1$ and therefore $1 = [N, G_i'] = [N, \mathrm{dc}(G_i')] = [N, G_i]$. So N and G_i commute. It follows that for $H = \prod_{i \notin I} G_i$, the group G is a central product of H and N. Hence $N \leq C_G(H) = Z(G)\prod_{i \in I} G_i$, so $\mathrm{dc}(N')$ equals $\prod_{i \in I} G_i$ and must be definably semi-simple. Furthermore, G is the central product of H and $\mathrm{dc}(N')$ and $N = \mathrm{dc}(N')C_N(G)$. This also proves the uniqueness of the decomposition of G: a minimal $\bigwedge$-definable normal non-abelian subgroup of G must be one of the G_i.

If M is an $\bigwedge$-definable normal subgroup of N, then by semi-simplicity of $\mathrm{dc}(N')$ we get that $\mathrm{dc}(M')$ is definably semi-simple and equals a finite product of some of the G_i. In particular it is normal in G. Furthermore M commutes with all the other G_i contained in $\mathrm{dc}(N')$. As N commutes with H, we get that M commutes modulo $\mathrm{dc}(M')$ with all G_i. Hence M itself is normal in G, and any subnormal $\bigwedge$-definable subgroup of G is normal.

Finally, let S be a soluble normal subgroup of G. By Theorem 1.1.10 we may assume that S is actually $\bigwedge$-definable. As S cannot contain any definably quasi-simple group, it must commute with all the G_i. Therefore S is central in G. $\square$

Definition 1.1.6 Let G be a group.

- The *definable socle* $S(G)$ of G is the group generated by all minimal normal $\bigwedge$-definable subgroups of G.

- The *definable layer* $E(G)$ is the group generated by all $\bigwedge$-definable definably semi-simple normal subgroups.

In Definition 0.1.13 we had defined $S(G)$ and $E(G)$ to stand for the socle and the layer of G, respectively, without definability assumptions. It will be clear from the context which version we shall mean. Since the central product of two normal definably semi-simple groups is again definably semi-simple (see

Lemma 0.1.27), the definable layer is definably semi-simple iff there is a finite bound on the number of factors.

Lemma 1.1.17 *Let G be a stable $\bigwedge$-definable group without abelian normal subgroups. Then the socle $S(G)$ is $\bigwedge$-definable and definably semi-simple. In fact, $S(G)$ decomposes as a direct product of non-abelian definably simple groups, and will contain all definably semi-simple normal subgroups.*

Proof: Note first that by the $|T|^+$cc every $\bigwedge$-definable normal subgroup contains a minimal $\bigwedge$-definable normal subgroup (which is non-trivial in a $|T|^+$-saturated model). As minimal $\bigwedge$-definable normal subgroups are non-abelian by assumption, there are only finitely many of them by Corollary 1.0.6, and $S(G)$ is their direct product.

Let N be a minimal normal $\bigwedge$-definable subgroup. If N is infinite, it is connected, lies in the connected component G^0, and contains a minimal $\bigwedge$-definable G^0-invariant subgroup. On the other hand, if N is finite, by Proposition 1.1.3 it is centralized by G^0. But as there are no abelian normal subgroups, $C_G(G^0)$ must intersect G^0 trivially. So $C_G(G^0)$ is a finite normal subgroup of G and possesses a maximal direct product S_0 of minimal normal subgroups of G. Furthermore S_0 decomposes as a finite direct product of simple groups, and $S(G) = S(G)^0 \times S_0$.

Consider a minimal $\bigwedge$-definable normal subgroup N of G^0. The G-conjugates of an abelian normal subgroup of N would generate a G-invariant soluble group by Theorem 1.1.12, contradicting our assumption on G. Hence the socle $S(N)$ of N, being characteristic in N and thus normal in G^0, must equal N by minimality of N. Therefore N decomposes into finitely many minimal $\bigwedge$-definable N-invariant subgroups, which are permuted by G, hence normalized by G^0. Again by minimality N must be definably simple. It follows that $S(G^0)$ is definable and definably semi-simple. But then $S(G^0)$ is G-invariant and the unique decomposition into definably quasi-simple G^0-invariant subgroups may be coarsened to a decomposition into minimal $\bigwedge$-definable G-invariant subgroups: $S(G^0) = S(G)^0$.

Let now N be a definably semi-simple normal subgroup of G. If N were not contained in the socle $S(G)$, then consider the product M of all definably quasi-simple normal subgroups of N which are not contained in $S(G)$. Then M is normal in G. Now $M \cap S(G)$ is central in M, but the centre $Z(M)$ is a normal abelian subgroup of G and must be trivial. It follows that $M \leq C_G(S(G))$, and the centralizer $C_G(S(G))$ contains a minimal $\bigwedge$-definable normal subgroup $\bar{M}$, which must be contained in the socle. Thus $\bar{M}$ is abelian, a contradiction. $\square$

Lemma 1.1.18 *A definably semi-simple normal subgroup N of a stable group G commutes with every normal soluble subgroup S. In particular, N is centralized by the Fitting subgroup $F(G)$.*

Proof: Consider $[N, S]$. This must be a normal soluble subgroup of N and hence contained in the centre $Z(N)$. Therefore $[N, S, N] = \{1\}$, whence $[N', S] = \{1\}$ and N' is centralized by S. It follows from definable semi-simplicity that N is centralized by S. $\square$

Theorem 1.1.19 *Let G be an $\bigwedge$-definable stable group. Then the definable layer $E(G)$ is $\bigwedge$-definable. There are only finitely many definably quasi-simple $\bigwedge$-definable subnormal subgroups of G. They generate $E(G)$, and are normalized by G^0.*

Proof: Let F be the Fitting subgroup of G. Then $C_G(F)$ and $Z(F)$ are both relatively definable, and by Lemma 0.1.9 any soluble subgroup of $C_G(F)$ is contained in $Z(F)$. Lemma 1.1.17 now yields an $\bigwedge$-definable, definably semi-simple group $\bar{E}/Z(F) := S(C_G(F)/Z(F))$. We put $E := \mathrm{dc}(\bar{E}')$. Then $Z(F)$ is central in $\bar{E}$, and in fact $Z(F) = Z(\bar{E})$ is the (unique) maximal normal soluble subgroup of $\bar{E}$. Definable semi-simplicity of $\bar{E}/Z(F)$ yields $\bar{E} = \mathrm{dc}(\bar{E}')Z(F) = EZ(F)$, whence

$$\mathrm{dc}(E') = \mathrm{dc}((EZ(F))') = \mathrm{dc}(\bar{E}') = E.$$

Furthermore $\bar{E}/Z(F)$ is isomorphic to $E/(E \cap Z(F))$; as $E \cap Z(F) = E \cap Z(\bar{E}) \leq Z(E)$ is normal abelian, we get equality, and E is definably semi-simple.

Now let N be a definably semi-simple $\bigwedge$-definable normal subgroup of G. We claim that N is contained in $\bar{E}$. So suppose not. Then $N \cap \bar{E}$ is normal in N, so there is a non-trivial definably semi-simple $\bigwedge$-definable group H such that N is the central product of $N \cap \bar{E}$ and H. In particular $[H, \bar{E}]$ must be central in H, whence by the Three Subgroup Lemma $[H', \bar{E}] = 1$. It follows that $H = \mathrm{dc}(H')$ is centralized by $\bar{E}$. Hence the centralizer of $\bar{E}$ in $C_G(F)/Z(F)$ is a non-trivial normal subgroup and contains a minimal $\bigwedge$-definable normal subgroup $\bar{H}/Z(F)$, which must be contained in the socle $\bar{E}/Z(F)$. But then $\bar{H}$ is abelian and must be contained in the soluble radical of $\bar{E}$, which is $Z(F)$, a contradiction. Therefore N is contained in $\bar{E}$, and by semi-simplicity even in E, as $N = \mathrm{dc}(N') \leq \mathrm{dc}(\bar{E}') = E$. Hence E is the definable layer $E(G)$, and is $\bigwedge$-definable.

If N is a subnormal definably quasi-simple $\bigwedge$-definable subgroup of G, then there is a sequence of $\bigwedge$-definable groups $G = G_0 \trianglerighteq G_1 \trianglerighteq \cdots \trianglerighteq G_n = N$. Now $N \leq E(G_n) \leq E(G_{n-1}) \leq \cdots \leq E(G_0) = E(G)$.

Finally, as E is definably semi-simple, it decomposes uniquely as a central product of finitely many definably quasi-simple $\bigwedge$-definable normal subgroups; these are obviously subnormal in G and normalized by G^0, and every $\bigwedge$-definable definably quasi-simple subgroup must equal one of them. $\square$

Definition 1.1.7 The *definable generalized Fitting subgroup* $F^*(G)$ of a group G is the central product of the Fitting subgroup and the definable layer.

Again, whether we shall require $\bigwedge$-definability or not for $F^*(G)$ will be clear from the context. We shall prove in the next section that the Fitting subgroup is nilpotent in any group with chain condition on centralizers. Hence it is relatively definable, and we get

Theorem 1.1.20 *Let G be an $\bigwedge$-definable stable group. Then the definable generalized Fitting subgroup $F^*(G)$ is type-definable. The centralizer of the definable generalized Fitting subgroup equals the centre of the Fitting subgroup $F(G)$. Hence $G/Z(F(G))$ embeds into $\mathrm{Aut}(F^*(G))$.*

Proof: By Theorem 1.2.11 and Theorem 1.1.19 the definable generalized Fitting subgroup is $\bigwedge$-definable. By Lemma 1.1.18 the centre of $F(G)$ centralizes the whole of $F^*(G)$. On the other hand, if $C_G(F^*(G))/Z(F(G))$ were non-trivial, it would have to contain some minimal type-definable normal subgroup N, which by definition lies in the socle $S(C_G(F(G))/Z(F(G)))$. However, this socle equals the layer $E(G)$ modulo $Z(F(G))$, hence N is contained in $F^*(G)$. So N is soluble and therefore contained in $Z(F(G))$, contradicting our assumptions. $\square$

1.2 $\mathfrak{M}_c$-groups and Nilpotency

Definition 1.2.1 A group G is

- $\mathfrak{M}_c$, if it satisfies the descending chain condition for centralizers, and

- $\mathfrak{Z}_f$, if for every iterated centre $Z_i(G)$ the centre of $G/Z_i(G)$ is the centralizer of finitely many elements.

Note that $C_G(C_G(C_G(H))) = C_G(H)$, so the descending and the ascending chain conditions on centralizers are equivalent. Furthermore, a subgroup of an $\mathfrak{M}_c$-group is obviously again $\mathfrak{M}_c$. However, the class $\mathfrak{M}_c$ is not closed under quotients; in fact, Bryant [25] constructs an $\mathfrak{M}_c$-group G such that $G/Z(G)$ is not $\mathfrak{M}_c$. Similarly, although the quotient of a $\mathfrak{Z}_f$-group by one of its iterated centres is again $\mathfrak{Z}_f$, a subgroup of a $\mathfrak{Z}_f$-group or an arbitrary quotient of a $\mathfrak{Z}_f$-group need not be $\mathfrak{Z}_f$.

Theorem 1.2.1 *An $\mathfrak{M}_c$-group is $\mathfrak{Z}_f$.*

Proof: Given any n, by the chain condition on centralizers there is some finite subset $\bar{g}_n \in G^n$ such that $C_G(G^n) = C_G(\bar{g}_n)$. Furthermore, there is a finite subset $\bar{k}_n \in G$ such that $\bar{g}_n \in \left\langle \bar{k}_n \right\rangle^n$. Consider $K = \left\langle \bar{k}_0, \dots, \bar{k}_n \right\rangle$. Then $K \leq G$, and for all $i \leq n$ we have $\bar{g}_i \in K^i$, i.e. $C_G(K^i) = C_G(G^i)$. It follows from Lemma 0.1.12 that $C_G^n(K) = C_G^n(G) = Z_n(G)$. In particular, for $\bar{G} = G/Z_{n-1}(G)$, we have $Z(\bar{G}) = C_{\bar{G}}(K) = C_{\bar{G}}(\bar{k})$, where $\bar{k}$ is the image in $\bar{G}$ of the finite generating set $\{\bar{k}_0, \dots, \bar{k}_n\}$ of K. $\square$

Definition 1.2.2 Let P and Q be some properties of groups. We call a group *P-by-Q* if it has a normal subgroup with property P, such that the quotient has property Q. It is *locally P* if every finitely generated subgroup is contained in a group with property P. If the property has an associated parameter n (like the derived length for solubility, or the class for nilpotency), a group is *uniformly locally P* if every finitely generated group is contained in a group which has property P with parameter n, where n only depends on the number, not on the choice of the generators. Finally, the group is *bi-P*, if every 2-generated subgroup is contained in a group with property P, and *uniformly bi-P*, if we can fix a value of the parameter n which works independently of the particular choice of the two generators.

Lemma 1.2.2 *A subgroup K of a finitely generated nilpotent group H is finitely generated.*

Proof: If H is generated by $\{h_1, \dots, h_k\}$ and is nilpotent of class n, then it is easy to see that H^{n-1} is generated by conjugates of commutators $[h_{i_1}, \dots, h_{i_n}]$, with $1 \leq i_j \leq k$. But these elements are central and equal to their conjugates, so H^{n-1} is finitely generated. As H^{n-2}/H^{n-1} is finitely generated for the same reason, H^{n-2} is finitely generated, and so is H^i for $0 \leq i \leq n$.

Now inductively we may assume that $K \cap H'$ is finitely generated. But H/H' is abelian, and a subgroup of a finitely generated abelian group – which is a finite direct product of cyclic groups – is clearly finitely generated. Hence $KH'/H' \cong K/(K \cap H')$ is finitely generated, and so is K. $\square$

Lemma 1.2.3 *Let M and N be two locally nilpotent normal subgroups of G. Then MN is locally nilpotent.*

Proof: Let $X = \langle m_1, \dots, m_s \rangle$ be a finitely generated subgroup of M, and $Y = \langle n_1, \dots, n_t \rangle$ be a finitely generated subgroup of N. We have to show that $\langle m_1, \dots, m_s, n_1, \dots, n_t \rangle$ is nilpotent. By Lemma 1.2.2 the group $C := \langle [m_i, n_j] : 1 \leq i \leq s, 1 \leq j \leq t \rangle$ is a finitely generated subgroup of $M \cap N$ and therefore nilpotent. Hence $\langle X, C \rangle$ is a finitely generated nilpotent subgroup of M which contains the finitely generated subgroup $D := \langle [X, n_j] : 1 \leq j \leq t \rangle$, which is contained in $M \cap N$. Therefore $\langle D, Y \rangle$ is a finitely generated nilpotent subgroup of N which contains $[X, Y]$. It follows that both $X[X, Y]$ and

$Y[X,Y]$ are finitely generated nilpotent groups which are normal in $\langle X,Y \rangle$. By Lemma 0.1.4 they generate a nilpotent group, namely $\langle X,Y \rangle$. $\square$

Definition 1.2.3 Let G be a group.

- The *locally soluble radical* $R_{ls}(G)$ of G is the maximal locally soluble normal subgroup of G.

- The *Hirsch-Plotkin radical* $HP(G)$ of G is the maximal locally nilpotent normal subgroup of G.

Corollary 1.2.4 *Every group has a unique Hirsch-Plotkin radical.*

Proof: Let H be the union of all locally nilpotent normal subgroups of G. Then any finite tuple $\bar{h} \in H$ lies in the product of finitely many normal locally nilpotent subgroups of G, which is locally nilpotent by Lemma 1.2.3. So H is locally nilpotent, and must equal $HP(G)$. $\square$

There are examples of infinite groups without a unique locally soluble radical [100]; it is an open question whether the locally soluble radical is unique in every stable group.

Lemma 1.2.5 *Let G be a locally nilpotent $\mathfrak{M}_c$-group. Then G is nilpotent, or the series of iterated centres is properly ascending. Furthermore, G is soluble, and the derived length is bounded by the maximal length k of a chain of centralizers (if there is such k).*

Proof: G is also $\mathfrak{Z}_f$, so if G is not nilpotent of class i, the $(i+1)$-st centre is the centralizer modulo $Z_i(G)$ of a finite number of elements $\bar{g}$. These, however, generate a nilpotent group G_0, so $Z_{i+1}(G) = C_G(G_0/Z_i(G)) > Z_i(G)$.

For the second assertion, consider some element $z \in Z_2(G) - Z_1(G)$. As z is non-central, by the chain condition on centralizers we may assume that the theorem holds for $C_G(z)$. Note that if k is a bound for the length of a chain of centralizers in G, then $k-1$ is such a bound for $C_G(z)$. Now Lemma 0.1.11 implies that $[G', Z_2(G)] \leq Z_0(G) = \{1\}$, so $C_G(z)$ contains G' and is normal, and $G/C_G(z)$ is abelian. $\square$

Lemma 1.2.6 *Let the abelian group G act on the abelian group A, in such a way that there are finitely many elements $g_0, \ldots, g_k$ in G with $C_A(G) = C_A(g_0, \ldots, g_k)$. Suppose a is an element in A such that for every i there is $n_i < \omega$ such that $[a,_{n_i} g_i] = 1$. Then $a \in C_A^m(G)$, where $m = 1 + \sum_{i=0}^{k}(n_i - 1)$.*

Proof: We shall use group ring notation, i.e. consider the action of $\mathbb{Z}G$ on A. Then the commutator condition translates as $(g_i - 1)^{n_i} a = 0$. Put $m = 1 + \sum_{i=0}^{k} (n_i - 1)$. Then for any m indices $i_1, \ldots, i_m$ in $\{0, \ldots, k\}$ at least one of them must occur at least n_i times, so by commutativity of G we get

$$(g_{i_1} - 1) \cdots (g_{i_m} - 1)a = 0.$$

Hence $(g_{i_2} - 1) \cdots (g_{i_m} - 1)a$ is annihilated by $(g_i - 1)$ for all i in $\{0, \ldots, k\}$, and thus centralized by G. So for any g_1' in G we get

$$(g_1' - 1)(g_{i_2} - 1) \cdots (g_{i_m} - 1)a = 0.$$

But now by commutativity of the ring $\mathbb{Z}G$, for any $i \in \{0, \ldots, k\}$ the element

$$(g_1' - 1)(g_{i_3} - 1) \cdots (g_{i_m} - 1)a$$

is annihilated by $(g_i - 1)$, and hence by $(g_2' - 1)$ for any $g_2' \in G$. Repeating this exchange m times yields

$$(g_1' - 1)(g_2' - 1) \cdots (g_m' - 1)a = 0$$

for any m elements $g_1', \ldots, g_m'$ in G, whence $a \in C_A^m(G)$. $\square$

Definition 1.2.4 A *(left normalized) commutator condition* on $x_0, \ldots, x_k$ is a word $w(x_0, \ldots, x_k)$ of the form

$$[x_0, x_{i_1}, \ldots, x_{i_n}],$$

where $i_0, \ldots, i_n$ are some (not necessarily distinct) indices in $\{0, \ldots, k\}$ with $i_1 \neq 0$. A group G satisfies a commutator condition w if $w(\bar{g}) = 1$ for all tuples $\bar{g} \in G$.

Note that a commutator condition $w(x, y_1, \ldots, y_k)$ gives rise to one in two variables, namely $w'(x, y) = w(x, y, \ldots, y)$.

Proposition 1.2.7 *Let G be a soluble $\mathfrak{M}_c$-group such that for any $g \in G$ there is a commutator condition $w(x, y)$ such that $w(gh, g) = 1$ for all $h \in G'$. Then G is nilpotent.*

Proof: We use induction on the derived length of G, the case of an abelian group being obvious. So we may assume that G' is nilpotent, and it is sufficient to show inductively that all iterated centres of G' lie in some iterated centre of G.

Trivially, $\{1\} = Z_0(G')$ is contained in $Z_0(G)$. Now suppose that $Z_i(G')$ is contained in $Z_k(G)$, and consider the action of G on $A := Z_{i+1}(G')/Z_k(G)$ by conjugation. A is centralized by G', so we get an action of $\mathbb{Z}(G/G')$ on A.

As G is 3_f, there are finitely many elements $g_1, \ldots, g_n$ in G such that $C_A(G) = C_A(g_1, \ldots, g_n)$. For any g_i there is a commutator condition w_i such that $w_i(g_ih, g_i) = 1$ for all $h \in G'$. But $[g_ih, g_i] = [g_i, g_i]^h[h, g_i] = [h, g_i]$, and for $h \in G'$ the actions of g_i and g_ih on A agree. Hence in group ring notation the identity $w_i(g_ih, g_i) = 1$ for $a = hZ_k(G) \in A$ becomes $(g_i - 1)^{s_i} a = 0$, where s_i is the number of commutators in w_i.

By Lemma 1.2.6 we get $a = hZ_k(G) \in C_A^m(G)$ for $m = 1 + \sum_{i=1}^n (s_i - 1)$, i.e. $h \in Z_{k+m}(G)$. Hence $Z_{i+1}(G') \leq Z_{k+m}(G)$, finishing the proof. $\square$

Corollary 1.2.8 *Let G be a soluble $\mathfrak{M}_c$-group satisfying some commutator condition. Then G is nilpotent.*

Proof: If G satisfies an identity $w(x, x_1, \ldots, x_n) = 1$, then it satisfies $w'(x, y) = w(x, y, \ldots, y) = 1$, and we may use Proposition 1.2.7. $\square$

Corollary 1.2.9 *A uniformly bi-nilpotent soluble $\mathfrak{M}_c$-group is nilpotent.*

Proof: If G is uniformly bi-nilpotent, it satisfies some commutator condition. $\square$

Corollary 1.2.10 *A uniformly locally nilpotent $\mathfrak{M}_c$-group G is nilpotent.*

Note that the local nilpotency is in particular uniform if the group has finite exponent.

Proof: G is soluble by Lemma 1.2.5 and satisfies a commutator condition. $\square$

Theorem 1.2.11 *Let G be an $\mathfrak{M}_c$-group. Then the Fitting subgroup of G is nilpotent.*

Proof: Clearly $F := F(G)$ is locally nilpotent and hence must be soluble by Lemma 1.2.5. Consider some element $f \in F$. It lies in a normal nilpotent subgroup K of class c, say. So if h is any element of F, then $[h, f] \in K$, and $[fh, _{c+1} f] = [h, _{c+1} f] = 1$. By Proposition 1.2.7 we get that F is nilpotent. $\square$

We now trivially get

Corollary 1.2.12 *Let G be $\mathfrak{M}_c$. Then any family of normal nilpotent subgroups generates a normal nilpotent subgroup. Any subnormal nilpotent subgroup lies in a normal nilpotent subgroup.*

Proof: Any normal nilpotent subgroup of G lies in $F(G)$. As for the second assertion, if $G_0 \trianglerighteq G_1 \trianglerighteq \cdots \trianglerighteq G_n$ and G_n is nilpotent, then $G_n \leq F(G_{n-1})$ by normality in G_{n-1}. But for any i, the Fitting subgroup of G_{i+1} is characteristic in G_{i+1} and hence normal in G_i, and must be contained in $F(G_i)$ by nilpotency. Therefore G_n is contained in the Fitting subgroup of G_0. $\square$

A final application yields

Theorem 1.2.13 *Let G be an ω-saturated $\mathfrak{M}_c$-group. Then $HP(G)$ is nilpotent.*

Proof: $HP(G)$ is soluble by local nilpotency. Suppose that $HP(G)$ properly contains the Fitting subgroup $F(G)$. Then we can find a characteristic subgroup F of $HP(G)$ such that $F/F(G)$ is abelian; we claim that F is itself nilpotent. This will contradict the maximality of the Fitting subgroup.

By local nilpotency, for every $f \in F$ and $g \in F(G)$ there is n_g such that $[fg, _{n_g} f] = 1$. Suppose that the n_g were not bounded (fixing f and varying g). If c is the nilpotency class of the Fitting subgroup, then the following set of formulæ would be consistent:

1. $\forall x_0, \ldots, x_c \, [x, x^{x_0}, x^{x_1}, \ldots, x^{x_c}] = 1$, together with

2. $[fx, _m f] \neq 1$, for all $m < \omega$.

For condition 1. just says that the group $\langle x^G \rangle$ is nilpotent of class c, and this is satisfied by any element g in $F(G)$. By ω-saturation there is some g_0 in G satisfying all these formulæ together. But then g_0 lies in a normal nilpotent subgroup of class c by 1. and must be contained in the Hirsch-Plotkin radical. However, $\langle g, f \rangle$ is not nilpotent by 2., contradicting local nilpotency of the Hirsch-Plotkin radical.

Hence there is a uniform bound (depending only on f), and F is nilpotent by Proposition 1.2.7. $\square$

For the rest of this section, we shall be dealing with periodic $\mathfrak{M}_c$-groups.

Lemma 1.2.14 *Let G be a periodic $\mathfrak{M}_c$-group. Then for every i there is some n_i such that for all $z \in Z_i(G)$ we have $z^{n_i} \in Z(G)$.*

Proof: By 3_f there are finitely many elements $\bar{g}$ in G such that $Z_i(G) = C_G(\bar{g}_i / Z_{i-1}(G))$. Let n be the least common multiple of the orders of the elements g_j from $\bar{g}_i$. Then modulo $Z_{i-1}(G)$ and for $g \in Z_{i+1}$ we get

$$[g_j, g^n] = [g_j^n, g] = 1,$$

whence $g^n \in Z_i(G)$. The lemma now follows by induction on i. $\square$

One might wonder what happens in Corollary 1.2.10 if the local nilpotency is not uniform. For a periodic group, we still get

Theorem 1.2.15 *A periodic locally nilpotent $\mathfrak{M}_c$-group G is nilpotent-by-finite.*

Proof: By the chain condition on centralizers, we may assume that every proper centralizer is nilpotent-by-finite. In particular, taking an element $z \in Z_2(G)$ of order n modulo the centre $Z(G)$ – and such a z must exist as in the proof of Lemma 1.2.5 – we get that $C_G(z)$ contains G', is nilpotent-by-finite, and for every $g \in G$ we have $[g^n, z] = [g, z^n] = 1$. So $G/C_G(z)$ has exponent n. But if N is a nilpotent G-invariant subgroup of finite index in $C_G(z)$, then by Lemma 1.2.14 the exponent of $N/Z(N)$ is finite. Therefore $G/Z(N)$ must have finite exponent.

Claim. Let M be an elementary abelian p-group of order p^2. Then there are $n := \frac{1}{2}(p^2 + p)$ non-trivial elements $m_1, \ldots, m_n$ of M such that in $\mathbb{Z}M$

$$(m_1 - 1) \cdots (m_n - 1) = 0.$$

Proof of Claim: Take $p + 1$ distinct non-trivial elements $g_0, \ldots, g_p$ in M. Then

$$\prod_{i<j}(g_i - g_j) = \det(g_i^j)_{i,j} = 0,$$

as the first and last columns of the Vandermonde matrix are equal. Now we take $g_i g_j^{-1}$ as elements m_k. $\square$

Claim. Let $A = C_{Z(N)}(\bar{g})$ for some $\bar{g} \in G$ be a subgroup of G such that $N_G(A)$ has finite index in G. Then there is a subgroup G_0 of finite index in G such that $A \leq C_G^k(G_0)$ for some $k < \omega$.

Proof of Claim: We may assume that the claim holds for all $A_1 < A$ satisfying the hypotheses. Now $N \leq C_G(A) \leq N_G(A) \leq G$. If $C_G(A)$ has finite index in $N_G(A)$, we can take $G_0 = C_G(A)$ and are done. Otherwise, since $M := N_G(A)/C_G(A)$ is finite-by-abelian of finite exponent, there is an element $v \in M$ of maximal order, and $C_M(v)$ has finite index in M. In particular, $C_M(v)$ is infinite, and we may take another element $w \in C_M(v) - \langle v \rangle$. So $\langle v, w \rangle$ is non-cyclic and abelian, and must contain an elementary abelian p-group of order p^2, for some prime p (using the maximality of the order of v). By the first claim, there are $n = \frac{1}{2}(p^2 + p)$ elements $m_1, \ldots, m_n$ in M with $\prod_i(m_i - 1) = 0$. As these elements are non-trivial, $C_A(m_i) < A$, and there are subgroups $G_1, \ldots, G_n$ of finite index in G and indices k_i with $C_A(m_i) \leq C_G^{k_i}(G_i)$. Take $G_0 := \bigcap_i C_{G,}(m_1, \ldots, m_n/C_G(A))$ and $k = \sum_i k_i$. As any m_i must have a centralizer of finite index in M, the index of G_0 in G is finite. We claim that $A \leq C_G^k(G_0)$. Indeed, since

$$(m_2 - 1) \cdots (m_n - 1)A \leq C_A(m_1) \leq C_G^{k_1}(G_1),$$

we get for any k elements $g_1, \ldots, g_k$ of G_0

$$(g_{k_1} - 1) \cdots (g_1 - 1)(m_2 - 1) \cdots (m_n - 1)A = 0.$$

Therefore

$$(g_{k_1} - 1) \cdots (g_1 - 1)(m_3 - 1) \cdots (m_n - 1)A \leq C_A(m_2) \leq C_G^{k_2}(G_2),$$

that is,
$$(g_{k_2+k_1} - 1) \cdots (g_1 - 1)(m_3 - 1) \cdots (m_n - 1)A = 0,$$
leading finally to
$$(g_k - 1) \cdots (g_1 - 1)A = 0. \; \square$$
But this implies that $Z(N) \leq C_G^k(G_0)$ for some $k < \omega$ and some subgroup G_0 of finite index in G. As $G_0/Z(N)$ has finite exponent, it is uniformly locally nilpotent, and so is G_0 itself. By Corollary 1.2.10, G_0 is nilpotent. $\square$

A similar theorem holds for locally soluble groups by Bryant and Hartley [26]:

Fact 1.2.16 *A locally soluble periodic $\mathfrak{M}_c$-group is nilpotent-by-(abelian of finite Prüfer rank)-by-finite. In particular, if the exponent of the group is finite, then the group is nilpotent-by-finite.*

In fact, in [12] Baudisch and Wilson have even shown

Fact 1.2.17 *Let G be a substable soluble group with a normal nilpotent subgroup N such that G/N is periodic. Then G is nilpotent-by-(abelian of finite Prüfer rank)-by-finite.*

(They state their theorem for the stable case, but their proof works just as well for a substable group.)

Corollary 1.2.18 *Let G be an $\bigwedge$-definable stable group with a normal nilpotent subgroup N such that G/N has finite exponent. Then G is nilpotent-by-finite.*

Proof: As normal nilpotent subgroups are contained in definable ones, we may assume that G is ω_1-saturated. So there is a normal nilpotent relatively definable subgroup M such that G/M is periodic and abelian of finite Prüfer rank. But G/M is ω_1-saturated, which implies that it must have finite exponent and hence be finite. $\square$

We can now describe the structure of locally nilpotent periodic $\mathfrak{M}_c$-groups.

Lemma 1.2.19 *Let G be a locally nilpotent periodic $\mathfrak{M}_c$-group. Then G has a unique maximal nilpotent subgroup N of finite index. Furthermore, N is normal, and for sufficiently large n we have $N = C_G(g^n : g \in G)$. Any normal nilpotent subgroup of G is contained in N.*

Proof: Let N_i be the centralizer of the $i!$-th powers of the elements of G. The sequence $N_1, N_2, \ldots$ is ascending; the chain condition on centralizers yields a maximal element $N = N_n$. Then $N/Z(N)$ has finite exponent (at most $n!$) and N must be nilpotent by Corollary 1.2.10. Clearly N is normal in G.

On the other hand, G has a nilpotent subgroup S of finite index by Theorem 1.2.15. For any such S, Lemma 1.2.14 implies that for big i the i-th powers of the elements of S, and hence the $(i \cdot |G : S|)$-th powers of the elements of G, are centralized by S. Therefore S must be contained in N.

Finally, if M is a normal nilpotent subgroup, MN is again nilpotent and of finite index in G, and must be contained in N. $\square$

Lemma 1.2.20 *Let N be a nilpotent $\mathfrak{M}_c$ p-group. Then there is a number $k(N)$ such that for all $k \geq k(N)$ the elements of order at most p^k form a subgroup B_k of N, which contains N'.*

Proof: Let B_k be the set of elements of order at most p^k. We show by induction on i that $B_{k(i)}$ contains $[Z_i(N), N]$ for some $k(i) < \omega$. This is trivial for $i = 1$.

By Lemma 1.2.19 the group $N/Z(N)$ has finite exponent p^j. So consider some x in $Z_{i+1}(N)$ and y in N. Modulo $[Z_i(N), N]$, the commutator $[x, y]^{p^j} = [x^{p^j}, y]$ is trivial. Hence the order of $[x, y]$ divides $p^{k(i)+j}$ and $[Z_{i+1}(N), N]$ is contained in $B_{k(i)+j}$.

So for some $k_0 = k(N)$ we get that B_{k_0} contains N', and for all $k \geq k(N)$ the elements in B_k form a subgroup. $\square$

Proposition 1.2.21 *Let G be a periodic $\mathfrak{M}_c$-group. Then the Hirsch-Plotkin radical is nilpotent-by-finite and the locally soluble radical exists and is soluble. If the exponent of G is finite, the Fitting subgroup equals the Hirsch-Plotkin radical and has finite index in the soluble radical.*

In particular, the (locally) soluble and the Hirsch-Plotkin radical are all definable. (The formula for the latter will, of course, not in general define a locally nilpotent subgroup in an elementary extension.)

Proof: By Lemma 1.2.19 the Hirsch-Plotkin radical $HP(G)$ of G contains a characteristic maximal nilpotent normal subgroup of finite index; this is the Fitting subgroup $F(G)$.

By Fact 1.2.16, a locally soluble normal subgroup is soluble. As the product of two normal soluble subgroups is again soluble, the locally soluble radical of G exists. By Fact 1.2.16 again, $R_{ls}(G)$ is soluble, and hence equals the soluble radical $R_s(G)$.

Finally, if G has finite exponent, then by Fact 1.2.16 the locally soluble radical is nilpotent-by-finite, and by Corollary 1.2.10 the Hirsch-Plotkin radical is nilpotent. $\square$

At this point we should mention the theorem of Kegel which states that a locally finite $\mathfrak{M}_c$-group of finite exponent is nilpotent-by-finite. Although a proof of the full theorem would lead us too far astray, in the next section we shall prove at any rate the substable case.

Definition 1.2.5 A group G satisfies the *normalizer condition*, if no subgroup is self-normalizing, i.e. $N_G(H) > H$ for any proper subgroup $H < G$.

Lemma 1.2.22 *Let G be locally nilpotent and nilpotent-by-finite. Then G is hypercentral.*

Proof: Let N be a normal nilpotent subgroup of finite index and A a system of representatives of the cosets of N in G. For every non-trivial $z \in Z(N)$ the conjugacy class z^G is finite, and the group F generated by A and z^G is nilpotent. Now $\langle z^G \rangle$ is normal in F and must contain a non-trivial element z' central in F. Then z' is centralized by A, and as $z^G \subseteq Z(N)$, it is also centralized by N. Since $G = NA$, the centre $Z(G)$ must be non-trivial. As $G/Z(G)$ is again nilpotent-by-finite and locally nilpotent, we may repeat this argument transfinitely often and exhaust the whole of G. Hence G is hypercentral. $\square$

Corollary 1.2.23 *Let G be a periodic $\mathfrak{M}_c$-group. Then the following are equivalent:*

1. *G is locally nilpotent.*

2. *G satisfies the normalizer condition.*

3. *G is hypercentral.*

Proof: If G is locally nilpotent, then G is nilpotent-by-finite by Theorem 1.2.15, and hypercentral by Lemma 1.2.22.

Suppose now that G is hypercentral, and $H < G$. If we choose β minimal such that $Z_\beta(G)$ is not contained in H, then β must be a successor, say $\beta = \gamma + 1$, and $Z_\beta(G)$ centralizes H modulo $Z_\gamma(G)$, which is contained in H. So $Z_\beta(G)$ normalizes H, and $N_G(H) \geq H Z_\beta(G) > H$.

Finally, if G satisfies the normalizer condition, for any $g \in G$ consider a maximal locally nilpotent subgroup H_g of G containing g, which exists by Zorn's Lemma. Now if all H_g are normal in G, they must all equal the Hirsch-Plotkin radical and G is locally nilpotent. Otherwise, by the normalizer condition, there is some g such that $N_G(N_G(H_g)) > N_G(H_g)$. But by maximality, H_g must be the Hirsch-Plotkin radical of $N_G(H_g)$ and characteristic therein; in particular, it is normal in $N_G(N_G(H_g))$, a contradiction. $\square$

1.3   Substability and Local Conditions

We shall now prove the substable case of Kegel's Theorem. Note, however, that the proof makes use of the Classification Theorem for finite simple groups.

Proposition 1.3.1 *A locally finite substable group G of finite exponent is nilpotent-by-finite.*

Proof: The soluble radical of G exists by Fact 1.2.16; it is relatively definable by Theorem 1.1.10. Again by Fact 1.2.16 it is sufficient to show that G is soluble-by-finite. Replacing G by $G/R_s(G)$, we may assume that G has no non-trivial abelian normal subgroup. By the chain condition on centralizers, we may suppose that any proper centralizer has infinite index. We aim to prove that G is trivial, so we may replace G by a countable elementary submodel. Therefore there is an ascending sequence $\{G_i : i < \omega\}$ of finite subgroups with $\bigcup_{i<\omega} G_i = G$. The classification of finite simple groups tells us that there are only finitely many simple groups of given exponent, so we may assume that the G_i have non-trivial proper normal subgroups A_i.

Case 1: The A_i may be chosen abelian.

We replace A_i by $A_i' := Z(C_G(A_i))$, a uniformly (in i) relatively definable abelian subgroup of G normalized by G_i. But now $N_j := \bigcap_{i>j} N_G(A_i')$ is increasing with j and uniformly relatively definable; it is therefore eventually stationary from some j_0 onwards. As N_j contains G_j, we get that N_{j_0} must contain G_j for all $j \geq j_0$, and $N_{j_0} = G$. Thus A_{j_0}' is a normal abelian subgroup of G, a contradiction.

Case 2: Almost all G_i have no non-trivial abelian normal subgroup.

If we choose the A_i to be minimal normal subgroups of the G_i, then they must be the direct products of simple non-abelian subgroups. However, their number is bounded by Corollary 1.0.6, so that the A_i and thus also the $N_G(A_i)$ are uniformly relatively definable, and there must be a j_0 such that $\bigcap_{i\geq j_0} N_G(A_i)$ is maximal (and hence the whole of G): A_{j_0} is a non-trivial normal subgroup. However, as G has no proper centralizer of finite index, any finite normal subgroup must be central and hence abelian, a contradiction.

   This finishes the proof. $\square$

A particularly easy case is the following:

Lemma 1.3.2 *Let A and B be substable abelian groups of finite, coprime exponent, such that B acts on A. Then a subgroup of B of finite index stabilizes A.*

Proof: Let $C \leq A$ be a minimal non-trivial intersection of kernels of endomorphisms, every one of which is generated by at most $e(B)+1$ elements from B. If $b_0, \dots, b_k \in B$ are $e(B)+1$ elements, then an endomorphism generated by them may be written in the form $\sum_{0 \leq j_0, \dots, j_k < e(B)} n_{j_0 \dots j_k} b_0^{j_0} \cdots b_k^{j_k}$ for some elements $0 \leq n_{j_0 \dots j_k} < e(A)$. Hence there are only finitely many ways in which such an endomorphism may be generated, and C is relatively definable by the icc. As B is abelian, C is B-invariant, and for any $e(B)+1$ elements $\bar{b} \in B$ any endomorphism of C generated by $\bar{b}$ is either zero or injective. Hence these $\bar{b}$-generated endomorphisms form a finite integral domain F, i.e. a finite field. But the number of elements of order $e(B)$ in F is at most $e(B)$, so two elements in $\bar{b}$ must have the same action on C. As $\bar{b}$ was arbitrary, there are at most $e(B)$ different actions by elements of B on C, and a subgroup B_1 of finite index fixes C.

Now consider $C_2 \leq A$, a maximal centralizer of some subgroup $B_2 \leq B$ of finite index, and suppose $C_2 < A$. As above, a subgroup B_3 of finite index in B_2 stabilizes some set $C_3/C_2 \leq A/C_2$. But then we get $(B_3 - 1)^2 C_3 = 0$, whence $C_3 \rtimes B_3$ is nilpotent of class 3. But the exponent of C_3 is coprime to the exponent of B_3, so C_3 and B_3 commute, contradicting the maximality of C_2. $\square$

Compare this with an – admittedly rather special – case of a group action on an abelian group of infinite exponent:

Proposition 1.3.3 *Let M be a group acting with finitely many orbits on some group G of infinite exponent. If $G \rtimes M$ is stable, then there is an $\bigwedge$-definable subgroup A of G carrying a definable field structure, such that multiplication on A gives the additive operation, and $N_M(A)/C_M(A)$ injects into the multiplicative group, where it has finite index.*

Proof: As two conjugate elements have the same order, the torsion of G is bounded by some finite $t < \omega$. We choose a minimal $\bigwedge$-definable subgroup A of G containing an element of infinite order. Then A is abelian and torsion-free, as A is contained in $Z(C_G(a))$ for any $a \in A$ of infinite order, and $t!A = A$. If there are a in A and g in G such that $ga \in A$ as well, then $A \cap gA$ is infinite and hence equals A. So the action of $N_M(A)$ on A has only finitely many orbits.

Let R be the ring of definable endomorphisms of A. By minimality of A, kernels and images of these endomorphisms are either trivial or the whole of A, so R is a division ring. Clearly its characteristic is zero, so in particular R has an infinite centre Z (the prime field consists of definable endomorphisms). Suppose now $a \in A$ is non-trivial. If $ra = r'a$ for any definable endomorphisms r, r', then $r - r'$ annihilates a and must be identically zero.

Since $N_M(A)$ has only finitely many orbits on A, for any non-trivial $a \in A$ the subgroup $\langle N_M(A)a \rangle$ is $\bigwedge$-definable as a finite union of orbits, and must equal A by minimality. Furthermore, there is a finite $n < \omega$ such that for any $a' \in A$ there are $r_1, \ldots, r_n \in N_M(A)$ with $\sum_{i=1}^{n} r_i a = a'$. Taking $a' = ra$ for some definable endomorphism r yields $r = \sum_{i=1}^{n} r_i$, and R is $\bigwedge$-definable as $N_M(A)^n$ modulo the equivalence relation E given by $\bar{r} E \bar{r}'$ iff $\sum r_i a = \sum r_i' a$ (for any non-zero $a \in A$).

But now the centre Z of R is $\bigwedge$-definable as well, as is the subgroup Za. But this equals A by minimality, and for every $r \in R$ there is $z \in Z$ with

$ra = za$, whence $r = z$. So R is commutative, and $N_M(A)/C_M(A)$ has finite index in R, bounded by the number of orbits. $\square$

As we shall see in the next chapter, the multiplicative group $K^\times$ of a stable field is connected as well, so in the context above $N_M(A)/C_M(A)$ must be isomorphic to it.

We now return to our investigation of nilpotency in substable groups. We first prove an analogue of Corollary 1.2.23.

Theorem 1.3.4 *Let G be a substable group. Then the following are equivalent:*

1. *G is locally nilpotent.*

2. *G satisfies the normalizer condition.*

3. *G is definably hypercentral.*

Proof: That hypercentrality implies the normalizer condition, which in turn implies local nilpotency, follows as in the proof of Corollary 1.2.23. It remains to show that a locally nilpotent substable group G is definably hypercentral.

By Lemma 1.2.5, the group must be definably soluble, and we may use induction on the derived length. Hence there is a relatively definable normal subgroup H such that H has smaller derived length, G/H is abelian, and H is definably hypercentral, i.e. there is an ascending sequence $(H_i : i \leq \alpha)$ of relatively definable subgroups H_i of H, with $H_0 = \{1\}$, $H_\alpha = H$, and $H_\lambda \leq \bigcup_{i<\lambda} H_i$ for limit ordinals λ, such that the quotient H_{i+1}/H_i is central in H/H_i for every $i < \alpha$. Suppose that $h \in H$ and $h' \in \bigcap_{g \in G} H_{i+1}^g$, then $[h, h'] \in H_i$, and by G-invariance $[h, h'] \in H_i^g$ as well. It follows that the series $\left(\bigcap_{g \in G} H_i^g : i \leq \alpha \right)$ is again a central series; as it is also relatively definable by the icc, we may thus assume that $(H_i : i \leq \alpha)$ is actually G-invariant.

Clearly, it is now sufficient to show that $H_{i+1} \leq C_{H_{i+1}}^\omega(G/H_i)$ for all $i < \alpha$, as then G is hypercentral of length at most $\omega \cdot \alpha + 1$, and the groups $C_{H_{i+1}}^n(G/H_i)$ are relatively definable and G-invariant.

Put $A := H_{i+1}/H_i$; then G' centralizes A. Since H_i is definable relative to H and hence relative to G, the quotient G/H_i is substable and there are finitely many elements $g_1, \ldots, g_n$ in G such that $C_A(G) = C_A(g_1, \ldots, g_n)$. As every $a = hH_i \in A$ and every g_i generate a nilpotent group of class n_i, say, we get $[a,_{n_i} g_i] = 1$, so Lemma 1.2.6 yields $a \in C_A^\omega(G)$. But this means exactly $H_{i+1} \leq C_{H_{i+1}}^\omega(G/H_i)$. $\square$

Remark 1.3.1 We should note that if G is locally nilpotent of derived length s, this gives us a bound $\omega^{s-1} + \omega^{s-2} + \cdots + 1$ for the hypercentral length of G. In particular, it is strictly bounded by ω^ω.

It is unknown whether Theorem 1.3.4 holds for $\mathfrak{M}_c$-groups in general. The above proof clearly breaks down when taking the quotient by some normal relatively definable subgroup. In the nilpotent case of the Fitting subgroup we could use $\mathfrak{Z}_f$ to deduce that even in the quotient by some iterated centre the (new) centre is the centralizer of finitely many elements. However, in the hypercentral case we merely take the quotient by some (relatively definable) normal soluble subgroup.

Definition 1.3.1 A subgroup H of a group G is *ascendant* if there is a sequence $H = H_0 \lhd H_1 \lhd \cdots \lhd H_\alpha = G$ of subgroups H_i of G indexed by ordinals, with $H_\lambda = \bigcup_{\beta<\lambda} H_\beta$ at limit ordinals $\lambda \le \alpha$, such that every H_i is normal in its successor H_{i+1}.

Corollary 1.3.5 *Suppose H is a locally nilpotent subgroup of G. Then if H is ascendant in G, it is contained in $HP(G)$. If G is substable and H relatively definable, the converse holds as well.*

Proof: Let $H = H_0 \lhd \cdots \lhd H_i \lhd \cdots \lhd H_\alpha = G$ be an ascending sequence from H to G. We show by transfinite induction that H lies in $HP(H_i)$ for all $i \le \alpha$. This is clear for $i = 0$ and successor steps, as $HP(H_i)$ is characteristic in H_i and hence normal in H_{i+1}. So consider a limit ordinal $\lambda \le \alpha$. We claim that $N := \bigcup_{i<\lambda} HP(H_i)$ is normal in H_λ; as it is locally nilpotent, it must lie in $HP(H_\lambda)$. But for any $n \in N$ and $h \in H_\lambda$ there is some $i < \lambda$ with $h \in H_i$ and $n \in HP(H_i)$, whence $n^h \in HP(H_i) \le N$ and we are done.

Conversely, suppose that G is substable and H is a relatively definable subgroup with locally nilpotent quotient G/H. Then by Theorem 1.3.4 the quotient G/H satisfies the normalizer condition, and H must be ascendant in G. In particular, if $H \le HP(G)$, then H is ascendant in $HP(G)$ and hence in G. $\square$

For a group satisfying the ωacc^0, we can even prove an analogue of Theorem 1.2.15, removing the periodicity condition:

Theorem 1.3.6 *Suppose G satisfies the ωacc^0. Then a locally nilpotent subgroup H of G is nilpotent-by-finite.*

Proof: First, we shall show that if A is a relatively definable normal abelian subgroup of H centralized by H', then there are a subgroup H_0 of finite index in H and some $n < \omega$ such that $A \le C_H^n(H_0)$. So consider the series $C_A^i(H)$ of iterated centralizers. By Lemma 1.2.6 and local nilpotency, every element of A lies in some iterated centraliser $C_A^m(H)$ for some $m < \omega$, whence $C_A^\omega(H) = A$. However, the iterated centralizers $C_G^m(H)$ are definable, so by the ωacc^0 there must be some finite $n < \omega$ such that for $i > n$ all indices $|C_A^i(H) : C_A^n(H)|$ are finite. Hence $H_i := C_H(C_A^i(H)/C_A^n(H))$ has finite index in H. As $C_A^n(H)$ is definable relative to H, the quotient $H/C_A^n(H)$ is $\mathfrak{M}_c$ and there must be a smallest centralizer modulo $C_A^n(H)$, say H_{i_0}, of finite index in H. But then $A \leq C_H^{n+1}(H_{i_0})$.

We shall now use induction on the derived length of H. If H is abelian, we are done. So consider a relatively definable normal soluble subgroup S containing H' of the same derived length as H'. By the inductive hypothesis, it has a nilpotent subgroup N of finite index and class c, say, which we may take to be relatively definable and normal in H. Applying the first paragraph to the action of $H/Z_i(N)$ on $Z_{i+1}(N)/Z_i(N)$, we see that for every $i < c$ there are a subgroup H_i of finite index in H and some $n_i < \omega$ such that $Z_{i+1}(N)/Z_i(N) \leq C_{H/Z_i(N)}^{n_i}(H_i)$. Furthermore, there is a subgroup H_c of finite index in H such that H_c stabilizes the finite group S/N. Then $\bigcap_{i \leq c} H_i$ is a nilpotent subgroup of finite index in H, of class at most $2 + \sum_{i < c} n_i$. $\square$

As usual, solubility remains rather more elusive than nilpotency. At least we can show that a normal locally soluble subgroup gives rise to a normal abelian subgroup.

Proposition 1.3.7 *Let G be a locally soluble substable group. Then G contains a non-trivial abelian normal subgroup.*

Proof: By local solubility, for any finite subset X of G there is an abelian intersection of centralizers in the ambient group $\mathcal{G}$, say A_X, which intersects G non-trivially and is normalized by X, namely the centre of the centralizer of the last non-trivial commutator subgroup of $\langle X \rangle$. But then the groups $N_X = \bigcap_{X' \supseteq X} N_{\mathcal{G}}(A_{X'})$ are uniformly definable, so by the ucc there is a maximal element N_{X_0}. By maximality, $N_{X_0} = N_{X'}$ for every $X' \supseteq X_0$. However, for any X we have $X \subseteq N_X$. Therefore N_{X_0} must contain the whole of G, i.e. $A_{X_0} \cap G$ is a non-trivial abelian normal subgroup of G. $\square$

We shall conclude this section with a result on the definability of a field in a locally nilpotent group.

Proposition 1.3.8 *Let G be an $\mathfrak{M}_c$-group with $Z(G) < Z_2(G)$. Then there are $a, b \in G$ and a definable subgroup $H \leq G$ such that*

1. *H is nilpotent of class 2, contains a, b, and $[a, b] \neq 1$,*

2. *all proper centralizers in H are abelian, and*

3. *$H' = [a, C_H(b)] = [C_H(a), b] \leq Z(G)$.*

Proof: Consider any $a_0 \in Z_2(G) - Z(G)$ and $b_0 \in G$ with $[a_0, b_0] \neq 1$. Then $[a_0, b_0] \in Z(G)$ and we may put $G_1 = C_G(b_0/Z(G))$ and $G_2 = C_{G_1}(G_1/Z(G))$. Then G_2 contains a_0, b_0 and $G_2' \leq Z(G) \leq Z(G_2)$, so G_2 is nilpotent of class 2. As G_2 is still $\mathfrak{M}_c$, we may find a minimal non-abelian centralizer

H_0 in G_2; choose any $a, b \in H_0$ with $[a, b] \neq 1$. Let $A_0 := C_{H_0}(a)$ and $B_0 := C_{H_0}(b)$. Note that $[a, B_0]$ and $[A_0, b]$ are central subgroups of H_0, so $A_1 := C_{A_0}(b/[a, B_0])$ and $B_1 := C_{B_0}(a/[A_0, b])$ are subgroups of H_0 containing H_0', so $H_1 := A_1 B_1$ is a subgroup of H_0, and $a \in A_1 = C_{H_1}(a)$ and $b \in B_1 = C_{H_1}(b)$. All these subgroups are obviously definable; if G is merely type-definable, they are type-definable as well. (However, if G is substable, then $[a, B_0]$ and $[A_0, b]$ and the subsequent groups are not necessarily relatively definable.)

Claim. For $x \in A_1$ and $y \in B_1$, if $[a, y] = [x, b]$, then $[A_1, y] = [x, B_1]$.

Proof of Claim: By symmetry it is enough to show that $[A_1, y] \leq [x, B_1]$. So take any $u \in A_1$. Then $[u, b] \in [a, B_0]$, so there is $v \in B_0$ with $[u, b] = [a, v]$; this implies $[a, v] \in [A_0, b]$, and $v \in B_1$.

As ab is non-central, $C_{H_1}(ab)$ is abelian. But $[uv, ab] = [u, a][v, b] = 1$; similarly $[xy, ab] = [x, b][y, a] = 1$ by assumption. Hence $1 = [uv, xy] = [u, y][v, x]$ and $[u, y] = [x, v] \in [x, B_1]$. $\square$

Now $[a, B_1]$ and $[A_1, b]$ are again central subgroups; we may put $A_2 := C_{A_1}(B_1/[a, B_1])$ and $B_2 := C_{B_1}(A_1/[A_1, b])$, and $H := A_2 B_2$. Then again A_1 and B_1 contain H_1' and H is a subgroup of H_1 with $a \in A_2 = C_H(a)$ and $b \in B_2 = C_H(b)$. Clearly (H, a, b) satisfies conditions 1. and 2.

Claim. $[a, B_2] = [A_2, b] = [A_2, B_2] = H'$.

Proof of Claim: By symmetry and since A_2 and B_2 are abelian, we only have to show $[A_2, B_2] \leq [A_2, b]$. So let $x \in A_2$ and $y \in B_2$. Then $[x, y] \in [A_1, y] \leq [A_1, b]$, so there is $u \in A_1$ with $[x, y] = [u, b]$. It remains to show that $u \in A_2$, i.e. $[u, B_1] \leq [a, B_1]$.

Let $v \in B_1$. By definition of A_2 and the first claim we get $[A_2, B_1] \leq [a, B_1] = [A_1, b]$, so there is $z \in A_1$ with $[x, v] = [z, b]$. But xb is not central, so $C_H(xb)$ is abelian, and it contains vz and uy. Therefore $1 = [uy, vz] = [u, v][y, z]$, and $[u, v] = [z, y] \in [A_1, B_2] \leq [A_1, b] = [a, B_1]$. $\square$

But this proves the proposition. Again, if G is type-definable, so is H. $\square$

Note that $C_H(a) \cap C_H(b) = Z(H)$. We now use a correspondence due to Mal'cev:

Theorem 1.3.9 *Suppose G is a nilpotent group of class 2 with elements a and b such that*

1. *$C_G(a)$ and $C_G(b)$ are abelian,*

2. *$C_G(a) \cap C_G(b) = Z(G)$, and*

3. *$[a, C_G(b)] = [C_G(a), b] = G'$.*

Then (G, a, b) interprets a ring R (but multiplication may be non-associative). If all proper centralizers are abelian, then R is an integral domain.

Proof: The underlying domain of R is G'; we put $x \oplus y = xy$ for $x, y \in G'$, and $x \otimes y = [x', y']$ for any $x' \in C_G(a)$ and $y' \in C_G(b)$ with $x = [x', b]$ and $y = [a, y']$. Then multiplication is well-defined by conditions 2. and 3. Let $1_R = [a, b]$ and $0_R = 1$, the group unit. Clearly $(R, 0_R, \oplus)$ is an abelian group, $0_R \otimes x = 0_R = x \otimes 0_R$ for any x (consider $0_r = 1 = [a, 1] = [1, b]$), and distributivity follows from bi-linearity of the commutator map: if in addition $y_1 = [a, y_1']$ and $z_1 = [x', y_1']$ for some $y_1' \in C_G(b)$, then

$$x \otimes y \oplus x \otimes y_1 = z \oplus z_1 = zz_1 = [x', y'][x', y_1'] = [x', y'y_1']$$
$$= x \otimes [a, y'y_1'] = x \otimes ([a, y'][a, y_1']) = x \otimes (y \oplus y_1).$$

Now suppose all proper centralizers are abelian. Consider $x'' \in C_G(b)$ and $y'' \in C_G(a)$ with $x = [x', b] = [a, x'']$ and $y = [a, y'] = [y'', b]$. So $x'x'', y'y'' \in C_G(ab)$, which is abelian, and multiplication is commutative:

$$x \otimes y = [x', y'] = [y'', x''] = y \otimes x.$$

If $z = [z', b]$ for some $z' \in C_G(a)$, then by commutativity

$$(x \otimes y) \otimes z = [z', b] \otimes [x', y']$$

and

$$x \otimes (y \otimes z) = [x', b] \otimes [z', y'].$$

Now there are $x'', z'' \in C_G(b)$ with $[x', y'] = [a, x'']$ and $[z', y'] = [a, z'']$. Either $y = 0_R$, in which case associativity is trivial, or $y \neq 0_R$. Then ay' is not central and $C_G(ay')$ is abelian and contains $x'x''$ and $z'z''$. Therefore $[z', x''] = [x', z'']$ and associativity follows:

$$(x \otimes y) \otimes z = [z', b] \otimes [a, x''] = [z', x''] = [x', z''] = [x', b] \otimes [a, z''] = x \otimes (y \otimes z).$$

Finally suppose $x \otimes y = [x', y'] = 1 = 0_R$, but $x \neq 0_R$, whence $x' \notin Z(G)$. So $C_G(x')$ is abelian and contains a and y'. Hence $y = [a, y'] = 1 = 0_R$, and R has no zero-divisors. $\square$

Corollary 1.3.10 *Let G be a type-definable stable non-abelian locally nilpotent group such that G' is torsion-free. Then a field of characteristic zero is interpretable in G.*

Proof: By Lemma 1.2.5 the second centre properly contains the first centre. By Proposition 1.3.8 we can find a type-definable subgroup H which satisfies the assumptions of Theorem 1.3.9, and type-interprets an integral domain R. By Lemma 1.0.1 the ring must have inverses. As its domain is additively a subgroup of $Z(G) \cap G'$, the characteristic is zero. $\square$

1.4   Engel Conditions

Definition 1.4.1 The *n-th Engel condition* is the condition $[x,_n y] = 1$.

An element $g \in G$ is *left Engel* if $[x,_n g] = 1$ for all $x \in G$, where n may depend on x. It is *bounded left Engel* or *left n-Engel* if n is independent of x. The set of left Engel elements is denoted by $L(G)$, and the set of bounded left Engel elements by $\bar{L}(G)$.

An element $g \in G$ is *right Engel* if $[g,_n x] = 1$ for all $x \in G$. It is *bounded right Engel* or *right n-Engel* if n is independent of x. The set of right Engel elements is denoted by $R(G)$, and the set of bounded right Engel elements by $\bar{R}(G)$.

An element is *Engel* if it is left or right Engel. It is *bounded Engel* if it is bounded left or right Engel.

Note that if g is left (right) Engel, then also all conjugates of g are left (right) Engel. Thus, if g^{-1} is right n-Engel and $y \in G$, we have $[y,_{n+1} g] = [g^{-y}g,_n g] = [g^{-y},_n g]^g = 1$ and g is left $(n + 1)$-Engel. Engel conditions are closely related to nilpotency, and also arise in the context of groups of small exponent. For example, a group of exponent 3 satisfies the second Engel condition. They also play a rôle in the analysis of the Restricted Burnside Conjecture. A finite group satisfying an Engel condition is nilpotent; more generally, the set of left Engel elements in a finite group is the Fitting subgroup, and the set of right Engel elements equals the hypercentre. In general it is not known whether the sets $L(G)$, $\bar{L}(G)$, $R(G)$, and $\bar{R}(G)$ even form subgroups (although Gruenberg [42] has shown that in a soluble group this is indeed the case).

Remark 1.4.1 Engel conditions are commutator conditions, so by Corollary 1.2.8 a soluble $\mathfrak{M}_c$-group satisfying an Engel condition is nilpotent. Moreover, it is easily seen that $L(G)$ contains the Hirsch-Plotkin radical, $\bar{L}(G)$ contains the Fitting subgroup, $R(G)$ contains the hypercentre $Z_\infty(G)$, and $\bar{R}(G)$ contains the ω-centre $Z_\omega(G)$.

Lemma 1.4.1 *Let G be a soluble $\mathfrak{M}_c$-group. Then $\bar{L}(G) = F(G)$.*

Proof: We shall prove by induction on the derived length of G and without $\mathfrak{M}_c$, that a left n-Engel element g generates a subnormal subgroup $\langle g \rangle$ of G. As it is also cyclic and thus nilpotent, g must be contained in the Fitting subgroup by Corollary 1.2.12.

For abelian G there is nothing to show, so we can assume that there is an abelian normal subgroup A such that $\langle gA \rangle$ is subnormal in G/A. But clearly $\langle g, A \rangle$ is nilpotent of class $n + 1$, so $\langle g \rangle$ is subnormal in $\langle g, A \rangle$. Hence $\langle g \rangle$ is subnormal in G. This finishes the induction; the reverse inclusion is obvious. $\square$

Lemma 1.4.2 *Let G be a soluble group. Then $L(G) = HP(G)$.*

Proof: Again we use induction on the derived length, showing that an un-bounded left Engel element g generates an ascendant subgroup of G; by Corollary 1.3.5 we get $g \in HP(G)$. The abelian case being trivial, we may assume that there is an abelian subgroup A such that $\langle gA \rangle$ is ascendant in G. But clearly $\langle g, A \rangle$ is hypercentral: for any normal N and any $a \in A$, the last iterated commutator $[a,_s g]$ not contained in N is central in $\langle g, A \rangle / N$ and the ascending central series can only stop by exhausting the whole group. Therefore $\langle g \rangle$ is ascendant in G. The reverse inclusion is again obvious. $\square$

It follows that in a soluble group an element g generates an ascendant sub-group iff it lies in the Hirsch-Plotkin radical.

Theorem 1.4.3 *Let G be a substable group. Then $\bar{L}(G)$ equals the Fitting subgroup of G, and $L(G)$ equals the Hirsch-Plotkin radical.*

Proof: Suppose, for a contradiction, that the group generated by $L(G)$ were not soluble. By Lemma 1.4.2 any soluble subgroup generated by $L(G)$ is locally nilpotent, and thus k-soluble for some fixed $k < \omega$ by Lemma 1.2.5, where k is the maximal length of a chain of centralizers in G. By Proposition 1.0.11 there are two definable k-soluble subgroups S and T such that $S \cap L(G)$ and $T \cap L(G)$ are different and maximal and such that the intersection $S \cap T \cap L(G) =: I$ is maximal. Put $S_1 := \langle S \cap L(G) \rangle$ and $T_1 := \langle T \cap L(G) \rangle$. Note that S_1 and T_1 are both contained in G and are locally nilpotent.

By Lemma 0.1.18 there are $x \in N_{S \cap L(G)}(I) - I$ and $y \in N_{T_1 \cap L(G)}(I) - I$. As x is left Engel, there is a maximal commutator $u = [y, x, \dots, x] \notin S_1$, such that $x^{-u}x = [u, x] \in S_1$, that is $x^u \in S_1 \cap S_1^u$. However, both x and y are in $N_G(I)$ and so is u, whence $x^u \notin I$ and $S_1 \cap S_1^u \cap L(G) > I$. By maximality we must have $S_1 = S_1^u$, so $u \in N_G(S_1)$.

We now distinguish two cases. If $u = y \in L(G)$, then $\langle S_1, y \rangle$ is soluble and $y \in S_1$, a contradiction. Otherwise there is $v = [y, x, \dots, x] \in G$ with $u = [v, x] = x^{-v}x$, whence $x^v \in N_G(S_1)$. But x^v is also left Engel; since $\langle S_1, x^v \rangle$ is soluble, we get $x^v \in S_1$. This implies $u = [v, x] \in S_1$, contradicting the choice of u.

Hence $L(G)$ generates a characteristic soluble group S. But a left Engel element in G is left Engel in S, so it follows from Lemmas 1.4.2 and 1.4.1 that $L(G) \subseteq HP(S) \leq HP(G)$ and $\bar{L}(G) \subseteq F(S) \leq F(G)$. As $HP(G) \subseteq L(G)$ and $F(G) \subseteq \bar{L}(G)$, the result follows. $\square$

In particular, the set of (bounded) left Engel elements is closed under inver-sion and products. To prove this directly would have been rather difficult! Furthermore, a bounded left Engel element is left $(n + 1)$-Engel, where n is the nilpotency class of $F(G)$. We may also improve Corollary 1.2.9 in the case of a substable group by omitting the solubility condition:

Corollary 1.4.4 *A substable group is nilpotent iff it is uniformly bi-nilpotent.*

Proof: One direction is obvious. Suppose now that any two elements generate a nilpotent subgroup of bounded class. Then the group satisfies some Engel condition, and is nilpotent by Theorem 1.4.3. $\square$

The situation is more difficult in the case of right Engel elements. One problem arises from the fact that while in an $\mathfrak{M}_c$-group the Fitting subgroup is nilpotent and there must be a bound on left Engelness, the example of the Prüfer 2-group acted upon by an inversion shows that the hypercentre even of a group of Morley rank 1 (and degree 2) need not be definable in general (go to an elementary extension!); in particular it need not equal some finite iterated centre, and there is no bound on right Engelness. More serious, however, is the lack of a suitable result for 2-generated subgroups: if N is right n-Engel in G and $g, h \in G$, then it may happen that N is not hypercentral in $\langle N, g, h \rangle$ (consider an infinite Burnside group B of exponent p and generated by g and h acting on the group ring $N := \mathbb{Z}_p B$, with $G = N \rtimes B$).

Clearly, any element in $Z_n(G)$ is right n-Engel. We shall establish the reverse inclusion $\bar{R}(G) \subseteq Z_\omega(G)$ in a number of special cases. First, we need a result of Gruenberg, which is proved by commutator calculus:

Fact 1.4.5 [42] *Let H be normalized by g and suppose that H has a generating set Y such that $[Y, _n g] = 1$ holds. If H is nilpotent of class c, then $\langle H, g \rangle$ is nilpotent of class at most $m(m-1)$, where $m = max\{n, c+1\}$.*

Lemma 1.4.6 *The right n-Engel elements in a substable group G are contained in a normal subgroup N such that $[N, _k G] = 1$ holds for some $k < \omega$.*

Proof: As bounded right Engel elements are bounded left Engel, they are all contained in the Fitting subgroup F of G. So consider the group N generated by the set Y of all right n-Engel elements. Note that N is normal in G. Now let $g \in G$ be arbitrary. Then $[Y, _n g] = 1$ must hold, so by Fact 1.4.5 the group $\langle N, g \rangle$ is nilpotent of class at most $m(m-1)$ with $m = max\{n, \text{class}(F)+1\}$. Hence $[N, _{m(m-1)} g] = 1$ holds, but as $g \in G$ was arbitrary and m does not depend on g, the assertion is true for $k = m(m-1)$. $\square$

Note that the proof also shows that if a and b are two unbounded right Engel elements in the Fitting subgroup, then ab^{-1} is again an (unbounded) right Engel element (take H to be the group generated by the set $Y := \{y \in R(G) \cap F(G) : [y, _n g] = 1\}$ for sufficiently big n). So $F(G) \cap R(G)$ forms a normal subgroup of G.

Lemma 1.4.7 *The set of unbounded right Engel elements in a substable group G forms a subgroup of $HP(G)$.*

Proof: As a right Engel element has an inverse which is left Engel, by Theorem 1.4.3 right Engel elements must lie in $HP(G)$. We have to show that $R(G)$ forms a subgroup. So consider $x, y \in R(G)$ and $g \in G$. Put $Y := \{[x,_n g], [y,_n g] : n < \omega\}$ and $H := \langle Y \rangle$. Then H is a finitely generated subgroup of $HP(G)$ and hence nilpotent; in addition there is some $n < \omega$ such that $[Y,_n g] = 1$ holds. Furthermore, $Y^g \subset H$, whence $H^g \leq H$.

For $z \in Y$ we have

$$[z, g^{-1}] = z^{-1} g z g^{-1} = (g^{-1} z^{-1} g z)^{g^{-1}} = [g, z]^{g^{-1}}$$
$$= ([z, g]^{g^{-1}})^{-1} = ([z, g][z, g, g^{-1}])^{-1};$$

so $[z, g]^{g^{-1}} \in H$ implies $z^{g^{-1}} \in H$. As $[z, g]$ is again in Y and $[z,_n g] = 1$ for sufficiently big n, we see inductively that $z^{g^{-1}} \in H$ for all $z \in Y$, whence $Y^{g^{-1}} \subset H$ and $H^{g^{-1}} \leq H$. Therefore $H^g = H$. By Fact 1.4.5, the group $\langle H, g \rangle$ is nilpotent, and $H \subset R(G)$. $\square$

Note that this shows that in any group the set $R(G) \cap HP(G)$ forms a subgroup.

Theorem 1.4.8 *Let R be a G-invariant set of right Engel elements in a substable group G, such that $G/C_G(R)$ is soluble. Then $R \subseteq Z_\infty(G)$.*

In particular, the set of right Engel elements in a soluble substable group equals the hypercentre.

Proof: By Lemma 1.4.7, we know that R is a subgroup of $HP(G)$. We shall show inductively that there is a transfinite series $\{1\} = R_0 \leq R_1 \leq \cdots \leq R_\alpha = R$ of relatively definable G-invariant subgroups of R such that for every $i < \alpha$ the quotient R_{i+1}/R_i is centralized by $RG^{(j)}$, for j ranging from the derived length of $G/C_G(R)$ to zero. If j is the derived length of G, we first get a relatively definable hypercentral series $(R_i)_i$ for R by hypercentrality of the Hirsch-Plotkin radical (Theorem 1.3.4; by G-invariance of R we may intersect every R_i with its G-conjugates and obtain a G-invariant hypercentral series for R, which is relatively definable by Proposition 1.0.11).

Now suppose $(R_i : i \leq \alpha)$ is a relatively definable G-invariant series with successive quotients central in $RG^{(j+1)}$, and consider the action of $RG^{(j)}$ on a quotient R_{i+1}/R_i by conjugation. Since R_i is definable relative to R and G-invariant, there is a relatively definable normal subgroup N of $RG^{(j)}$ with $N \cap R = R_i$. Then $RG^{(j)}/N$ is substable and $\mathfrak{M}_c$; in particular $Z(RG^{(j)}/N) = C_{RG^{(j)}/N}(\bar{g})$ for a finite subset $\bar{g} \in RG^{(j)}$. But for any $r \in R$ and $g \in RG^{(j)}$ the commutator $[r, g]$ lies in N if and only if it lies in $N \cap R = R_i$, and it follows that $C_{R_{i+1}/R_i}(RG^{(j)}) = C_{R_{i+1}/R_i}(\bar{g})$.

If $\bar{g} = \{g_0, \ldots, g_k\}$, then for any $r \in R_{i+1}$ there are $n_0, \ldots, n_k < \omega$ such that $[r_{,n_s} g_s] = 1$ holds for all $0 \leq s \leq k$. So by Lemma 1.2.6 for $RG^{(j)}/RG^{(j-1)}$ acting on R_{i+1}/R_i and $m = 1 + \sum_{s \leq k}(n_s - 1)$, we get $r \in C^m_{R_{i+1}/R_i}(RG^{(j)})$, whence $R_{i+1}/R_i \leq C^m_R(RG^{(j)})$. Clearly, the iterated centralizers $C^n_R(RG^{(j)})$ are relatively definable G-invariant subgroups of R, and $R \leq C^{\omega \cdot \alpha + 1}_R(RG^{(j)})$. This proves the result. $\square$

An inspection of the proof shows that the hypercentrality length is strictly less than ω^ω.

Corollary 1.4.9 *Let G be substable. Then $Z_\infty(G) = Z_\alpha(G)$ for some $\alpha < \omega^\omega$.*

Proof: Let β be minimal such that $G/C_G(Z_\beta(G))$ is not soluble. As $C_G(Z_\beta(G))$ is the centralizer of a finite number of elements, β must be a successor ordinal. But now for $z \in Z_\beta$ the map $\varphi_z : x \mapsto [z, x]$ is a homomorphism from $C_G(Z_{\beta-1}(G))$ to $Z_{\beta-1}(G) \cap C_G(Z_{\beta-1}(G))$, which is abelian. It follows that

$$C_G(Z_\beta(G)) = \bigcap_{z \in Z_\beta(G)} \ker(\varphi_z) \geq C_G(Z_{\beta-1}(G))'.$$

Hence $G/C_G(Z_\infty(G))$ is soluble. We may now apply Theorem 1.4.8 to $R := Z_\infty(G)$ and are done. $\square$

Theorem 1.4.10 *Let N be a normal subgroup of right n-Engel elements of the $\mathfrak{M}_c$-group G, such that $G/C_G(N)$ is soluble. Then N lies in some iterated centre of G.*

In particular, the bounded right Engel elements of a soluble $\mathfrak{M}_c$-group lie in the ω-centre.

Proof: Consider a sequence $C_G(N) = G_0 < G_1 < G_2 < \cdots < G_s = G$ of normal subgroups of G such that every quotient G_i/G_{i-1} is abelian. We shall prove inductively on i that N lies in some iterated centre of NG_i. For $i = 0$ this follows from nilpotency of N.

So suppose now that the assertion is true for $i - 1$. By induction, we may assume that there is k such that $C^{j-1}_N(NG_{i-1}) \leq Z_k(NG_i)$. We shall consider the action of G_i on $A := C^j_N(NG_{i-1})/Z_k(NG_i)$ by conjugation. Note that G_{i-1} centralizes A, so this action induces a commutative ring of endomorphisms. Furthermore, NG_i is $\mathfrak{3}_f$, so there are elements $g_1, \ldots, g_m$ in NG_i such that $C_A(g_1, \ldots, g_m) = C_A(NG_i)$. But any $h \in N$ is n-Engel, so $[a_{,n} g_i] = 1$ for any $a = hZ_k(NG_i) \in A$. Lemma 1.2.6 implies that $A \leq C^{1+m(n-1)}_A(NG_i)$, and hence $C^j_N(NG_{i-1}) \leq Z_{k+mn-m+1}(NG_i)$. As $N = C^j_N(NG_{i-1})$ for big j by the inductive hypothesis, this finishes the induction and hence the proof. $\square$

If G is not soluble, then the problem remains open even in the case of a group of finite Morley rank. We can, however, obtain partial information. For that, we shall need a theorem due to Heineken:

Fact 1.4.11 [46] *Let A be a normal abelian subgroup of G of right n-Engel elements. Suppose that A either has odd prime exponent $p > (n - 1)^2$ or is divisible and torsion-free. Then $G/C_G(A)$ is locally nilpotent.*

Definition 1.4.2 Let G be a group of automorphisms of some group A. Then G *stabilizes a chain in* A, if there is a chain $\{1\} = A_0 \leq A_1 \leq \cdots \leq A_n = A$ such that G stabilizes A_{i+1}/A_i for all $0 \leq i < n$, i.e. for any $g \in G$ and $a \in A_{i+1}$ we have $(aA_i)^g = aA_i$. If G stabilizes a chain in A, we call G a *stability group* for A. If in addition all the A_i in a chain stabilized by G are normal subgroups of A, then G *stabilizes a normal chain*.

Lemma 1.4.12 *Let G stabilize a normal chain in A of length m. Then G is nilpotent of class at most $m - 1$.*

Proof: We consider the group $H := A \rtimes G$. By normality of the chain stabilized, $A \leq C_H^m(G)$. Hence $[G^{m-1}, A] \leq C_H^0(G) = 1$ by Lemma 0.1.11, so G^{m-1} stabilizes A. But this means that G^{m-1} is trivial, as G is a group of automorphisms and $C_G(A) = \{1\}$. $\square$

It has been shown by Philip Hall [45] that any stability group is nilpotent, regardless of the normality of the chain stabilized.

Proposition 1.4.13 *Let G be a substable group, and A a normal abelian subgroup of right n-Engel elements. Suppose that either A has exponent 2 or exponent p for some prime $p > (n - 1)^2$, or else A is definable, A^0 is divisible and has only finitely many elements of every finite order. Then A is contained in some iterated centre of G.*

Proof: Suppose first that A has exponent p. Then for any $g \in G$, we have $(g - 1)^n A = 0$, whence $(g - 1)^q A = 0$ for any power q of p greater than or equal to n. But for any $a \in A$ we have $0 = (g-1)^q a = g^q a - a$, so $g^q \in C_G(A)$. It follows that $G/C_G(A)$ is a substable p-group. In case $p = 2$ it follows from Theorem 1.5.1 that $G/C_G(A)$ is locally nilpotent (the present theorem is not used in the proof there); for $p > (n - 1)^2$ local nilpotency follows from Fact 1.4.11. By Lemma 1.2.5 this quotient is soluble, so Theorem 1.4.10 proves the assertion.

In order to be able to apply Heineken's Theorem in the third case, we go to a saturated model and replace A by $\bar{A} := A^0/t(A)$, where $t(A)$ is the torsion subgroup of A. Then $\bar{A}$ is divisible and torsion-free, whence $G/C_G(\bar{A})$ is locally nilpotent by Fact 1.4.11. On the other hand, for every finite n

the group $A^0[n]$ is finite and G-invariant and must be centralized by the centralizer-connected component G^{cc}. So G^{cc} centralizes $t(A^0)$. But G/G^{cc} is finite, and so is $\langle a^G \rangle \rtimes G/G^{cc}$ for every $a \in t(A^0)$. Let $g \in G/G^{cc}$ be an element of prime power order p^s. If $a \in t(A^0)$ has order coprime to p, then g must act trivially on a by nilpotency of $\langle t(A^0), g \rangle$. Hence the action of G on the p-part A_p^0 of A^0 reduces to the action of a finite p-Sylow subgroup S_p of G/G^{cc}; since S_p is soluble, there is k_p such that $A_p^0 \leq Z_{k_p}(G)$ by Theorem 1.4.10. Taking k to be the maximum of all the k_p (only finitely many primes occur), we obtain $t(A^0) \leq Z_k(G)$. And now $C_G(\bar{A}) \leq C_G(A^0/Z_k(G))$, so $G/C_G(A^0/Z_k(G))$ is substable and locally nilpotent, hence soluble. Therefore A^0 is contained in $Z_i(G)$ for some $i < \omega$ by Theorem 1.4.10. Now $A/Z_i(G)$ is a finite subgroup of bounded right Engel elements; it must be centralized by the centralizer-connected component of $G/Z_i(G)$. Considering the finite group $(A/Z_i(G)) \rtimes \big((G/Z_i(G))/(G/Z_i(G))^{cc} \big)$, we see that A must lie in some iterated centre of G. $\square$

Theorem 1.4.14 *A right n-Engel element g of finite order s lies in the ω-centre, provided every prime factor of s is either 2 or greater than $(m-1)^2$, where $m = \max\{n(n-1), c(c-1)\}$ and c is the nilpotency class of the Fitting subgroup of G.*

Proof: A bounded right Engel element is bounded left Engel and must lie in the Fitting subgroup, which is nilpotent. By Lemma 1.4.6 the set of right n-Engel elements is contained in a G-invariant subgroup N of right m-Engel elements, whose p-part N_p is a characteristic subgroup by nilpotency, for every prime p. So it is sufficient to show the corollary for prime power exponent. In fact, we shall show that N_p lies in some iterated centre of G.

Let $\bar{n}$ be a finite subset of N_p such that $C_G(\bar{n}) = C_G(N_p)$. By Lemma 1.2.20 there is a characteristic subgroup H of N_p of finite exponent which contains $\bar{n}$. As H is nilpotent, there is a G-invariant sequence $\{1\} = H_0 < H_1 = Z(H)[p] < \cdots < H_t = H$ such that every quotient H_{i+1}/H_i has exponent p. By Proposition 1.4.13 every quotient lies in some iterated centre of G, and so does H. Therefore G stabilizes a normal chain in H, and $G/C_G(H)$ is nilpotent. But $C_G(H) = C_G(N_p)$, so N_p is contained in some iterated centre of G by Theorem 1.4.10. $\square$

The case of bounded right Engel elements of small odd order is still open. For superstable groups, slight improvements are possible: If N is a subgroup of bounded right Engel elements of Lascar rank $U(N) = \omega^{\alpha_1}n_1 + \cdots + \omega^{\alpha_k}n_k$, with $\alpha_1 > \cdots > \alpha_k$ and $n_1, \ldots, n_k < \omega$, then N has a series $N = N_0 > N_1 > \cdots > N_k = \{1\}$ of characteristic definable subgroups such that every quotient has rank $U(N_{i-1}/N_i) = \omega^{\alpha_i}n_i$. It can be shown that $[N_{i-1}/N_{i,n_i}, G] = 1$ must hold, so any bounded right Engel element in N of order coprime to max$\{n_1, \ldots, n_k\}$! must be ω-central in G. Furthermore, if N is abelian of monomial rank, the sequence of subgroups $i!N^0$ must become stationary at some i_0! by the dcc^0, and this can only happen if $i_0!N^0$ is divisible and its generic type is foreign to $i_0!N^0[k]$ for every $k < \omega$ (generic types will be introduced in the next chapter, and foreignness in chapter 3). Then by Corollary 3.0.3 the torsion part $t(i_0!N^0)$ is centralized by G^{cc}, and we see by the same argument as in Proposition 1.4.13 that $i_0!N^0$ is contained in some iterated centre of G. It follows that any bounded right Engel element has an ω-central power.

An inspection of the proofs above shows that the only obstacle to a hypercentrality theorem for bounded right Engel elements is the possible existence of a *very bad group:* a stable p-group G of finite exponent with non-trivial soluble radical R_s which is not contained in the hypercentre (note that every element of R_s is bounded right Engel). In the case of finite Morley rank we may simplify further and assume that G/R_s is a *bad group:* a simple non-soluble connected group whose proper connected subgroups are nilpotent. Groups whose maximal soluble subgroups are infinite, cyclic and conjugate (and hence self-normalizing and covering the whole) have been constructed by Ivanov and Ol'shanski [61] and it seems probable that one may also obtain similar examples with divisible maximal soluble subgroups; however, the examples found so far are not even superstable. Of course, even if there is a bad group of finite Morley rank, whether one can in addition make it act non-trivially on an elementary abelian p-group (and retain finite Morley rank) is another question. Our results at least imply that p cannot be greater than the square of the rank. We shall come back to bad groups in Section 3.8.

1.5 Sylow Theory

In this section we shall treat the Sylow theory. However, *a priori* it is not clear how a p-Sylow subgroup should be defined: whether it should be a maximal p-subgroup or a maximal locally nilpotent p-subgroup. Furthermore Sylow's Theorem does not automatically hold: the semi-direct product $\mathbb{Z} \rtimes \langle i \rangle$ of $\mathbb{Z}$ with an involution acting by inversion has two non-conjugate 2-Sylow subgroups, and is even superstable of rank one (and in particular stable). The following theorem gives a criterion for the local nilpotency of maximal substable p-subgroups:

Theorem 1.5.1 *A substable p-group G is locally finite iff it is bi-finite. A substable 2-group is locally finite.*

By Theorem 1.2.15 locally finite p-groups are locally nilpotent, and nilpotent-by-finite.

Proof: By Lemma 1.2.5 a maximal locally finite subgroup S of G is soluble of derived length k for some fixed $k < \omega$, and thus contained in a uniformly definable soluble subgroup S^* (of the same derived length). But $S^* \cap G$ is again locally finite, whence $S^* \cap G = S$ and maximal locally finite subgroups of G are uniformly relatively definable.

Now if G were not locally finite, then there would be two distinct maximal locally finite subgroups S and T. Their intersection $I = S \cap T = S^* \cap T^* \cap G$ is again uniformly relatively definable; by the icc we may assume that it is maximal. By Lemma 0.1.18 there are elements x in $N_S(I) - I$ and y in $N_T(I) - I$ of order p modulo I. For $p \neq 2$ the group $P = \langle x, y \rangle$ is finite by assumption; for $p = 2$ we obtain a dihedral subgroup PI/I of $N_G(I)/I$, hence a finite 2-group.

Therefore PI is contained in a third maximal locally finite subgroup U, and both $U \cap S$ and $U \cap T$ properly contain I. The maximality of I now implies $S = U = T$, a contradiction. $\Box$

If the p-exponent of G is finite (i.e. all p-elements have bounded order), then the maximal p-subgroups are nilpotent and definable. Otherwise all we know is that there is a definable nilpotent-by-finite group H^* containing H.

Before formulating the main theorem of this section, we have to rule out cases like $\mathbb{Z} \rtimes \langle i \rangle$.

Definition 1.5.1 A group has property ABD if every relatively definable abelian section A decomposes as a sum of a divisible group D with one of bounded exponent B.

Clearly, $\mathbb{Z}$ is not ABD.

Lemma 1.5.2 *If G satisfies the ωdcc, then G is ABD.*

Proof: The sequence of subgroups $n!A$ is descending and must contain a minimal element $D = n_0!A$. Then D is divisible. Put $B = A[n_0!]$, and consider a in A. Since $n_0!a$ lies in D, there is d in D with $n_0!d = n_0!a$, whence $n_0!(d - a) = 0$ and $d - a$ lies in B. Therefore $A = B + D$. $\Box$

As we shall see in chapter 4, other important groups also satisfy ABD. If $A = B + D$, we may choose B and D characteristic in A: if $n = e(B)$, then necessarily $nA = nD = D$ is characteristic, and we may replace B by the characteristic supergroup $A[n]$.

We shall need a fact from elementary cohomology theory:

Fact 1.5.3 SCHUR-ZASSENHAUS, GASCHÜTZ *Let A be an abelian normal subgroup of G, and B a subgroup of finite index k in G which contains A. Suppose that the map $a \mapsto a^k$ is invertible on A. Then:*

1. *If A has a complement in B, it has a complement in G.*

2. *If all complements of A in B are conjugate, then all complements of A in G are conjugate.*

Theorem 1.5.4 *Let G be a substable group.*

1. *If $p = 2$, and G is ABD, or*

2. *if G is soluble-by-finite and ABD, or*

3. *if $p = 2$ and G is periodic, or*

4. *if any two p-elements of G generate a finite subgroup,*

then the maximal p-subgroups of G are all conjugate.

Remark 1.5.1 • By Theorem 1.5.1 these assumptions imply in particular that the maximal p-subgroups of G are locally finite.

• Conditions 1. and 2. are elementary (at least if the ABD-property is uniform!), but not 3. and 4., so we have to remain in a fixed model G. The question of how conjugacy of the maximal p-subgroups behaves under elementary equivalence is still open.

• In the following proof *definable* will mean *definable relative to G*.

Proof: We shall give the proofs in four steps.

Step 1: *If G is substable, soluble-by-finite and ABD, then any two finite p-subgroups of G have conjugates which together generate a p-group.*
Proof: Consider two finite p-subgroups U and V of G and a normal definable soluble subgroup H of finite index in G. As Sylow's Theorem holds in G/H, we may conjugate V, say, and assume that modulo H the group generated by U and V is a finite p-group and thus soluble. Therefore it is sufficient to prove the assertion for soluble G.

We use induction on the derived length of G. The assertion is trivial for abelian G. So consider a definable abelian normal subgroup A of G which contains the last non-trivial commutator subgroup. By the inductive hypothesis we may assume that, modulo conjugation, G/A is a finite p-group. By assumption A decomposes as $B + D = B_p + B_{p'} + D$, so A/A_p is uniquely divisible by p. Fact 1.5.3 implies that A/A_p has a complement P/A_p in G/A_p, and any two complements are conjugate. Now the maximal p-subgroups of G are exactly the pre-images of those complements; they are all conjugate. □

As the assumptions in 4. do not carry over to quotient groups, we have to single out in any definable section H/K those elements which come from the p-elements in G: we shall call them *special*. A special subgroup of H/K is then a maximal subgroup consisting of special elements.

Obviously in cases 1. and 2. all p-elements are special: if $h \in H$ is a p-element modulo K, then $Z(C_K(h))$ is the product of a p-divisible group D with a p-group B of finite exponent. So if $h^q \in K$ for some power q of p, then clearly $h^q \in DB$, so by divisibility there is $k \in D \le C_K(h)$ such that $k^q B = h^q B$, i.e. $(k^{-1}h)^q \in B$. Thus h is congruent to the p-element $k^{-1}h$ modulo K. In case 3. any 2-element is special as well: if $h \in H$ is a 2-element modulo K, then $\langle h \rangle$ is a finite group and $\langle h \rangle/(\langle h \rangle \cap K)$ is a nilpotent 2-group. Hence any 2-Sylow subgroup of $\langle h \rangle$ is a supplement for $\langle h \rangle \cap K$; in particular h is equivalent modulo K to some 2-element.

In case 4. any two special elements of any section generate a finite subgroup. Furthermore the special elements of a locally nilpotent section H/K form a subgroup, because if two special elements $\bar{x}$ and $\bar{y}$ from H/K generate a nilpotent group $\bar{P}$, then $\bar{P}$ is the image of any p-Sylow subgroup of the finite group $\langle x, y \rangle$ generated by their pre-images x and y, and hence special.

Step 2: *If all special subgroups of G are nilpotent of class at most m for some fixed $m < \omega$, then they are conjugate.*
Proof: For any special subgroup S of G, we consider a definable nilpotent supergroup S^* of S of class m, such that the family of S^*, with S special, is uniformly definable. Then S is the set of special elements of S^*. Now if there were two non-conjugate special subgroups, we could choose two such that the set I of special elements of $S^* \cap T^*$ is maximal. Then $I = S \cap T$, and S and T have maximal intersection.

As S and T are locally nilpotent and nilpotent-by-finite, by Corollary 1.2.23 there are elements x in $N_S(I) - I$ and y in $N_T(I) - I$ of order p modulo I. We claim that the assumptions imply that there is some s in $N_G(I)$ such that x and y^s generate a finite p-group P modulo I. In cases 3. and 4. this follows from the finiteness of $\langle x, y \rangle/I$ and Sylow's Theorem; in cases 1. and 2. we use step 1 and the fact that $N_G(I)$ is definable as $N_G(\bigcap\{U^* : U \ge I \text{ special}\})$ and, in case 1., that two involutions always generate a soluble group.

Now PI is contained in a third special subgroup U. Both $U \cap S$ and $U \cap T^s$ properly contain I; by maximality of I we get that S is conjugate to U and U is conjugate to T^s, whence S is conjugate to T, a contradiction. $\square$

Step 3: *If G is soluble, then the special subgroups are all conjugate.*
Proof: We use induction on the derived length, the case of an abelian group being trivial. So consider a definable abelian normal subgroup A of G containing the last non-trivial commutator subgroup. Let S and T be two special subgroups of G, with maximal nilpotent subgroups M and N, respectively, of finite index.

By the inductive hypothesis we may assume that S and T are contained in a special subgroup modulo A, which is nilpotent-by-finite. Hence there is a definable subgroup H of G containing A, M and N, such that H/A is nilpotent of class k, say.

Let P be the set of special elements of A. Then P is contained in all special subgroups of G. By Lemma 1.2.19 there is some $m = p^n$ such that P^m is centralized both by M and by N. Put $B = A[m]$; then P/B is centralized by M/B and N/B, as $[x, s]^m = [x^m, s] = 1$ for any x in P and s in M. We shall now consider $K := C_H(P/B)$. The special subgroups of K/B are nilpotent of class $k + 1$, as their intersection with A/B is central. Hence they are all conjugate by step 2. But the pre-image of a special subgroup of K/B is a locally finite p-group and thus locally nilpotent; its special elements form a subgroup. Therefore the special subgroups of K are conjugate, and we can find a third special subgroup which contains a conjugate of M and a conjugate of N.

If S and T are not conjugate, there must be (possibly after replacing S by T) a special non-conjugate subgroup U such that the index $|S : S \cap U|$ is minimal among all conjugates of U by elements in $N_G(M)$. (For example, the third special group at the end of the last paragraph would do except for minimality of the – finite – index.) Note that we do not yet know that $I := S \cap U$ has finite index in U, so the situation is asymmetrical.

Again by Corollary 1.2.23 there are $x \in N_S(I) - I$ and $y \in N_U(I) - I$ of order p modulo I. Note that M is characteristic in I as its unique maximal nilpotent subgroup of finite index, so $N_G(I) \leq N_G(M)$. We claim that there is $s \in N_G(M)$ such that x, y^s, I and I^s are contained in a special subgroup V. In cases 3. and 4. this follows from Sylow's Theorem for $\langle x, y \rangle / I$. In cases 1. and 2. we first notice that $N_G(M)$ is definable as the normalizer of $\bigcap_{n \in N_G(M)} \bar{M}^n$, where $\bar{M}$ is a definable nilpotent supergroup of M; for a finite system X of representatives of I/M we now find by step 1 some $s \in N_G(M)$ such that x, X, y^s and X^s generate a finite p-group.

By minimality of $|S : I|$ we get that $V = S^t$ for some element t in $N_G(M)$. On the other hand $S^t \cap U^s$ properly contains $(S \cap U)^s = I^s$, so the index of M in $S \cap U^{st^{-1}}$ is bigger than its index in $S \cap U = I$. Therefore the index of $S \cap U^{st^{-1}}$ in S is smaller than the index of I in S; by maximality again, there is some u normalizing M such that $S^u = U^{st^{-1}}$, a contradiction. $\square$

Step 4: *The general case.*

As the induction on the derived length has been dealt with, we need no longer refer to special groups. By local nilpotency, any maximal p-subgroup S is soluble of class k for some fixed $k < \omega$, and S is contained in a soluble definable supergroup S^* of derived length k, such that the family of these S^*, for varying S, is again uniformly definable.

Consider pairs of non-conjugate maximal p-subgroups S_i and T_i, and let I_i be a maximal p-subgroup of $S_i^* \cap T_i^*$. If we could find such a sequence, for $i < \omega$, with $(I_i)_{i<\omega}$ properly ascending, this would contradict Proposition 1.0.11 for the intersections of $S_i^* \cap T_i^*$ with $\bigcup I_i$. Hence there is a maximal such intersection $I = I_n$, which is contained in a conjugate S of S_n and in a conjugate T of T_n by step 3. Thus we have found two non-conjugate maximal p-subgroups with maximal intersection.

In cases 3. and 4. this immediately leads to a contradiction, as we can pick $x \in N_S(I) - I$ and $y \in N_T(I) - I$ of order p. Then modulo I these elements generate a finite group, so they can be conjugated in $N_G(I)$ to generate a p-group, which is in turn contained in a third maximal p-subgroup U. By maximality of the intersection, U must be conjugate both to S and to T, whence S and T are conjugate.

In cases 1. and 2. we would like to use step 1; however, $N_G(I)$ need not be definable. Choose maximal (normal) nilpotent subgroups S_0 in S and T_0 in T, and definable nilpotent supergroups S_1 of S_0 and T_1 of T_0. So S_0 is the set of p-elements in S_1, and T_0 the corresponding set in T_1. Then $N := \bigcap_{g \in N_G(I)} (S_1 \cap T_1)^g$ is a definable nilpotent group whose p-elements form a characteristic subgroup I_0 of finite index in I, and which is normalized by $N_G(I)$. If we choose $x \in N_S(I) - I$ and $y \in N_T(I) - I$ of order p, we can find a definable soluble-by-finite subgroup H of $N_G(N)$ containing I, x and y. Note that H must then normalize I_0. Let X be a system of representatives of I/I_0. By step 1. there is some $s \in H$ such that x, X, y^s and X^s generate a finite p-group P. Then PI is contained in a maximal p-subgroup U of G, which by maximality of I is conjugate to both S and T, a contradiction. $\square$

As usual, a conjugacy theorem implies the existence of supplements:

Corollary 1.5.5 *Let G be a substable group satisfying assumption 1., 2., or 3. of Theorem 1.5.4. Suppose N is a normal relatively definable subgroup of G. Then the p-Sylow subgroups of G/N are the images of the p-Sylow subgroups of G.*

Proof: We shall first prove the assertion for a relatively definable p-group P/N. It is soluble by Lemma 1.2.5, and we may use induction on the derived length of P/N. So suppose first it is abelian.
Claim. Any element $\bar{x}$ in P/N has a pre-image which is a p-element.
Proof of Claim: Let x be a pre-image of $\bar{x}$. If G is periodic, then it is easy to see that $\bar{x}$ must be equivalent to a p-element $y \in xN$, as any power coprime to p generates the same cyclic group modulo N. In the other cases, we consider the relatively definable abelian group $A := Z(C_G(x))$. Then also $N \cap A$ is relatively definable, whence ABD, so $(N \cap A)/(N \cap A)_p$ is p-divisible. But there is $k < \omega$ such that $x^{p^k} \in N \cap A$, and by divisibility there is $y \in N \cap A$ with $y^{p^k} = x^{p^k}$ modulo $(N \cap A)_p$. Hence xy^{-1} is a p-element in xN. $\square$

So for any $\bar{x}$ in P/N there is a p-pre-image x in P, which lies in some p-Sylow subgroup of P. Now any element in P/N remains unchanged under conjugation. But P is a relatively definable subgroup of G and inherits the assumptions, so the p-Sylow subgroups of P are all conjugate, and $\bar{x}$ lies in the image of any one of them. As this holds for all elements in P/N, a p-Sylow subgroup of P projects onto P/N.

Now suppose the assertion holds for derived length n and consider a p-Sylow subgroup P/N of derived length $n + 1$. There is a P-invariant soluble subgroup H/N of P/N of derived length n containing P'/N, which is definable relative to G. Then H inherits the assumptions from G, so by the inductive hypothesis H/N lies in the image of any p-Sylow subgroup of H. On the other hand, we may replace G by $N_G(H)$ and N by H – the assumptions of the corollary still hold – and the abelian case implies that P/H lies in the image of any p-Sylow subgroup of P. Therefore for any $\bar{x} \in P/N$ and any p-Sylow subgroup S in P there is $y \in S$ with $yH = xH$. Now $S \cap H$ may be extended to a p-Sylow subgroup S_1 of H, which is itself contained in a p-Sylow subgroup S_2 of P. As S and S_2 are conjugate in P and H is P-invariant, we get that $S \cap H$ and $S_2 \cap H$ are conjugate, whence $S \cap H = S_1$ is a p-Sylow subgroup of H. Therefore there is $y' \in S \cap H$ with $y^{-1}xN = y'N$, whence $xN = yy'N \in S/N$.

This proves the relatively definable case. Now by Theorem 1.2.15 and Lemma 1.2.20 a general p-group P/N has a normal nilpotent subgroup of finite index, whose elements of order p^k form a subgroup for all sufficiently big $k < \omega$. If $\bar{x}$ is a system of representatives for the cosets of P/N modulo this nilpotent subgroup of finite index, then the groups $P_k/N := B_k\langle \bar{x}\rangle/N$ form an ascending series whose union is P/N. Replacing G by P_k, it follows from the above that P_k/N is the image of any p-Sylow subgroup of P_k. But any p-Sylow subgroup of P_k can be extended to a maximal p-subgroup of P_{k+1}, so there is a p-Sylow subgroup S of P with image P/N. Clearly S must be a maximal p-subgroup of G, since P/N is maximal in G/N.

Finally consider another p-Sylow subgroup T of P. Then S and T are conjugate in G, so $P/N = NS/N$ and NT/N are conjugate in G/N. But then NT/N is again a p-Sylow subgroup, and contained in P/N, whence equal. $\square$

We now want to generalize the Baer-Suzuki Theorem.

Theorem 1.5.6 *Let C be a G-invariant subset of a substable group G such that any two elements of C generate a finite p-group. Then C generates a locally finite p-group.*

Proof: We shall show first that the conclusion holds in general for soluble groups, even without assuming substability. So consider a counter-example G of minimal derived length, and $x_1, \dots, x_n \in C$ such that $F = \langle x_1, \dots, x_n \rangle$ is not a finite p-group. Let A be the intersection of F with the last non-trivial derived subgroup of G. By the minimality of the solubility class, F/A must be a finite p-group. But F is finitely generated, and so is any subgroup of

finite index, in particular A. On the other hand, A is abelian and there is
a finite k such that A/A^k is not a p-group. Hence F/A^k is not a p-group,
contradicting Theorem 0.1.19. This proves the assertion for soluble groups.

Returning to our substable group, this implies that any subset X of C
generates a locally finite p-group if and only if it generates a soluble group
of derived length k, where k is necessarily fixed by the length of a chain of
centralizers. So if the theorem did not hold, we could find two uniformly
relatively definable soluble groups H and K of derived length k, with distinct
maximal $A = H \cap C$ and $B = K \cap C$, such that the intersection $D = A \cap B =
H \cap K \cap C$ is maximal (and it cannot equal A or B).

By Lemma 0.1.18 there are $x \in N_A(D) - D$ and $y \in N_B(D) - D$. Then
x, y and D generate a locally finite p-group, contradicting the maximality of
D. $\square$

Theorem 1.5.7 *Let C be a G-invariant subset of a substable group G such
that any two elements of C generate a finite p-group of bounded exponent.
Then C generates a nilpotent p-group.*

Proof: By Theorem 1.5.6 the group generated by C is a locally finite p-group,
whose maximal nilpotent subgroup of finite index we shall denote by N. Let
p^k be a bound for the exponent of the groups generated by two elements of
C. For any x in C and n in N the order of the commutator $[x, n] = x^{-1}x^n$ is
bounded by p^k; furthermore this commutator is an element of N. By Lemma
1.2.20 there is some $k'(N) < \omega$ such that the set of elements of order at most
p^n form a subgroup B_n of N for all $n \geq k'(N)$. If m is the maximum of k
and $k'(N)$, then xB_m contains all conjugates of x by elements of N. Now
B_m and x generate a p-group $P = B_m \cdot \langle x \rangle$ of finite exponent. But P is
nilpotent, normalizes N and is normalized by N. By maximality of N, it
must be contained in N, whence x lies in N and C generates the nilpotent
group N. $\square$

Finally we shall have a closer look at the structure of the 2-Sylow subgroups.
First notice that the Frattini argument remains valid in our context:

Remark 1.5.2 Let G be a substable periodic group. Then any subgroup
H of G which contains the normalizer of a 2-Sylow subgroup S is self-
normalizing; if N is normal in G and P a 2-Sylow subgroup of N, then
$G = NN_G(P)$. This is also true in stable ABD groups, provided H and N
are definable.

Proof: Consider $n \in N_G(H)$. The conjugate S^n is another 2-Sylow subgroup
of H; by Theorem 1.5.4 there is some h in H with $S^n = S^h$, whence $h^{-1}n \in
N_G(S) \leq H$. So n lies in H.

For the second part consider any g in G. By normality, P^g is another 2-Sylow subgroup of N, so there is $n \in N$ with $P^g = P^n$, and $n^{-1}g$ normalizes P. $\square$

Theorem 1.5.8 *Let S and T be 2-Sylow subgroups of some substable group G of finite exponent, such that their intersection I is maximal subject to $S^{co} \neq T^{co}$. Then $N = N_G(S^{co}) \cap N_G(T^{co}) \cap N_G(I)$ acts transitively on the infinite group $A = (N_S(I)/I)[2]$. Furthermore, N/I has odd exponent and any abelian subgroup of $N/C_N(A)$ is finite.*

Proof: First note that both S and T are relatively definable as maximal nilpotent 2-groups of finite exponent. Secondly, if I had finite index in S, then it would contain S^{co}, and $S^{co} = I^{co}$ by conjugacy of S and T. Since $S^{co} \neq T^{co}$, the index of I in S or in T must be infinite. But S^{co} and T^{co} are conjugate and definable relative to G, so if $I^{co} = T^{co}$ (say), we could conjugate S^{co} to the proper subset T^{co} by some element g, and the sequence $(S^{co})^{g^i}$ (with $i < \omega$) would contradict Proposition 1.0.11. Thus I has infinite index in both S and T.

By Lemma 1.1.9 first $N_S(I)/I$, and then $N_{S^{co}}(I)/I$, are infinite. Fix an involution j in $N_{T^{co}}(I)/I$ and consider any involution i in $N_S(I)/I$. If the order of ij were even, both i and j would commute with a common third involution k in $N_G(I)/I$ and we could find maximal 2-subgroups $S' \supset I \cup \{i,k\}$ and $T' \supset I \cup \{k,j\}$. But then the maximality of I would imply $S^{co} = (S')^{co} = (T')^{co} = T^{co}$, a contradiction. Hence the order is odd and there is an involution $k_i = ij \cdots i = ji \cdots j$ in $N_G(I)/I$ (with $o(ij)$ factors) such that $i^{k_i} = j$ and $j^{k_i} = i$. Since I is maximal, we must have $(S^{co})^{k_i} = T^{co}$ and $(T^{co})^{k_i} = S^{co}$; in particular $j \in T^{co}$ implies $i \in S^{co}$. But then, if $i' \in N_S(I)/I$ is another involution, $k_i k_{i'}^{-1}$ lies in $N_G(S^{co}) \cap N_G(T^{co}) \cap N_G(I) = N$ and conjugates i to i'.

As we have seen above, $X := N_{S^{co}}(I)/I$ is an infinite nilpotent 2-group, and must contain a central involution i. But all other involutions of $N_S(I)/I$ are conjugate to i in N; they must also lie in S^{co} and be central. Hence the involutions form a central subgroup A of X. By Lemma 1.1.7 there is a subgroup Y of finite index in X such that $Z(Y)$ is infinite; as it is abelian of finite exponent, it must contain infinitely many involutions: A is infinite.

Suppose now that there were an involution $n \in N/I$. Then we should find 2-Sylow subgroups $S' \supset S^{co} \cup nI$ and $T' \supset T^{co} \cup nI$, and $S' \cap T' > I$. Hence by the maximality of I we get $S^{co} = (S')^{co} = (T')^{co} = T^{co}$, a contradiction.

As for the last statement, this follows from Lemma 1.3.2. $\square$

1.6   Historical and Bibliographical Remarks

The chain condition in its various guises appears in Macintyre [70], Baldwin and Saxl [4] (in particular Theorem 1.0.5), Felgner [41] and Baur, Cherlin and Macintyre [14]. The argument for Theorem 1.0.3 comes from Reineke [99].

The definability of solubility and nilpotency was proven by Berline and Lascar [18]; Theorems 1.1.12 and 1.1.13 come from [119] (the former generalizes a result in [115], the latter a result of Borovik and Thomas [23]). The existence of the Hirsch-Plotkin radical (Corollary 1.2.4) was proven independently by Hirsch [49] and Plotkin [89]. Variants of simplicity and the socle are dealt with by Nesin [79], and generalized in [122].

The group theory of section 1.2 (in particular Theorem 1.2.15) is taken from Bryant [25], Fact 1.2.16 from Bryant and Hartley [26]; the nilpotency of the Fitting subgroup and related results were proven in Derakhshan and Wagner [39]. One should also refer to Kegel [62], where Proposition 1.3.1 is shown generally for locally finite groups of finite exponent and with the chain condition on centralizers. Lemma 1.3.2 comes from [119]. Equivalence of 1. and 2. in Theorem 1.3.4 is shown in [117]; the full result had to wait until [39]. The construction of the field from Proposition 1.3.8 to Corollary 1.3.10 is taken from Grünenwald and Haug [44]; the Mal'cev correspondence itself appeared in Mal'cev [71].

Lemmas 1.4.1 and 1.4.2 are due to Gruenberg [42], who in fact showed that the set of bounded left Engel elements in any soluble group equals the Baer radical. The Engel condition is analysed in [117] and [39]; one should compare this with Boffa and Point [19, 90], where a variant of Corollary 1.2.10 is proven for linear groups and finitely generated soluble groups. Section 1.5 on Sylow theory and the Baer-Suzuki Theorem originated in [22]; the present version appeared co-authored with Poizat [97]. Fact 1.5.3 is due to Schur, Zassenhaus and Gaschütz, and can be found e.g. in [60], Satz I.17.4; the proof uses elementary cohomology theory. Finally, Remark 1.5.2 and Theorem 1.5.8 are taken from [119].

Chapter 2

Groups & Genericity

In this chapter we shall develop the second main tool for analysing stable groups, the method of generic types. In section 0 we shall develop the relevant notions for the case of a superstable group before introducing them in full generality in section 1, which culminates in a general existence and definability theorem for stable structures: any semi-group with a certain cancellation condition is contained in a definable group. The next, short, section generalizes the results obtained for groups to transitive group actions. We shall see the first simple applications in section 3, where we look at fields and prove equality of the additive and the multiplicative generic types. Furthermore we shall transfer the definability results to fields.

In section 4 we shall look at generic properties. This is inspired by algebraic group theory: if some equation is satisfied generically in an algebraic group, then it is satisfied by the whole group, as it defines a Zariski closed set of maximal dimension. For stable groups similar questions are mostly open; the only settled cases deal with nilpotency, solubility, and exponent 2 or 3. If the group is soluble-by-finite, then generically finite exponent implies finite exponent, and generically prime exponent implies that the group is nilpotent-by-finite.

2.0 The Superstable Case

Before we start with the general theory, we shall illuminate the relevant concepts in the superstable case. Readers familiar with this part of the theory may safely continue with section 1. Throughout this section G will denote a superstable group.

We shall call a 1-type *generic* if it has maximal Shelah rank; as types of maximal Shelah rank always exist in a superstable theory, there are generic types. Clearly, an extension of a generic type is generic iff it is non-forking. We call an element g *generic over* A if $\operatorname{tp}(g/A)$ is generic.

Claim. If g is generic over G and $h \in G$, then $\mathrm{tp}(g^{-1}/G)$ and $\mathrm{tp}(gh/G)$ are generic, and do not fork over $\emptyset$.

Proof of Claim: $RC(gh/G) = RC(g/G) = RC(g^{-1}/G)$ by invariance of Shelah rank under definable bijections, and this is maximal possible. But this implies that these types cannot be forking extensions. $\square$

In particular, if g and h are independent over A and $\mathrm{tp}(g/A)$ is generic, then $\mathrm{tp}(g/h, A)$ is generic, and there is a model G containing A, h and independent of g over A, h. So $\mathrm{tp}(g/G)$ is generic, as is $\mathrm{tp}(gh/G)$, and hence $\mathrm{tp}(gh/A, h)$. It follows that gh and h are independent over A.

Claim. There are only boundedly many generic types (over any model).

Proof of Claim: A generic type does not fork over $\emptyset$, and types over $\emptyset$ have only boundedly many non-forking extensions. $\square$

Claim. Every element is the product of two generic elements.

Proof of Claim: If $g \in G$ and h is generic over G, then so is gh^{-1}, and clearly $(gh^{-1})h = g$. $\square$

We call a formula *generic* if it lies in some generic type.

Claim. Finitely many translates of a generic formula cover G.

Proof of Claim: Suppose $\varphi(x)$ is generic, but for any $g_0, \ldots g_{n-1}$ there is some $g \in G - \bigcup_{i<n} g_i \varphi^G$. Then $\{\neg\varphi(gx) : g \in G\}$ is consistent, and realized by some element h. Let now g be generic over G, h and such that $\varphi(g)$ holds. Then $gh^{-1} \underset{G}{\downarrow} h$, and $\mathrm{tp}(h/G, gh^{-1})$ is an heir of $\mathrm{tp}(h/G)$. As $\varphi(gh^{-1}x) \in \mathrm{tp}(h/G, gh^{-1})$, there must be some $g' \in G$ with $\varphi(g'x) \in \mathrm{tp}(h/G)$, a contradiction. $\square$

We might equally well have taken translates on the right rather than on the left.

Claim. If G is $|T|^+$-saturated, then for any two generic types p and p' there is $g \in G$ with $gp = p'$ (where $gp = \{\varphi(x) : \varphi(gx) \in p\}$).

Proof of Claim: Let G_0 be an elementary substructure of G of cardinality $|T|$. For any $\varphi(x) \in p'\lceil G_0$ there is $g \in G$ with $\varphi(gx) \in p$, since translates of φ cover G. So $G \models \exists x \, d_p y \, \varphi(xy)$, and this sentence must be true in G_0 as well (it has parameters in G_0); by saturation there is a $g \in G$ such that $d_p x \, \varphi(gx)$ for all formulæ $\varphi \in p'\lceil G_0$. Hence for $h \models p$ we have $\varphi(gh)$ for all $\varphi \in p'\lceil G_0$; as furthermore $\mathrm{tp}(gh/G)$ is again generic and hence the unique non-forking extension of $p'\lceil G_0$, it follows that $gh \models p'$, whence $p' = gp$. $\square$

Claim. For any formula $\varphi(x, y)$ there is some $n < \omega$ such that $\varphi(x, m)$ is generic iff n translates of φ cover G.

Proof of Claim: Clearly, if finitely many translates of $\varphi(x, m)$ cover G, then $RC(\varphi(x, m)) = RC(G)$ and φ must contain generic types. On the other hand, if p is *any* generic type, then $\varphi(x, m)$ is generic iff $\exists x \, d_p y \, \varphi(xy, m)$ holds (as any generic type is a translate of p) iff finitely many translates of φ cover

G. By compactness, there must be the required bound n on the number of translates required. □

Define the φ-*stabilizer* of p in G as

$$\operatorname{stab}_G(p, \varphi) := \{g \in G : \forall y, z \, d_p x \, [\varphi(gyx, z) \leftrightarrow \varphi(yx, z)]\},$$

and the *stabilizer* of p in G as $\operatorname{stab}_G(p) := \bigcap_\varphi \operatorname{stab}_G(p, \varphi)$. Then $\operatorname{stab}_G(p, \varphi)$ is definable and $\operatorname{stab}_G(p)$ is type-definable over $\emptyset$.

Claim. p is generic iff $\operatorname{stab}_G(p) = G^0$.

In particular, $\operatorname{stab}_G(p, \varphi)$ has finite index in G for generic p, and there are only finitely many generic φ-types.

Proof of Claim: Let p be generic. As the number of generic types is bounded and the translate of a generic type is again generic, $\operatorname{stab}(p)$ must have bounded index in any model of G, and hence contain G^0. Conversely, p must specify its coset modulo every subgroup of finite index, whence $\operatorname{stab}(p) \le G^0$.

Now suppose $\operatorname{stab}(p) = G^0$. Then $RC(\operatorname{stab}(p)) = RC(G)$ and we can find a generic type q inside $\operatorname{stab}(p)$. If $g \models p$ and $h \models q$ with $g \mathop{\smile}\limits_G h$, then $\operatorname{tp}(g/G, h) = \operatorname{tp}(hg/G, h)$. Therefore $\operatorname{tp}(g/G) = \operatorname{tp}(hg/G)$, but the latter is generic. □

Example 2.0.1 Let $G = (\mathbb{Q}, +)$. Then there is only one non-algebraic type over $\emptyset$, and this must be the generic type; $G = G^0$.

Example 2.0.2 Let $G = (\mathbb{Z}, +)$. Then $G^0 = \bigcap nG$ (in a saturated model!); G^0 has a unique generic type p which says that x is divisible by n for all $n < \omega$. In general a type is generic iff it is non-algebraic (as $RC(G) = 1$), and it is determined by its class modulo nG, for all $n < \omega$. Note that these classes are all definable over the single parameter 1.

Example 2.0.3 Let K be a field. Then K^+ is connected, so K has just one additive generic type p. Hence $kp = p$ for any $k \in K^\times$, but this implies that p is also the unique multiplicative generic type, and $K^\times$ is connected as well. If the field is algebraically closed, then a type is generic iff it is transcendental.

Example 2.0.4 Let G be an algebraic group over an algebraically closed field K. Then g is generic over K iff it is a generic point for the variety G in the sense of algebraic geometry.

2.1 Generic Types

We shall now consider the general case of a semi-group living on some set in a stable structure. Hence we assume that G is a set interpretable in a stable structure (and we may as well suppose that G is the domain of that structure) together with an interpretable binary function $* : G \times G \to G$, such that there is some (not necessarily definable) infinite subset M of G on which $*$ induces a semi-group structure. We shall write $g * h$ or even gh instead of $*(g, h)$. If $*$ is only partially defined, we may always extend it definably to the whole of G; modifying it definably (outside M), we may assume that in addition $g1 = 1g = g$ for all $g \in G$ and the unit element $1 \in M$.

Important special cases are that of an interpretable group ($M = G$) and that of a group given by some infinite conjunction (M is the set of realizations of some partial type in a sufficiently saturated model). As the arguments are considerably easier in this context, readers unfamiliar with the theory might restrict their attention to it at a first reading; we shall indicate the appropriate simplifications at every step.

The main idea is to distinguish a special class of types in our theory, the *types generic for M*.

Definition 2.1.1 A definable subset φ of G is *generic for M* if there are finitely many elements $g_1, h_1, \ldots, g_n, h_n$ of M such that for any $m \in M$ the disjunction $\bigvee_{i=1}^{n} \varphi(g_i m h_i)$ is satisfied (in G).

So a set is generic if finitely many translates of it cover the whole — but note that these are "translates" by g_i^{-1} and h_i^{-1}, so to speak, and these inverses need not actually exist. More precisely, therefore, a set is generic if it contains one of finitely many translates of every element. Furthermore, it is clear that if $\varphi(gxh)$ is generic, so is $\varphi(x)$, but the converse need not hold (just consider the "multiplication" $xy := x$; the formula $g_0 = x$ is generic, but $g_0 = gxh$ is not for $g_0 \neq g$).

Lemma 2.1.1 *Let φ be a definable subset of G. Then φ or $\neg\varphi$ is generic for M.*

Proof: We suppose the lemma to be false and find inductively elements g_i and h_i in M, for $i < \omega$, such that φ is satisfied by $g_i h_j$ iff $i > j$. This will then contradict stability.

We choose any g_1 in M. Suppose we have already found $g_1, \ldots, g_{i+1}$ and $h_1, \ldots, h_i$ in M. As φ is not generic for M, there is an element h_{i+1} in M not satisfying $\bigvee_{j=1}^{i+1} \varphi(g_j x)$.

If we have already found $g_1, \ldots, g_i$ and $h_1, \ldots, h_i$ in M, then by the non-genericity of $\neg\varphi$ there is some g_{i+1} in M not satisfying $\bigvee_{j=1}^{i} \neg\varphi(x h_j)$. □

Lemma 2.1.2 *If φ and ψ are two non-generic subsets of G, then $\varphi \vee \psi$ is not generic for M either.*

Proof: Suppose otherwise. Then M contains elements $g_1, h_1, \ldots, g_n, h_n$ such that any m in M satisfies the disjunction $\bigvee_{i=1}^{n} \varphi(g_i m h_i) \vee \psi(g_i m h_i)$.

We put $\varphi_1(x) := \bigvee_{i=1}^n \varphi(g_i x h_i)$ and $\psi_1(x) := \bigvee_{i=1}^n \psi(g_i x h_i)$. Then by non-genericity of φ and ψ neither φ_1 nor ψ_1 is generic for M, and $(\varphi_1 \vee \psi_1)(m)$ holds for any m in M. Now by non-genericity of φ_1, its negation $\neg\varphi_1$ must be generic for M, so there are elements $g_1', h_1', \ldots, g_m', h_m'$ of M such that any $m \in M$ satisfies $\bigvee_{i=1}^m \left(\neg\varphi_1(g_i' m h_i')\right)$. But any $m \in M$ satisfying $\neg\varphi_1$ must satisfy ψ_1, whence any m in M satisfies $\bigvee_{i=1}^m \psi_1(g_i' m h_i')$, contradicting the non-genericity of ψ_1. $\square$

From now on let A be a superset of M.

Definition 2.1.2 A type $p \in S_1(A)$ is *generic for M* if all formulæ in p are generic for M. An element g is *generic* (for M) over A if $\mathrm{tp}(g/A)$ is generic for M.

Corollary 2.1.3 *A set π of formulæ generic for M which is closed under finite intersections can be completed to a type generic for M.*

In particular, the set of types over A generic for M is a closed subset of $S_1(A)$ (extending the empty partial 1-type).

Proof: Let $\pi' = \pi \cup \{\varphi(x) \in \mathfrak{L}(A) : \neg\varphi \text{ not generic for } M\}$. Since π is closed under finite intersections, π' must be consistent by Lemma 2.1.2, and any completion of it is a generic type for M. $\square$

Definition 2.1.3 $S_1^M(A) = \{p \in S_1(A) : p \text{ is finitely satisfiable in } M\}$.

As any type in $S_1^M(A)$ is finitely satisfiable in M, it cannot fork over M. And certainly any type generic for M is finitely satisfiable in M. If E is some A-definable equivalence relation intersecting M in a finite number of classes $Em_1, \ldots, Em_n$ (for $m_1, \ldots m_n \in M$), and if p is some type in $S_1^M(A)$, then p cannot contain the formula $\bigwedge_{i=1}^n \neg x E m_i$ by finite satisfiability. Therefore it must contain one of the formulæ $x E m_i$, for $1 \leq i \leq n$. By the Finite Equivalence Relation Theorem 0.2.22, any type in $S_1^M(A)$ must be stationary. Finally, we should remark that $S_1^M(A)$ is also a closed subset of $S_1(A)$.

On types in S_1^M we expect a behaviour similar to that of elements in M; in particular we shall try to extend the multiplicative structure on M to realizations of types in S_1^M. As we do not know the properties of $*$ outside of M (it may not even be associative), some such restriction is clearly necessary in general. However, if M is already a (type-)definable stable group and $\mathcal{G}$ is an elementary extension of G, then the elements of $\mathcal{G}$ whose types over M lie in $S_1^M(M)$ are exactly the elements of $M^{\mathcal{G}}$.

Lemma 2.1.4 *Let π be a set of formulæ over A which is finitely satisfiable in M. Then π can be completed to a type in $S_1^M(A)$.*

Proof: By Zorn's Lemma it is enough to show that for any such π and any formula $\varphi(x)$ with parameters in A either $\pi \cup \{\varphi\}$ or $\pi \cup \{\neg\varphi\}$ is finitely satisfiable in M. However, if there were formulæ $\varphi_1, \ldots, \varphi_n, \psi_1, \ldots, \psi_m$ in π such that both $\varphi_1 \wedge \cdots \wedge \varphi_n \wedge \varphi$ and $\psi_1 \wedge \cdots \wedge \psi_m \wedge \neg\varphi$ were not satisfiable in M, then also $\varphi_1 \wedge \cdots \wedge \varphi_n \wedge \psi_1 \wedge \cdots \wedge \psi_m$ would not be satisfiable in M, a contradiction. $\square$

Corollary 2.1.5 *Let B be a superset of A. Then any type p in $S_1^M(A)$ has exactly one extension in $S_1^M(B)$, namely its (unique) non-forking extension.*

Proof: The existence has just been shown, and the uniqueness follows from the fact that any type in $S_1^M(B)$ does not even fork over M and has a stationary restriction to M. $\square$

Corollary 2.1.6 *The following are equivalent:*

1. $\mathrm{tp}(\bar{a}/A)$ and $\mathrm{tp}(\bar{b}/A\bar{a})$ are finitely satisfiable in M.

2. $\mathrm{tp}(\bar{a}\bar{b}/A)$ is finitely satisfiable in M and $\bar{a} \underset{A}{\smile} \bar{b}$.

3. $\mathrm{tp}(\bar{a}/A)$ and $\mathrm{tp}(\bar{b}/A)$ are finitely satisfiable in M and $\bar{a} \underset{A}{\smile} \bar{b}$.

Proof: (1. $\Rightarrow$ 2.) Consider $\varphi(\bar{x}, \bar{y}) \in \mathrm{tp}(\bar{a}\bar{b}/A)$. Since $\mathrm{tp}(\bar{b}/A\bar{a})$ is finitely satisfiable in M, there is $\bar{b}' \in M$ with $\models \varphi(\bar{a}, \bar{b}')$, and now we can find $\bar{a}' \in M$ with $\models \varphi(\bar{a}', \bar{b}')$ by finite satisfiability of $\mathrm{tp}(\bar{a}/A)$ in M. The independence follows from Corollary 2.1.5, as $\mathrm{tp}(\bar{b}/A\bar{a})$ may not even fork over M.

(2. $\Rightarrow$ 3.) is immediate, and (3. $\Rightarrow$ 1.) follows immediately from Corollary 2.1.5. $\square$

Note that in the above we did not use the semi-group structure of M; the last three results also hold if M is just an arbitrary subset of A. On the other hand, generic types for M are obviously finitely satisfiable in M. So if $p \in S(A)$ is generic for M, it can be completed to a type q over B generic for M. But q may not fork over M, hence not over A, and must be the unique non-forking extension of p to B.

However, we shall also want to deal with forking extensions of types in $S_1^M(A)$. In order to do this, we shall have to consider another closed set of types.

Definition 2.1.4 $S_1^*(A)$ is the set of types over A whose restriction to M is finitely satisfiable in M.

Any extension of a type in $S_1^*(A)$ to a superset B of A is in $S_1^*(B)$. Furthermore, if $\mathrm{tp}(a/A) \in S_1^*(A)$, $\mathrm{tp}(\bar{g}/A)$ is finitely satisfiable in M, and a and $\bar{g}$ are independent over A, then any product of a and elements in the tuple $\bar{g}$ is associative, as $\text{tp}(\bar{g}/A, a)$ is finitely satisfiable in M, and non-associativity is a first-order formula.

If M is a (type-)definable group, then S_1^* is just the collection of types containing the formula(e) "$x \in M$".

Definition 2.1.5 Let $p \in S_1^*(A)$ and $g, h \in A$ be such that $p(x)$ proves that $\text{tp}(g, h/M, x)$ is finitely satisfiable in M. We define the *translate of p by g and h* as $gph = \{\varphi(x) : \varphi(gxh) \in p\}$.

The finite satisfiability condition is equivalent to saying that $\text{tp}(g, h/M)$ is finitely satisfiable in M and p proves $x \underset{M}{\downarrow} g, h$. It implies in particular that the product gph is associative, and hence well-defined. Note that if a realizes p, then gah realizes gph. Therefore this is a consistent set of formulæ; either $\varphi(gxh)$ or $\neg\varphi(gxh)$ lies in p for any $\varphi(x)$, it is also complete and gph is a type over A. Furthermore, it is again in $S_1^*(A)$: if $\psi(gah)$ is a true formula with parameters in M, then we find first $g', h' \in M$ such that $\psi(g'ah')$ holds, and then $a' \in M$ such that $\psi(g'a'h')$ holds. Obviously $g'a'h' \in M$.

Clearly, if p is finitely satisfiable in M and $g, h \in M$, then gph is finitely satisfiable in M: if $\varphi(x) \in gph$, then $\varphi(gxh) \in p$, so by finite satisfiability of p there is $m \in M$ such that $\models \varphi(gmh)$, and $gmh \in M$ as well. Furthermore, if p is generic for M, so is gph: if $\varphi(x) \in gph$, then $\varphi(gxh) \in p$, and genericity of the latter implies genericity of the former (using associativity of the product).

We shall now define a suitable rank on types, which characterizes forking and is translation-invariant (i.e. the ranks of a type p and a translate gph should be equal). In the (type-)definable case, we may just take $R_\varphi(p)$, where we only consider *stratified* formulæ $\varphi(uxv, \bar{y})$. In general, however, we have to take care of the possible lack of associativity and inverses. The main problem is that the rank might conceivably go down when multiplying by some elements in M, unless we have both inverses and associativity of multiplication. In order to remedy this problem, we shall *stratify* our rank:

Definition 2.1.6 Let $p \in S_1^*(A)$ and φ be a formula. Then the *stratified φ-rank $SR_\varphi^M(p)$* is the smallest n such that there is a non-forking extension q of p over some superset B of A, and $g, h \in B$ with $\text{tp}(g, h/A) \in S_2^M(A)$, such that $R_\varphi(gqh) = n$.

The conditions in particular imply that gqh is a translate of q.

Remark 2.1.1 Note that if $SR_\varphi^M(p) \leq n$ and $q \in S(B)$ is a non-forking extension of $p \in S_1^*(A)$ with translate gqh and such that $SR_\varphi^M(p) = R_\varphi(gqh) \leq n$, then there is a formula $\psi'(x, \bar{b})$ in gqh restricting the φ-rank, and $\psi'(gxh) \in q$. Hence $\psi(x) := d_{\text{tp}(\bar{b}/A)}\bar{y}\,\psi'(x, \bar{y})$ is a formula over $\text{acl}(A)$ restricting the stratified φ-rank, and the union of its finitely many $\text{acl}(A)$-conjugates is a formula in p restricting the stratified φ-rank. In other words, $SR_\varphi^M(q) \leq n$ for all q containing ψ, and "$SR_\varphi^M(p) \leq n$" is an open condition.

Remark 2.1.2 Suppose $q \in S_1^*(B)$ is an extension of $p \in S_1(A)$. Consider a non-forking extension p' of p over B' with translate $gp'h$ such that $SR_\varphi^M(p) = R_\varphi(gp'h)$, and realize q by a and p' by a'. As $\operatorname{tp}(a/A) = p = \operatorname{tp}(a'/A)$, we may assume that $a = a'$, and furthermore $B \downarrow_{Aa} B'$. Since $a \downarrow_A B'$ we get $B' \downarrow_A aB$, whence $a \downarrow_B B'$. Put $q' = \operatorname{tp}(a/BB')$. Note that $g, h \downarrow_A aB$, so $\operatorname{tp}(g, h/B)$ is finitely satisfiable in M, and the translate $gq'h$ is well-defined and realized by gah. Now

$$SR_\varphi^M(p) = R_\varphi(gp'h) = R_\varphi(gah/B) \geq R_\varphi(gah/BB') = R_\varphi(gq'h) \geq SR_\varphi^M(q),$$

and the stratified rank is decreasing.

If now q is a non-forking extension of p, then $SR_\varphi^M(q) \geq SR_\varphi^M(p)$, since the set of translates of non-forking extensions of q by suitable elements is a subset of the corresponding set for p. Hence $SR_\varphi^M(p) = SR_\varphi^M(q)$.

The next lemma deals with the behaviour of stratified rank under translation.

Lemma 2.1.7 Let $p \in S_1^*(A)$ and q be a non-forking extension of p to some superset B of A. Suppose $g, h \in B$ satisfy $\operatorname{tp}(g, h/A) \in S_2^M(A)$. Then $SR_\varphi^M(p) \leq SR_\varphi^M(gqh)$ for any formula φ.

Proof: Consider a superset B' of B and a non-forking extension q' of gqh over B'. If a realizes q, then gah realizes gqh, and we may conjugate B' over B so that $gah \models q'$ and $a \downarrow_B B'$. Put $p' = \operatorname{tp}(a/B')$, so $q' = gp'h$; note that $a \downarrow_A B'$.

If $g', h' \in B'$ are such that $\operatorname{tp}(g', h'/B)$ is finitely satisfiable in M, then $\operatorname{tp}(g', h', g, h/A)$ is finitely satisfiable in M by Corollary 2.1.6; as $a \downarrow_A g', h', g, h$ this is also true of $\operatorname{tp}(g', h', g, h/A, a)$ by Corollary 2.1.5, and hence also of $\operatorname{tp}(g'g, hh'/A, a)$. Because $p \in S_1^*(A)$, the product $g'gahh'$ is associative, and $(g'g)p'(hh') = g'(gp'h)h' = g'q'h'$. Therefore

$$SR_\varphi^M(p) \leq R_\varphi((g'g)p'(hh')) = R_\varphi(g'q'h');$$

as B' (and g', h') was arbitrary, $SR_\varphi^M(p) \leq SR_\varphi^M(gqh)$. $\square$

All we have done until now would also work for the constant multiplication $xy = c$, or the projections $xy = x$ or $xy = y$. We shall hence have to make an additional assumption on $\langle M, * \rangle$ in order to proceed to our construction of a group. In Lemma 1.0.1 of chapter 1 we have seen that (left and right) cancellation is an assumption which in a stable context yields the existence of inverses.

Definition 2.1.7 Let $\langle M, * \rangle$ be a semi-group. M has *generic cancellation* if whenever a, a', g, h, A are such that $\{g, h\}$ is independent over A of a and of a', $\operatorname{tp}(a/A)$ and $\operatorname{tp}(a'/A)$ are in $S_1^*(A)$, $\operatorname{tp}(g, h/A) \in S_2^M(A)$, and

$\text{tp}(gah/A, g, h) = \text{tp}(ga'h/A, g, h)$, then $\text{tp}(a/A) = \text{tp}(a'/A)$. In other words, M has generic cancellation if any translate $\text{tp}(gah/A, g, h)$ determines the original type $\text{tp}(a/A)$.

Proposition 2.1.8 *Suppose M is a group. Then M has generic cancellation.*

Proof: Suppose otherwise and let this be witnessed by A, a, a', g, h. We may choose a' independent of a over A, g, h, so that the triple $a, a', (g, h)$ is independent over A; furthermore we may assume that $\text{stp}(gah/A, g, h) = \text{stp}(ga'h/A, g, h)$. As $\text{tp}(g, h/A)$ is finitely satisfiable in M, there are elements g', h' such that $\text{tp}(g, h, g', h'/A) \in S_4^M(A)$ and $g'g = 1 = hh'$. We may choose g', h' independent of a, a' over A, g, h, so the products $g'gahh'$ and $g'ga'hh'$ are associative, and both gah and $ga'h$ are independent of g', h' over A, g, h. This implies $\text{tp}(gah/A, g, h, g', h') = \text{tp}(ga'h/A, g, h, g', h')$, whence

$$\begin{aligned} \text{tp}(a/A, g, h, g', h') &= \text{tp}(g'gahh'/A, g, h, g', h') \\ &= \text{tp}(g'ga'hh'/A, g, h, g', h') = \text{tp}(a'/A, g, h, g', h'), \end{aligned}$$

and in particular $\text{tp}(a/A) = \text{tp}(a'/A)$, a contradiction. $\square$

From now on we assume that M has generic cancellation.

Proposition 2.1.9 *Suppose $q \in S_1(B)$ is an extension of $p \in S_1^*(A)$. Then q does not fork over A iff $SR_\varphi^M(q) = SR_\varphi^M(p)$ for every formula $\varphi(x, \bar{y})$.*

Proof: Remark 2.1.2 already gives us the implication from left to right. So suppose that q does fork over A. We may assume that A is algebraically closed and B is a big saturated model (again using Remark 2.1.2 to replace q by a non-forking extension). Let C be a canonical base for q. Then C is not algebraic over A and has a distinct A-conjugate C', which is a base for a distinct A-conjugate q' of q. Consider a completion of the partial type

$$\{\forall \bar{y} [d_q x \, \psi(gxh, \bar{y}) \leftrightarrow d_{q'} x \, \psi(gxh, \bar{y})] : \psi \in \mathfrak{L}\} \cup S_2^M(B)$$

in variables g, h over B. By generic cancellation, this is inconsistent, so there is already some inconsistent finite part π of it. Let X be a small subset of B containing A, C, C', and all parameters mentioned in π. Let $\{\psi_i : i \leq k\}$ be all the formulæ occuring in π as $d_q \psi_i \leftrightarrow d_{q'} \psi_i$, and consider the formula

$$\psi(x, \bar{y}, z_0, \dots, z_k) := \bigvee_{i \leq k} \Big(\psi_i(x, \bar{y}) \wedge z_i = 1 \wedge \bigwedge_{j < i} z_j \neq 1 \Big).$$

Clearly, all translates of q and q' by realizations of types in $S_2^M(X)$ differ by a ψ-formula.

Now take a superset A' of A with elements $g, h \in A'$ satisfying $\text{tp}(g, h/A) \in S_2^M(A)$, and a non-forking extension p' of p to A', such that $SR_\psi^M(p) =$

$R_\psi(gp'h)$. By the saturation of B, we may assume that A' is contained in B and independent over A of X. Now if a realizes $q\lceil_X$ independently of A' over X, then a realizes p and $A' \underset{A}{\smile} Xa$, so a realizes the unique non-forking extension of p to A', namely p'. It follows that q restricts to p' over A', and the same holds for q'. Furthermore, since $\mathrm{tp}(A'C/A) = \mathrm{tp}(A'C'/A)$, by saturation we can conjugate C to C' and hence q to q' over A'. Note that $g, h \underset{A}{\smile} X$ implies $\mathrm{tp}(g, h/X) \in S_2^M(X)$, and the translates gqh and $gq'h$ are well-defined.

Suppose $R_\psi(gqh) = n$. As q' is conjugate to q over A', this implies $R_\psi(gq'h) = n$ as well; since both gqh and $gq'h$ extend $gp'h$ and differ by a ψ-formula, this yields

$$SR_\psi^M(q) = SR_\psi^M(q\lceil X) \le R_\psi(gqh) < R_\psi(gp'h) = SR_\psi^M(p). \ \square$$

Lemma 2.1.10 *Let p be generic for M over A. Then for every formula $\varphi(x, \bar{y})$ the rank $SR_\varphi^M(p)$ is maximal among types in $S_1^*(A)$.*

Proof: Firstly, as generic types do not fork over M and have the same φ-ranks as their restrictions to M by Proposition 2.1.9, and as extensions in general cannot have greater φ-rank by Remark 2.1.2, it is sufficient to consider types over M. But now if $\psi(x)$ is a formula in the generic type $p \in S_1(M)$, then finitely many translates of ψ by elements in M cover M. So for any type $q \in S_1^M(M)$ there are g and h in M such that q contains $\psi(gxh)$, whence gqh contains $\psi(x)$. By Lemma 2.1.7

$$SR_\varphi^M(\psi) \ge SR_\varphi^M(gqh) \ge SR_\varphi^M(q).$$

As ψ was an arbitrary formula in p, the assertion follows from Remark 2.1.1. $\square$

Proposition 2.1.11 *Let $\mathrm{tp}(a/A)$ be generic for M and $\mathrm{tp}(g, h/A)$ finitely satisfiable in M, with a independent of g, h over A. Then $\mathrm{tp}(gah/A, g, h)$ is generic for M.*

Proof: Let $\varphi(x)$ be a formula in $\mathrm{tp}(gah/A)$. As $\mathrm{tp}(g, h/A, a)$ is finitely satisfiable in M and $\varphi(gah)$ holds, there are g', h' in M such that $\varphi(g'ah')$ holds. But then $\varphi(g'xh') \in \mathrm{tp}(a/A)$, so this formula is generic. Hence $\varphi(x)$ must be generic as well, and $\mathrm{tp}(gah/A)$ is generic.

We now want to prove that $gah \underset{A}{\smile} g, h$. So let $\varphi(x, \bar{y})$ be any formula. We have

$$SR_\varphi^M(a/A) \le SR_\varphi^M(gah/A \cup \{g, h\}) \le SR_\varphi^M(gah/A) \le SR_\varphi^M(a/A).$$

The first inequality holds by Lemma 2.1.7, the second one by Remark 2.1.2, and the last inequality is implied by Lemma 2.1.10, as $SR_\varphi^M(a/A)$ must be maximal.

Hence equality holds all the way through. So *gah* is independent of g, h over A by Proposition 2.1.9. $\square$

In particular, $\mathrm{tp}(ah/A, h)$ and $\mathrm{tp}(ga/A, g)$ are generic. Proposition 2.1.11 plays a pivotal rôle in our development: everything done so far leads to it, and the remainder of this section will rest on it. In particular, we shall make no more use of stratified rank.

If we have a generic type p for M and a type q finitely satisfiable in M, there is a product type $p*q$, namely the type of the product ab of a realization a of p and an independent realization b of q (remember that these types are stationary). By Proposition 2.1.11 this is again a generic type, and in fact ab is independent over A of b. If q is also generic, then ab is also independent over A of a (but, of course, not of both a and b at the same time). This product is clearly associative: $p * q * r$ is just the type of the product of independent realizations of p, q and r.

Remark 2.1.3 If p and q are both generic over A and $r \in S_1^M(A)$ with $p * r = q * r$, then $p = q$.

Proof: Let p, q, r be realized independently over A by a, b, c, respectively. Then $\mathrm{tp}(ac/A) = \mathrm{tp}(bc/A)$ by hypothesis, and $ac \underset{A}{\downarrow} c$ and $bc \underset{A}{\downarrow} c$ by Proposition 2.1.11. Hence both ac and bc realize the unique non-forking extension of $\mathrm{tp}(ac/A)$ to A, c, whence $\mathrm{tp}(ac/A, c) = \mathrm{tp}(bc/A, c)$ and $p = \mathrm{tp}(a/A) = \mathrm{tp}(b/A) = q$ by generic cancellation. $\square$

We should note at this point

Lemma 2.1.12 *If P is a set of types in $S_1^M(A)$ such that $S_1^M(A)$ acts on P transitively by multiplication, then P consists of generic types for M.*

Proof: Let p be in P and q generic for M. Then $p * q$ is generic for M and in P. So P contains a generic type for M. But if $p \in P$ is generic and $q \in P$ is arbitrary, then by transitivity there is some $r \in S_1^M(A)$ such that $q = p * r$, whence q is generic as well. $\square$

Lemma 2.1.13 *Let p and q be generic types over A. Then there are two types r and s in $S_1^M(A)$ such that $p = r * q * s$.*

Proof: For any formula $\varphi(x)$ in p there are g, h in M such that $\varphi(gxh) \in q$. Hence $d_q x \varphi(gxh)$ holds, and the set $\{d_q x \, \varphi(yxz) : \varphi(x) \in p\}$ is finitely satisfiable in M and can be extended to a type $t(y, z) \in S_2^M(A)$. Then for $(g, h) \models t$ and $a \models q$ independently of g, h over A, the product gah realizes p. But if q is generic for M, then $\mathrm{tp}(ga/A, g, h)$ is generic for M; in particular ga and h are independent over A. If $r = \mathrm{tp}(g/A)$ and $s = \mathrm{tp}(h/A)$, then $\mathrm{tp}(ga/A) = r * q$ and $\mathrm{tp}(gah/A) = (r * q) * s$. However, this equals p. It follows that every generic type is a two-sided translate of every other one. $\square$

Lemma 2.1.14 *For every formula $\varphi(x, \bar{y})$ there are only finitely many different generic φ-types, i.e. the set $\{p\restriction_\varphi : p$ generic for $M\}$ is finite.*

Proof: Let p be generic for M, and suppose first that MpM contains infinitely many different φ-types. Then we could find an infinite subset X of $M \times M$ such that for any $(g', h') \neq (g, h)$ in X we have

$$\models \exists \bar{y} \, \neg[d_p x \, \varphi(gxh, \bar{y}) \leftrightarrow d_p x \, \varphi(g'xh', \bar{y})].$$

By Lemma 2.1.4 there is a non-algebraic type q over A finitely satisfiable in X. Let $(g_i, h_i : i \in I)$ be a Morley sequence in q. So for $i \neq j$ we get

$$\models \exists \bar{y} \, \neg[d_p x \varphi(g_i x h_i, \bar{y}) \leftrightarrow d_p x \varphi(g_j x h_j, \bar{y})].$$

Now consider a realization a of p over A and independent of $\{g_i, h_i : i \in I\}$ over A. As $\{g_i, h_i : i \in I\}$ is independent over A, the type $\mathrm{tp}(g_i, h_i/A \cup \{g_j, h_j : j \neq i\})$ is finitely satisfiable in M. On the other hand, a is generic for M over $A \cup \{g_i, h_i : i \in I\}$, so by Proposition 2.1.11 the types $p_i = \mathrm{tp}(g_i a h_i/M \cup \{g_j, h_j : j \in I\})$ are also generic for M. But they have φ-definitions $d_{p_i} x \varphi(x, \bar{y}) = d_p x(g_i x h_i, \bar{y})$, hence they are all different. On the other hand they do not fork over M, which leads to a contradiction for big I.

Hence there are only finitely many φ-types among the two-sided translates of p by elements of M. So there are $g_1, h_1, \ldots, g_n, h_n$ in M such that

$$\bigvee_{i=1}^{n} \forall \bar{y} \, [d_p x \, \varphi(g_i x h_i, \bar{y}) \leftrightarrow d_p x \, \varphi(mxm', \bar{y})]$$

holds for all $m, m' \in M$ and must hence also be satisfied by any pair whose type over A is finitely satisfiable in M. It follows that there are only finitely many φ-types in $\{q_1 * p * q_2 : q_1, q_2 \in S_1^M(A)\}$, and all of them occur in MpM. The result now follows from Lemma 2.1.13. $\square$

Lemma 2.1.15 *Let p be finitely satisfiable in M and q generic. Then there is a generic r such that $q = r * p$. In particular, for every generic formula φ there is some $m \in M$ with $\varphi \in mp$. Furthermore there are finitely many elements $g_1, \ldots, g_n$ in M such that for any $m \in M$ we have $\bigvee_{i=1}^n \varphi(g_i m)$.*

Proof: Let $\varphi(x, \bar{y})$ be a formula, and say that two generic types p and q are *φ-equivalent* if for all generic types r the φ-type of $p * r$ equals the φ-type of $q * r$. Then if s is finitely satisfiable in M and p and q are φ-equivalent, for any generic r the type $s * r$ is again generic, and the φ-type of $(p * s) * r = p * (s * r)$ equals the φ-type of $(q * s) * r = q * (s * r)$. Hence φ-equivalence is preserved under multiplication on the right. On the other hand, there are only finitely many equivalence classes: there are only finitely many generic definitions

$\psi_i(x, \bar{y}) = d_i y\, \varphi(xy, \bar{y})$, and p and q are φ-equivalent iff they have the same ψ_i-definition for all i, and there are only finitely many generic definitions for those.

Now consider a type p finitely satisfiable in M. We can find some power $p^k = p * \cdots * p$ (again a type finitely satisfiable in M) such that the number of φ-equivalence classes C_φ in $\{q * p^k : q \text{ generic}\}$ is minimal. Then further multiplication by p just permutes these classes, and there is some power p^{k_φ} which fixes them. Note that right multiplication by some type $t(z) \in S_1^M(A)$ stabilizes the φ-equivalence class of q iff the formula

$$\forall \bar{y}\, d_q x \bigwedge_i \Big(d_i y\, \varphi(xy, \bar{y}) \leftrightarrow d_i y\, \varphi((xz)y, \bar{y}) \Big)$$

is contained in t (this uses Proposition 2.1.11), and t maps to C_φ iff t contains the formula

$$\bigvee_j \forall \bar{y}\, d_q x\, d_{q_j} x' \bigwedge_i \Big(d_i y\, \varphi((xz)y, \bar{y}) \leftrightarrow d_i y\, \varphi(x'y, \bar{y}) \Big),$$

where the disjunction runs over a set $\{q_j\}_j$ of representatives of the classes in C_φ. Note that if t stabilizes C_φ and maps to C_φ, it is necessarily idempotent, and maps onto C_φ.

If ψ is another formula, we likewise find a power p^{k_ψ} which fixes the ψ-equivalence classes C_ψ in $\{q * p^k : q \text{ generic}, k \text{ big}\}$, and $p^{k_\varphi k_\psi}$ stabilizes both C_φ and C_ψ. As the idempotency condition is first-order expressible, there is a type t finitely satisfiable in M, such that multiplication by t maps onto C_φ and stabilizes C_φ for all formulæ φ. Hence for any generic types q and r we get that $q * t * r$ and $q * t * t * r$ have the same φ-types for all φ, and thus are identical. By generic cancellation ($q * t$ and $q * t * t$ are both generic), we get $q * t = q$. In particular, C_φ must encompass all the possible classes, and multiplication by p permutes them.

Now for any generic type q the formula

$$\forall \bar{y}\, [d_p x \varphi(x, \bar{y}) \leftrightarrow d_q y\, \varphi(zy, \bar{y})]$$

says that $\mathrm{tp}(z) * p = q$; we have just seen that finite conjunctions (for various φ) are generic, and so there is a generic completion r with $r * p = q$.

Suppose now that $\varphi(x)$ is generic, and $p \in S_1^M(A)$. Take a generic type q containing φ, and r with $q = r * p$. Then for realizations $g \models r$ and $h \models p$, independent over A, we get $\models \varphi(gh)$. As $\mathrm{tp}(g/A, h)$ is finitely satisfiable in M, we find $m \in M$ with $\models \varphi(mh)$. But h realizes p, so $\varphi(mx) \in p$ and $\varphi(x) \in mp$.

Now if the last assertion were false, then the set $\{\neg\varphi(mx) : m \in M\}$ would be finitely satisfiable in M and could be completed to a type in $S_1^M(A)$, contradicting the previous paragraph. $\square$

Of course, we might just as well have translated on the other side.

Remark 2.1.4 We should pause at this point and consider the case of a definable, or even type-definable group G, given by a collection $\{\varphi_i(x) : i \in I\}$ of formulæ. We shall take a $|I|^+$-saturated model $\mathfrak{N}$ of the theory and the set M of realizations of $\bigwedge_{i \in I} \varphi_i$, and define a type to be generic for G if it is generic for M. The question is how genericity behaves under elementary extension, i.e. if $\mathfrak{N} \prec \mathfrak{N}'$ and M' is the set of realizations of $\bigwedge_{i \in I} \varphi_i$ in $\mathfrak{N}'$, are the generic types for M' exactly the non-forking extensions of those for M?

First, we should note that since there are only finitely many generic φ-types for all formulæ $\varphi(x, \bar{y})$, defining schemes for generic types for the type-definable group have only finitely many conjugates under automorphisms fixing the parameters of G. Hence the generic types do not fork over those parameters. Second, if $\varphi(x, \bar{m})$ is a generic formula for M with parameters in $\mathfrak{N}$, then by compactness and saturation finitely many translates of φ cover not only M, but even a finite subintersection of $\bigwedge_{i \in I} \varphi_i$, so being generic is an open condition on $\bar{m}$. On the other hand, if p is any generic type, then $\varphi(x, \bar{m})$ is generic iff there is $g \in G$ with $\models d_p x\, \varphi(gx, \bar{m})$, and this is a closed condition. By compactness, being generic is definable, and there must be a finite bound on the number of translates needed to cover M, depending only on the formula $\varphi(x, \bar{y})$. But this condition is preserved under elementary extension, and genericity for M equals genericity for M'.

The following important theorem of Hrushovski tells us how to reconstruct a group from a closed set of types and a generic multiplication. It is a model-theoretic version of Weil's Theorem on the reconstruction of a group from generic data.

Theorem 2.1.16 *Let P be a closed set of stationary types over $\emptyset$, and let $*$ be a partial $\emptyset$-definable operation defined on pairs of independent realizations of types in P, such that there are only finitely many different φ-types for the elements of P, where $\varphi(x, a, b, c, d)$ is the formula $[a * (b * x) = c * (d * x)]$. Suppose $*$ satisfies*

1. *(generic independence) for independent realizations $a \models p$ and $b \models q$ the product $a * b$ realizes a type in P and is independent of a and of b,*

2. *(generic associativity) for three independent realizations a, b and c of types in P we have $(a * b) * c = a * (b * c)$, and*

3. *(surjectivity) for any independent a and b realizing some types in P there are c, c' realizing types in P independently of $\{a, b\}$, with $a = b * c$ and $a = c' * b$.*

Then there are a type-definable group G, and a definable isomorphism between P and the collection of generic types of G, such that generically $*$ is mapped to the group multiplication. G is unique up to definable isomorphism.

Hrushovski calls the first condition *generic cancellation*.

Proof: We consider the set of definable functions f from P to P (i.e. realizations of types in P) such that if x realizes some $p \in P$ independently of the parameters used for f, then $f(x)$ realizes some $q \in P$ independently of those parameters. On this set we can define an equivalence relation by

$$f \equiv f' \iff \bigwedge_{p \in P} d_p x \, [f(x) = f'(x)].$$

The class of a function f modulo this equivalence relation is called the *germ* of f. If $f \equiv f'$ and $g \equiv g'$, and x satisfies the non-forking extension of some type in P over the parameters A used for f, f', g and g', then $g(x) = g'(x)$ satisfies some type in P independently of A, whence $f(g(x)) = f'(g'(x))$. Therefore $f \circ g \equiv f' \circ g'$, and composition of functions induces composition of germs.

We now consider functions of the form $f_a : x \mapsto a * x$, for any a realizing a type in P. If x realizes some type in P independently of a, then by generic independence, $f_a(x)$ realizes some type in P independently of a, so f_a is of the required type and we can consider the corresponding germs. We claim that for any a, b and c realizing types in P there are e and f also realizing types in P, such that $f_a \circ f_b \circ f_c \equiv f_d \circ f_e$. Indeed, by surjectivity we first find b' and b'' independently realizing some types in P and with $b = b' * b''$; and we may choose them in addition independent over b of a, c. Then b' is independent of b by generic independence, so b' is independent of a, and similarly b'' is independent of c. But now for any x realizing a type in P independently of everything else, generic independence implies that $c * x$ is independent of b', b'' and $b'' * c * x$ is independent of a, b'. Hence generic associativity yields

$$a*(b*(c*x)) = a*((b'*b'')*(c*x)) = a*(b'*(b''*(c*x))) = (a*b')*((b''*c)*x),$$

and $d = a * b'$ and $e = b'' * c$ realize types in P.

Therefore the set of germs of functions $f_a \circ f_b$, where a and b realize types in P, together with the identity, is closed under composition. Now the set of functions $f_a \circ f_b$ is uniformly definable; as there are only finitely many $[a * (b * x) = c * (d * x)]$-types among the types in P, the equivalence relation is definable as a finite conjunction, and the set G of germs is type-definable. Furthermore if $g \circ f \equiv h \circ f$ and x runs through realizations of all types in P independently of the parameters A needed for g, f and h, then by surjectivity $f(x)$ runs through realizations of all types in P independently of A, whence $g \equiv h$. So G has a unit and right cancellation, and must be a group by Lemma 1.0.1.

The map $\sigma : a \mapsto f_a / \equiv$ is injective: If $f_a \equiv f_b$, then for any realization c of some type in P independent of a and b, we first find some a_1 with $a = a_1 * c$, and some b_1 with $b = b_1 * c$ by surjectivity. But for x realizing a type in P independently of everything else,

$$a_1 * (c * x) = (a_1 * c) * x = a * x = b * x = (b_1 * c) * x = b_1 * (c * x).$$

However, by generic independence and surjectivity, $c * x$ realizes all the types in P independently of a_1, b_1 as x varies, so $f_{a_1} \equiv f_{b_1}$. We now choose some d such that $f_d \equiv f_{a_1} \equiv f_{b_1}$ and d is independent of a_1, b_1, c over this germ. Then c is generic over a_1, d and over b_1, d, so $a = a_1 * c = d * c = b_1 * c = b$.

The collection of the images $\sigma(p)$ of the types $p \in P$ is invariant under translation by any type q in S_1^G: if we realize q as the germ of the function $x \mapsto b * c * x$, where b and c realize types in P, and $\sigma(p)$ as the germ of the function $x \mapsto a * x$, with a independent of b, c, then the translate is the germ of $x \mapsto a * b * c * x$, and $a * b$ realizes a type in P independent of b, c, so $a * b * c$ realizes a type in P. Hence $\sigma(p)$ is generic for G; and injectivity of σ implies that $p \neq p'$ in P have different images $\sigma(p) \neq \sigma(p')$. Clearly, $*$ is mapped generically to the group multiplication. Furthermore, if a realizes some generic type q of G and b realizes $\sigma(p)$ for $p \in P$ independently of a, then ab^{-1} realizes a generic type q' of G independently of b. Hence $q' * p = q$ must lie in the image of P, so P maps surjectively onto the generic types of G.

Finally, G is determined uniquely up to isomorphism, as any group is isomorphic to the group of germs of its translations. $\square$

We shall now show that a type-definable group is actually a subgroup of a definable group.

Theorem 2.1.17 *Let G be a type-definable group. Then there is a definable supergroup for G.*

Proof: We work in a saturated model. By compactness there is a definable superset G_0 of G such that for all x, y and z in G_0 we have $(x * y) * z = x * (y * z)$ and $x * 1 = x = 1 * x$. Let G be type-defined by the set $\{\varphi_i : i \in I\}$, where the φ_i are closed under finite conjunctions, and put $\psi_i(x) = \bigwedge_j d_j y\, \varphi_i(x * y)$, where d_j runs through the finitely many different $\varphi_i(x * y)$-definitions for generic types of G. Then every ψ_i contains G. On the other hand, if $x \in G_0$ satisfies all the ψ_i and y is generic over x, then $x * y$ lies in G. Since $y \in G$, by associativity $x = x * y * y^{-1}$ also lies in G (note that y has an inverse, as G is a group). Thus $G = \bigcap \psi_i(G_0)$.

By compactness we find some subset $\psi_i = G_1$ of G_0 such that for all x and y in G_1 their product $x * y$ lies in G_0. But then for x in G_1 and y in G the product $x * y$ must lie in G_1, because firstly $x * y$ lies in G_0, and secondly for z generic over x and y the product $y * z$ is generic. As $x \in G_1$, we have $\models \varphi_i(x * (y * z))$, thus $\models \varphi_i((x * y) * z)$, and $x * y \in G_1$. We now put $G_2 = \{g \in G_1 : G_1 g \subseteq G_1\}$. This is a superset of G closed under multiplication: if $g, h \in G_2$, then $gh \in G_1 h \subseteq G_1$ and $G_1 gh \subseteq G_1 h \subseteq G_1$, whence $gh \in G_2$. Finally let $G_3 = \{g \in G_2 : \exists h \in G_2 \; gh = 1\}$ be the set of invertible elements of G_2. This is our definable supergroup. $\square$

Corollary 2.1.18 *A type-definable group G in a stable group is $\bigwedge$-definable.*

Proof: We use the notation of the proof of Theorem 2.1.17. So first we find a definable supergroup G_0 of G, and G is given as intersection of the definable subsets φ_i of G_0. We define ψ_i as above, and note that by the argument in the last paragraph, every ψ_i contains a definable supergroup G_i of G, and $\bigcap_i G_i = G$. $\square$

Note that we only needed parameters already used to type-define G. Putting the various pieces together, we get

Theorem 2.1.19 *Let X be a definable set in some stable structure, with a definable partial binary function $*$, such that $*$ induces a semi-group structure on some substable subset M of X. Suppose that $(M, *)$ has generic cancellation. Then M is contained in a definable group. The intersection $\mathrm{dc}(M)$ has the same generic types as M.*

Note that by Proposition 2.1.8 generic cancellation is satisfied if M is a group.
Proof: If M is finite, by generic cancellation, it must be a finite group and the result is trivial. We shall therefore assume that M is infinite. We may restrict X, possibly add a new element 1 to M, and complete $*$ to a total function from $X \times X$ to X, such that for all x in X the identity $x * 1 = 1 * x = x$ holds. (Note that if M already has a left or a right unit, then by generic cancellation this must actually be a unit, which we choose as 1.) This will preserve generic cancellation. We add the elements of M to the language, and let P be the set of types generic for M. Then all the hypotheses for Theorem 2.1.16 are satisfied, so the realizations of P embed into a type-definable group G_0, which in turn is contained in a definable group G, such that multiplication on M is preserved generically by the embedding σ of P into G.

Now given any element $m \in M$, we may choose any generic element x for M (over $\emptyset$) and define the image $\theta(m)$ of m in G to be $\theta(m) := \sigma(mx)\sigma(x)^{-1}$. This does not depend on the choice of our generic element, for if y is another one independent of x, then by Lemma 2.1.15 there is a third generic element z, independent of x and of y, with $xz = y$. So $\sigma(mxz) = \sigma(mx)\sigma(z)$ and $\sigma(xz) = \sigma(x)\sigma(z)$, whence

$$\sigma(my)\sigma(y)^{-1} = \sigma(mxz)\sigma(xz)^{-1} = \sigma(mx)\sigma(z)\sigma(z)^{-1}\sigma(x)^{-1} = \sigma(mx)\sigma(x)^{-1}.$$

As σ is injective, so is θ, and for $m, m' \in M$ we have

$$\begin{aligned}
\theta(mm') &= \sigma(mm'x)\sigma(x)^{-1} \\
&= \sigma(mm'x)\sigma(m'x)^{-1}\sigma(m'x)\sigma(x)^{-1} = \theta(m)\theta(m')
\end{aligned}$$

and θ is indeed an embedding of M into G.

Finally, it follows from Theorem 2.1.16 that the generic types of G_0 are in one-to-one correspondence with the generic types of M, and this correspondence is given by the embedding θ. We identify M with its image in G. Now clearly G_0 contains M, and hence $\mathrm{dc}(M)$ by Corollary 2.1.18. On the other hand, for every definable supergroup H of M the formula $x \in H$ contains M and is in all generic types for G_0. If $g \in G_0$ and h realizes a generic type of G_0 over h, then gh is generic. Thus both gh and h^{-1} lie in H, and so does g. Therefore $G_0 = \mathrm{dc}(M)$. $\square$

So we may assume without loss of generality that M is a semi-group contained in a stable group. In particular, we need no longer worry about associativity and the existence of inverses — possibly outside M, of course. Furthermore, M and $\langle M \rangle$ have the same generic types.

Definition 2.1.8 Let G be a substable group and p a 1-type over G.

- The *(left)* φ-*stabilizer of p in G,* denoted by $\mathrm{stab}_G(p, \varphi)$, is defined by the formula
$$\forall y, \bar{y} \, [d_p x \, \varphi(yx, \bar{y}) \leftrightarrow d_p x \, \varphi(ygx, \bar{y})].$$

- The *(left) stabilizer of p in G* is defined as
$$\mathrm{stab}_G(p) = \bigcap_{\varphi \in \mathfrak{L}} \mathrm{stab}_G(p, \varphi).$$

Right stabilizers are defined analogously.

Note that we have left-stratified the formulæ φ, so multiplication on the left transforms a $\varphi(yx, \bar{y})$-type into another $\varphi(yx, \bar{y})$-type. Therefore the φ-stabilizers are relatively definable subgroups of G; the stabilizers are relatively type-definable. As there are only finitely many different generic φ-types, the φ-stabilizers have finite index in G. If p is definable and stationary over some set A of parameters, then so are the stabilizers. In particular, if p is finitely satisfiable in A, then the stabilizers are relatively (type-)definable over A.

Proposition 2.1.20 *There is a one-to-one correspondence between the generic types for a substable group G and the directed system of cosets of relatively definable subgroups of finite index in G. In particular, a connected group has a unique generic type.*

Proof: We already know that the φ-stabilizers have finite index in G, and the stabilizer of a generic type contains the intersection of all relatively definable subgroups of finite index in G. Conversely, consider a relatively definable subgroup H of finite index in G. Then H induces an equivalence relation with a finite number of classes consistent with H — those cosets of H which intersect G. Therefore any p finitely satisfiable in G must determine its coset modulo H; hence the left stabilizer of p is contained in H.

Finally, suppose p and q are two generic types for G which lie in the same coset modulo all relatively definable subgroups of finite index. Let a realize p and b realize q with $a \underset{G}{\downarrow} b$. Then ab^{-1} stabilizes p and q, and is independent of b. Hence $p = \mathrm{tp}(a/G) = \mathrm{tp}(ab^{-1}b/G) = \mathrm{tp}(b/G) = q$. $\square$

In the case of a substable semi-group M, we have to consider relatively definable sub-semi-groups such that finitely many translates by elements of $M^{-1} := \{m^{-1} : m \in M\}$ cover M.

Definition 2.1.9 The *principal generic type* gen(G) for a substable group G is the unique generic type contained in all relatively definable subgroups of finite index.

Note that the intersection of all relatively definable subgroups of finite index in G may well be empty. So we have to be careful and regard them as a directed system, rather than simply as an intersection.

We shall now settle a question concerning relative definability in substable groups, which can be seen as an analogue of Corollary 2.1.18 for the substable case:

Theorem 2.1.21 *Let G be a substable group in $\mathcal{G}$, and X a definable subset such that $H := G \cap X$ forms a subgroup. Then H is relatively definable, i.e. there is a definable group intersecting G in H.*

Proof: If $\varphi(x, \bar{a})$ is the formula defining X, consider the φ-stabilizer S of gen(H) in $\mathcal{G}$, and put $\bar{S} = \bigcap_{g \in H} S^g$. Then $\bar{S}$ is normalized by H, so $\bar{S}H$ is a definable supergroup of H (in which $\bar{S}$ has finite index). We claim that $H = G \cap \bar{S}H$.

So consider $g \in G \cap \bar{S}$. As g stabilizes gen(H), which contains the formula $\varphi(x, \bar{a})$, the set $\{h \in H : gh \in X\}$ is generic for H. In particular, it is non-empty and contains an element h_0. But then $gh_0 \in X \cap G = H$, whence g must lie in H. The claim now follows. $\square$

We shall conclude this section with a further discussion of generic types.

Definition 2.1.10 Let p be a 1-type in G. The *inverse type* p^{-1} is defined as $\{\varphi(x^{-1}) : \varphi \in p\}$.

We should note that this obviously is a complete type, and for a realization g of p, the inverse g^{-1} satisfies p^{-1}.

Proposition 2.1.22 *Let p be a type generic for G. Then p^{-1} is generic for G. Any element of G is the product of two generic elements. Any type in $S_1^G(A)$ is a translate of an extension of a generic type (for $|T|^+$-saturated A).*

In particular, $\text{gen}(G)^{-1} = \text{gen}(G)$, as this is the unique generic type of $\text{dc}(G)^0$. Furthermore, generic types of substable semi-groups have inverses, as they are just the generic types of the definable hull.

Proof: Let $\varphi(x)$ be a formula in p^{-1}. As $\varphi(x^{-1})$ is a generic formula, there are elements $g_1, \ldots, g_n$ of G such that $\bigcup_i g_i\varphi(G)^{-1}$ covers G. But then $g_1^{-1}, \ldots, g_n^{-1}$ are also elements of G, and $\bigcup_i \varphi(G)g_i^{-1}$ covers $G^{-1} = G$, so φ is generic.

Now if g is an element of G, consider a realization h of any generic type for G. Then gh and h^{-1} realize generic types for G, and $g = ghh^{-1}$.

Finally, let $q \in S_1^G(A)$ be based on $A_0 \subseteq A$ and realized by g, and consider a generic element $h \in G \cap A$ over A_0, g. Then $\text{tp}(gh/A_0)$ is generic, and can be extended to gq. $\square$

Hence if the group is sufficiently saturated, the non-forking property characterizes the generic types:

Corollary 2.1.23 *Let G be a $|T|^+$-saturated type-definable stable group. Then any 1-type p over G is generic iff gp does not fork over $\emptyset$ for any g in G.*

In assertions like this, it shall be understood that the language has tacitly been expanded by the parameters needed to type-define G.

Proof: If p is generic, so is gp, and does not fork over $\emptyset$. On the other hand, p has a translate gp which is an extension of a generic type; if this extension is non-forking, then gp is generic, and so is p. $\square$

Proposition 2.1.24 *Let G be substable and $p \in S_1^G(A)$. Then p is generic for G iff for all formulæ $\varphi(x)$ the index $|G : \text{stab}_G(p, \varphi)|$ is finite.*

For sufficiently saturated G this means that p is generic for G iff $\text{stab}_G(p) = G^0$.

Proof: We have already seen the implication from left to right. For the other direction, consider the principal generic type q of G, and independent realizations b of q and a of p. The types $\text{tp}(a/A)$ and $\text{tp}(a/A, b)$ have the same stabilizer $\text{dc}(G)^0$, which contains q, so $\text{tp}(a/A, b) = \text{tp}(ab/A, b)$. But $\text{tp}(ab/A)$ is generic for G, and so is $\text{tp}(a/A)$. $\square$

Thus in a superstable theory the generic types are exactly those of maximal rank (U, RC, or, in the ω-stable case, RM). In particular, types of maximal U-rank exist, and generic types are invariant under definable bijections. However, in the merely stable case, they do depend on the group multiplication:

Example 2.1.1 Consider $G = \mathbb{Z}^\omega$. Then G is abelian and has a definable subgroup $2G$ of 2-divisible elements of infinite index. Hence $2G$ is a non-generic subset. However, the map defined by

$$x \mapsto \begin{cases} 2x & \text{for } x \in G - 2G, \\ x/2 & \text{for } x \in 2G - 4G, \\ x & \text{for } x \in 4G \end{cases}$$

exchanges the generic set $G - 2G$ with the non-generic set $2G - 4G$.

However, if the map is a (surjective) group homomorphism, then generic types are mapped (surjectively) to generic types.

Theorem 2.1.25 *Let G and H be type-definable stable groups. Then (g, h) is generic for $G \times H$ iff g is generic for G, h is generic for H, and g and h are independent. Furthermore $(G \times H)^0 = G^0 \times H^0$.*

Proof: Consider (a, b) in $G^0 \times H^0$ and let g be generic for G and h generic for H, independently over (a, b). Then $\text{tp}(ag/(a, b), h) = \text{tp}(g/(a, b), h)$ and $\text{tp}(bh/(a, b), g) = \text{tp}(h/(a, b), g)$. Therefore $G^0 \times H^0$ is contained in the stabilizer of $\text{tp}((g, h))$, which in turn must be contained in $(G \times H)^0$. But obviously $(G \times H)^0 \leq G^0 \times H^0$, so we must have equality and $\text{tp}((g, h))$ is generic for $G \times H$. Now $\text{gen}((G \times H)^0)$ is uniquely determined as $\text{gen}(G^0) \otimes \text{gen}(H^0)$ and all other generic types of $G \times H$ are translates of it, whence any generic type for $G \times H$ is of the form $p \otimes q$, with p generic for G and q generic for H. $\square$

2.2 Transitive Group Operations

If H is a definable normal subgroup of a stable group G, then we can form the quotient group G/H and consider its generic types. However, one would like to extend the technique also to the case when H is not normal, i.e. to coset spaces. In this section, we shall therefore look at a transitive action of a definable group on a definable set, and prove

Theorem 2.2.1 *Let G be a definable group which operates in a stable structure definably and transitively on a definable set X. Then in X there are generic types relative to that operation, namely the image of an arbitrarily chosen point $x_0 \in X$ under a generic element $g \in G$. They are maximal in the stratified fundamental order, have maximal local rank with respect to stratified formulæ, and only satisfy generic formulæ. In a $|T|^+$-saturated model a 1-type in X is generic iff for all group elements g the translate gp does not fork over $\emptyset$.*

For the rest of this section, G will be a definable group acting transitively on a definable set X, all in a stable structure. We shall (almost) only consider types p, q, etc. containing the formula $x \in X$. In analogy to the group case we first have to define "generic".

Definition 2.2.1 A formula φ defining a subset of X is *generic* if there are finitely many group elements $g_1, \ldots, g_n$ in G such that the union of the translates of φ covers X, i.e. $\models \forall x \in X \ \bigvee_{i=1}^n \varphi(g_i x)$.

Another route to generic types which we shall take in this case uses the "stratified fundamental order":

Definition 2.2.2   • The *stratified fundamental order* is the set

$$\Big\{ \{\varphi(gx, \bar{y}) : \varphi(hx, \overline{m}) \in p\} : \mathfrak{M} \models T, \ p \in S_1(\mathfrak{M}), \ h \in G^{\mathfrak{M}}, \ \overline{m} \in \mathfrak{M} \Big\},$$

where g is a variable ranging over group elements. It is partially ordered by reverse inclusion.

   • The *stratum* $\mathrm{str}(p)$ of a type p over a model $\mathfrak{M}$ is the set corresponding to p and $\mathfrak{M}$ in the stratified fundamental order. The stratum of some type p over an arbitrary set A of parameters is the stratum of any of its non-forking extensions to a model.

As all non-forking extensions of some given type to any model represent the same formulæ, the stratum is well-defined. By compactness (for theories with a predicate for an elementary substructure), the order is inductive. In particular, there are maximal strata. Clearly, if q extends p, then $\mathrm{str}(q) \leq \mathrm{str}(p)$.

Lemma 2.2.2 *A type p over a model $\mathfrak{M}$ has exactly one extension with the same stratum over any superset B of $\mathfrak{M}$, namely its heir.*

Proof: A non-forking extension has the same bound β, and hence the same stratum. Conversely, suppose there were an elementary superstructure $\mathfrak{M}'$ of $\mathfrak{M}$ and two extensions p_1 and p_2 of p over $\mathfrak{M}'$ of the same stratum. As $p_1 \neq p_2$, there are some formula φ and a parameter $\bar{b}$ in $\mathfrak{M}'$ such that $\varphi(x, \bar{b})$ lies in p_1, but $\neg\varphi(x, \bar{b})$ lies in p_2.

Claim. *If q is some type over a model $\mathfrak{N}$ of the same stratum as p, then there is some elementary extension $\mathfrak{N}'$ of $\mathfrak{N}$ such that q has two extensions to $\mathfrak{N}'$ of the same stratum which differ in some formula $\varphi(hx, \bar{b}')$, for some $h \in G^{\mathfrak{N}'}$ and $\bar{b}'$ in $\mathfrak{N}'$.*

Proof of Claim: Let P be a new one-place predicate. We add names for the elements of $\mathfrak{N}$ to the language and have to show that the following set of formulæ is consistent:

1. P is an elementary extension of $\mathfrak{N}$ (in the old language),

2. $P(h) \wedge P(\bar{b}') \wedge h \in G$,

3. $\mathrm{str}(a_1/P) = \mathrm{str}(a_2/P) = \mathrm{str}(p)$,

4. $\mathrm{tp}(a_1/\mathfrak{N}) = \mathrm{tp}(a_2/\mathfrak{N}) = q$, and finally

5. $\varphi(ha_1, \bar{b}') \wedge \neg\varphi(ha_2, \bar{b}')$.

But as p and q have the same stratum, for any finite part $\psi(x, \bar{d})$ of q there is some element h' in $G^{\mathfrak{M}}$ and $\bar{d}'$ in $\mathfrak{M}$, such that $\psi(h'x, \bar{d}')$ lies in p. Therefore we may interpret $\bar{b}'$ by $\bar{b}$ and h by h'^{-1}, $\bar{d}$ by $\bar{d}'$, P by $\mathfrak{M}'$, and a_1, a_2 by realizations of $h'p_1$ and $h'p_2$. $\square$

But this leads to an infinite binary $\varphi(ux, \bar{v})$-tree, contradicting stability. $\square$

Lemma 2.2.3 *Let x_0 be an element of $X^{\mathfrak{M}}$. If p is a type over $\mathfrak{M}$ of maximal stratum, then there is some group-generic element g such that gx_0 satisfies p.*

Proof: As G operates transitively on X, there must be some group element h (in an elementary extension) such that hx_0 realizes p. Let g be a group-generic element over $\mathfrak{M} \cup \{h\}$. Then

$$\mathrm{str}(ghx_0/\mathfrak{M}, g) = \mathrm{str}(hx_0/\mathfrak{M}, g) = \mathrm{str}(hx_0/\mathfrak{M})$$

(as hx_0 is independent of g over $\mathfrak{M}$), hence $\mathrm{str}(ghx_0/\mathfrak{M}, g)$ is maximal and therefore equals $\mathrm{str}(ghx_0/\mathfrak{M})$. So ghx_0 and g are independent over $\mathfrak{M}$; again by transitivity of the action there is h' independent of g over $\mathfrak{M}$ with $h'x_0 = ghx_0$. Now $g^{-1}h'$ is generic over $\mathfrak{M}$, and $(g^{-1}h')x_0 = hx_0$. $\square$

Definition 2.2.3 A type is *generic* if it has maximal stratum.

Lemma 2.2.4 *A type is generic iff it contains only generic formulæ.*

Proof: We shall show first that for every 1-type q over $\mathfrak{M}$ and every formula φ contained in some generic type p over $\mathfrak{M}$ there is some element a in $G^{\mathfrak{M}}$ with $\varphi \in aq$. So consider an element x_0 in $X^{\mathfrak{M}}$, a realization b of p with $b = gx_0$ for some group-generic element g, and let $c = hx_0$ realize q, where h is a group element independent of g over $\mathfrak{M}$. Then gh^{-1} is generic over $\mathfrak{M}, h$ and $\varphi(gh^{-1}c)$ holds. As gh^{-1} is independent of c over $\mathfrak{M}$, the type $\mathrm{tp}(gh^{-1}/\mathfrak{M}, c)$ is finitely satisfiable in $\mathfrak{M}$. Hence there is the required $a \in G^{\mathfrak{M}}$.

Now if φ were not generic, then $\{\neg a\varphi\}_{a \in G^{\mathfrak{M}}}$ would be consistent and could be completed to some type q over $\mathfrak{M}$, contradicting the first paragraph. Conversely, if p is generic, then so is gp for any g in $G^{\mathfrak{M}}$. Since for a generic formula there are finitely many elements $g_1, \ldots, g_n$ in $G^{\mathfrak{M}}$ such that $\forall x \in X \ \bigvee_{i=1}^{n} \varphi(g_i x)$ holds, the lemma follows. $\square$

As in the last section we conclude that for some $|T|^+$-saturated model $\mathfrak{M}$ a type p over $\mathfrak{M}$ is generic for X iff for all g in $G^{\mathfrak{M}}$ the translate gp does not fork over $\emptyset$. Hence a type is generic if its local ranks with respect to stratified formulæ are maximal; the reverse follows immediately from the fact that it only contains generic formulæ. This finishes the proof of Theorem 2.2.1. $\square$

Corollary 2.2.5 *If some definable subgroup H of G also operates transitively on X, then the generic types of X with respect to G and with respect to H are the same.*

Proof: H-generic formulæ are also G-generic, hence H-generic types are G-generic. The reverse follows from the characterization using $|T|^+$-saturated models. $\square$

Remark 2.2.1 The methods of this section work equally well in the type-definable context, i.e. for a type-definable group G acting transitively type-definably on a type-definable set X. In fact, by Corollary 2.1.18 the group G is the intersection of a family $\{G_i : i \in I\}$ of definable groups; picking any $x_0 \in X$, compactness yields some i_0 such that the action of G_{i_0} on x_0 is well-defined and definable. Clearly we may assume that $G_i \leq G_{i_0}$ for all $i \in I$; putting $X_i := G_i x_0$ yields a family of definable groups acting definably on definable sets, with $G = \bigcap_I G_i$ and $X = \bigcap_I X_i$.

The situation is similar for a transitive action of a substable group G on a substable set X: just take any $x_0 \in X$ and consider the action of $\mathrm{dc}(G)$ on $\mathrm{dc}(G)x_0$. As the generic types for G are exactly those for $\mathrm{dc}(G)$, a generic type for X over A (where $x_0 \in A$) will be the type $\mathrm{tp}(gx_0/A)$ for any group-generic g over A. We have already seen that the quotient of a substable group G by a relatively definable subgroup H is substable; as G acts transitively on the coset space, we obtain generic types for G/H, namely $\mathrm{tp}(gH/A)$ for any generic element g for G over A. Note that we have a distinguished point x_0, namely the coset H itself. In the case where H is normal, this definition agrees with the usual one; over a $|T|^+$-saturated model, a type is generic in either sense iff no translate forks over $\emptyset$.

   If H is substable in G and $g \in G$, then we may define a generic type for the coset gH to be just a translate of a generic type for H by g. Alternatively, we can consider gH as an H-space, where H acts transitively from the right by multiplication. Using Theorem 2.2.1, it is easy to see that again the two definitions of genericity agree. This time we do not have a distinguished point unless we specify g rather than merely its coset. However, if h is principal generic for H and g generic for G, with $h \perp g$, then $h \in H^0 \leq G^0$ and $\mathrm{tp}(g) = \mathrm{tp}(gh)$. If $\pi_H(g)$ denotes the imaginary element gH (and hence the canonical parameter for the coset gH), then $\pi_H(g) = \pi_H(gh)$; since $\mathrm{tp}(gh/\pi_H(g))$ is generic, so is $\mathrm{tp}(g/\pi_H(g))$. Often one does not distinguish between the coset gH as a set and as an imaginary element, but says that $\mathrm{tp}(g/gH)$ is generic for gH.

The most general case, however, is that of a quotient of a substable group by a relatively type-definable subgroup. Suppose $H = \bigcap_I H_i$ for some relatively definable subgroups $\{H_i : i \in I\}$ of a substable group G; if J denotes the collection of non-empty finite subsets of I, put $H_j = \bigcap_{i \in j} H_i$ for any $j \in J$. Then G/H will denote the directed system $(G/H_j : j \in J)$. If g is a generic element for G, put $p_j = \mathrm{tp}(gH_j)$ for $j \in J$. Then p_j is a generic type for G/H_j, all generic types for G/H_j arise in this fashion, and for $j' \supseteq j$ we have $gH_j \in \mathrm{dcl}(gH_{j'})$. By a generic type of G/H we shall mean the directed system of types $\{p_j : j \in J\}$; it is principal iff all the p_j are.

We have to choose our filtration $(H_i : i \in I)$ slightly more carefully when we want to take the canonical parameter of a type-definable group H or of a coset gH. For a definable supergroup H_i of H consider the intersection $\bar{H}_i$ of all model-theoretic conjugates of H_i which contain H. By the icc, this is a finite intersection, hence definable, and has a canonical parameter a_i. Then any automorphism fixing H must fix a_i; conversely, if it fixes a_i for all i it fixes $\bigcap_{i \in I} \bar{H}_i = H$. So $\{a_i : i \in I\}$ is the canonical parameter for H, and similarly the set of canonical parameters for $g\bar{H}_i$ is the canonical parameter for gH.

2.3   Fields

Using the technique of generic types, we may now prove connectivity of the multiplicative group in a stable division ring.

Lemma 2.3.1 *In an infinite stable type-definable division ring D the multiplicative group is connected; its generic type is the additive generic type.*

Proof: The additive group is connected, and for any d in $D^\times$ the additive automorphism $x \mapsto dx$ must map the additive generic type p to itself. Hence $dp = p$ for all d in $D^\times$, and p is the only multiplicative generic type. $\square$

Note that the lemma does not hold in general for fields with chain condition: $\mathbb{R}$ is o-minimal and hence satisfies the ωdcc. But $\mathbb{R}^+$ is connected, whereas $\mathbb{R}^\times$ has a definable subgroup $\mathbb{R}_{>0}^\times$ of index 2.

A type-definable division ring is an intersection of definable additive (or multiplicative) supergroups by Corollary 2.1.18. We may also get it as an intersection of definable division rings:

Proposition 2.3.2 *Let D be a type-definable division ring. Then D is an intersection of definable division rings.*

Proof: First we find some multiplicative supergroup M such that addition is defined on M (but may go outside of M), and such that the distributive laws hold on M. By Corollary 2.1.18 (and compactness) $M \cup \{0\}$ contains a definable additive supergroup A_0 of D^+. If p is the generic type of D, consider the set $A_1 := \{a \in A_0 : d_p x \, xa \in A_0\}$. By distributivity, A_1 is an additive subgroup of A_0 containing D.

Now put $R := \{m \in M \cup \{0\} : mA_1 \leq A_1\}$. If d lies in D, a in A_1, and x is generic for D over d, a, then xd is generic for D over a and $x(da) = (xd)a$ lies in A_0. Therefore R contains D. For elements m and n of R and a in A_1, we have $(m + n)a = ma + na \in A_1$ and $(mn)a = m(na) \in A_1$. As 1 lies in A_1, R must be contained in A_1, so $m + n$ and mn lie in $M \cup \{0\}$. Thus R is a ring. But $R^\times$ is contained in M and therefore without zero-divisors, so R must be a definable division ring containing D by Lemma 1.0.1.

If $D^\times = \bigcap M_i$, then for every i we can find a division ring R_i with $D \subseteq R_i \subseteq M_i \cup \{0\}$. Obviously $D = \bigcap R_i$. □

If T is a theory with the $\omega\mathrm{dcc}^0$, then D is even definable, as by connectivity every index $|R_i^+ : R_i^+ \cap R_j^+|$ must be zero or infinite.

For a substable ring D without zero-divisors, we have

Theorem 2.3.3 *Let D be a substable ring without zero-divisors. Then D may be embedded into a definable division ring; this maps the unique generic type of D to the generic type of some type-definable division ring.*

Proof: By Theorem 2.1.19 we may embed D^+ in a definable abelian group A. On $\mathrm{dc}(D)$ multiplication is defined on (additive) generic elements and is distributive for independent generics. But we can write any two elements a and b of $\mathrm{dc}(D)$ as $a = a_1 + a_2$ and $b = b_1 + b_2$, with a_i generic over b_1, b_2 and b_i generic over a_1, a_2. We define $ab = a_1 b_1 + a_1 b_2 + a_2 b_1 + a_2 b_2$. We have to check that this is well-defined. Clearly it suffices to show that for generic elements a_1 and a_2 and for independent generic elements a' and b' the equation $a_1 b' + a_2 b' = (a_1 + a')b' + (a_2 - a')b'$ holds. But this follows immediately from generic distributivity.

Thus $\mathrm{dc}(D)$ becomes a ring with unit. Then $\mathrm{dc}(D)$ is the intersection of definable additive groups $(D_i : i < \omega)$, and we may assume that on D_0 multiplication is defined, associative, and distributive (but may go outside). We consider the set $X := \{d \in D_0 : \bigwedge_i d_i x \, xd \in D_0\}$, where $d_i x$ runs over the finitely many defining schemes which are additively generic for D. Then X is an additive subgroup of D_0 containing D, and we may consider $R := \{x \in X : xX \leq X\}$. This is a definable super-ring of D.

Consider elements $0 \neq d, d' \in D$ such that $X := \{x \in R : dxd' = 0\}$ is maximal. Such d, d' exist by the ucc. Then dRd' is a subring of R, and

dRd' has no zero-divisors in $D \cap dRd'$ by maximality of X. Let Z be the multiplicative semi-group of non-zero-divisors of dRd'. This is definable, has left and right cancellation, and contains $D \cap dRd'$. By Lemma 1.0.1 this is a group, so in particular it contains 1, and both d and d' are invertible. Hence $dRd' = R$ contains D, and every element in $D - \{0\}$ is in Z and invertible: D extends to a division ring $D' \subseteq R$.

Proposition 2.1.8 implies that D has generic cancellation, and therefore the multiplicative generic types for D must be the same as those for $D' = \langle D^\times \rangle \cup \{0\}$. But when we look at $\mathrm{dc}(D')$ additively, we see as above for $\mathrm{dc}(D)$ that it carries a ring structure — and now the multiplication is generically invertible. Since every element a is the product of two generic elements (namely ab and b^{-1}, for any b generic over a), it must be invertible itself. Hence $\mathrm{dc}(D')$ is a type-definable division ring with a unique generic type, and it is contained in a definable division ring by Proposition 2.3.2. $\square$

It should be noted that the non-existence of zero-divisors in D seems to be weaker than generic cancellation in a semi-group, as generic cancellation must involve uniformity in order to survive to types finitely satisfiable in the semi-group. However, we do have the advantage of full associativity of multiplication on some definable set. Nevertheless, it might be interesting to see whether one might do with less than generic cancellation in Theorem 2.1.19.

2.4 Generic Properties

Both in the theory of finite groups and in the theory of algebraic groups one often considers the question of what properties of large subsets of the group carry over to the whole group. In finite group theory one might consider a subset to be large if it generates the whole group; in chapter 1 we have already encountered some theorems of this kind: a substable group which is generated by bounded Engel elements is nilpotent (Theorem 1.4.3), and a substable group is a (locally) nilpotent p-group iff it is generated by a normal subset whose two-element subsets all generate a finite p-group (of bounded exponent) (Theorems 1.5.6 and 1.5.7). In the theory of algebraic groups, large usually is taken to mean dense in the sense of the Zariski topology, and then any closed condition (i.e. equation) which holds on a dense subset must hold on the whole group.

The analogue of a large set in the stable context is a generic set. However, even in the case of finite Morley rank, it is not even known whether generic exponent n (i.e. any generic element has order n) implies exponent n, for general n.

Definition 2.4.1 Let $\varphi(\bar{x})$ be some formula. A stable group G satisfies φ

generically if $\varphi(\bar{g})$ holds for all tuples $\bar{g}$ of independent generic elements.

Problem 2.4.1 Cherlin's Conjecture claims that an infinite simple group of finite Morley rank is an algebraic group over an algebraically closed field. So if the conjecture held, it should be possible to recover the Zariski topology from the group structure, i.e. to distinguish between those formulae which are open and those which are closed.

In a stable group, one might try to define a formula φ to be *closed* if whenever a coset of some subgroup satisfies φ generically, φ must hold on the whole coset. (It does not matter whether we consider type-definable or arbitrary subgroups, since a subgroup and its definable hull share the same generic properties.) Clearly, the conjunction of closed formulæ is again closed, and the true and the false formulæare closed; due to uniqueness of the generic type for a connected group, the union of a finite number of closed formulæ is also closed, so this forms a topology. Under what conditions is it Noetherian? How does closedness behave under projections? And when is every formula equivalent to a Boolean combination of closed ones?

It is obvious that a generically commutative group ($xy = yx$ holds generically) is commutative, as every element is the product of two generic elements.

Theorem 2.4.1 *If G is stable and generically nilpotent of class n, then G is nilpotent of class n.*

Proof: We shall use induction on n, having just dealt with the case $n = 1$. So consider a group in which for any $n{+}1$ independent generic elements $g_0, \ldots, g_n$ the commutator $[g_0, g_1, \ldots, g_n]$ is trivial. Then $z := [g_0, g_1, \ldots, g_{n-1}]$ commutes with all generic elements and must lie in the centre $Z(G)$ (since any element is the product of two generic elements). Clearly $G/Z(G)$ is generically nilpotent of class $n - 1$, hence nilpotent of class $n - 1$ by the inductive hypothesis, and G is nilpotent of class n. $\square$

Theorem 2.4.2 *If G is stable and generically soluble of derived length n, then G is soluble of derived length n.*

Proof: Again we use induction on n, the case $n = 1$ being clear. So suppose the theorem holds for n and G is generically soluble of derived length $n + 1$. Define inductively a sequence of iterated commutators via $c_0(g) = g$ and

$$c_{i+1}(g_1, \ldots, g_{2^{i+1}}) = [c_i(g_1, \ldots, g_{2^i}), c_i(g_{2^i+1}, \ldots, g_{2^{i+1}})]$$

(so generically soluble of derived length $n + 1$ means $c_{n+1}(\bar{x}) = 1$ generically). Let X be a sequence of independent generic commutators $c_n(\bar{g})$, where all possible choices for generic strong types among the $\bar{g}$ occur infinitely often.

Then X is commutative by generic solubility of derived length $n+1$. Further-more, the centralizer of any such sequence is the centralizer of any sufficiently big finite subset which encompasses suitable choices for generic strong types. In particular, any two such sequences (containing all the strong types infinitely often) have the same centralizer — we may just choose a common prolongation and take the centralizer of that. Hence $C_G(X)$ is a definably characteristic subgroup, and $A := Z(C_G(X))$ is a definable abelian normal subgroup of G. Clearly G/A is generically soluble of derived length n, hence soluble of derived length n, and G is soluble of derived length $n + 1$. $\square$

We shall now consider equations of the form $x^n = 1$. The case of general n is unknown and complicated by the fact that such a group need not be locally finite. Indeed, Olshanskii and Ivanov have constructed countably infinite groups of prime exponent such that every two non-commuting elements generate the whole group. This is an extreme example of a *Tarski monster*, i.e. a countably infinite group whose proper subgroups are all finite (the question of the existence of these groups was first put forward by Tarski).

The case $n = 2$ is trivial, as such a group is abelian of exponent 2: if x and y are independent generic elements, then xy is generic and hence an involution. But $xy = (xy)^{-1} = y^{-1}x^{-1} = yx$, so the group is generically commutative. The rest follows immediately from the fact that every element is the product of two generic ones.

Theorem 2.4.3 *If G is a stable group of generic exponent 3, then G is nil-potent and has exponent 3.*

Proof: Let x be an element of G and g generic over x. Then gx and xg^{-1} are generic as well, so

$$x^{g^2} x^g x = gxgxgx = 1 = xg^{-1}xg^{-1}xg^{-1} = xx^g x^{g^2}.$$

Therefore $xx^g = (x^{g^2})^{-1} = x^g x$. But the centralizer of x^G is the centralizer of finitely many elements $x^{g_1}, \ldots, x^{g_n}$, and if g is an independent and generic element, $g_1 g, \ldots, g_n g$ are generic as well. Hence x commutes with all $x^{g_i g}$, so $x^{g^{-1}}$ commutes with all x^{g_i}, hence with x^G. Therefore $x \in C_G(x^G)^g = C_G(x^G)$. So x^G generates an abelian normal subgroup, and G must be nilpotent by Theorem 1.1.12.

The group generated by any two elements x and y must be nilpotent of class 2, as the commutator $[x, y] = x^{-1}x^y = y^{-x}y$ commutes both with x and with y. Now for arbitrary x and y generic over x we get that xy is generic, so

$$1 = (xy)^3 = x^3 y^3 [y, x]^3 = x^3 [y, x]^3.$$

But for a second independent generic z over x the product yz is also generic over x and

$$x^3 = [x, y]^3 = [x, z]^3 = [x, yz]^3 = ([x, y]^z [x, z])^3 = ([x, y]^z)^3 [x, z]^3$$

(note that all commutators lie in $\langle x^G \rangle$ and must commute), whence $x^3 = 1$. Therefore G has exponent 3. $\square$

For general n we have to assume solubility:

Theorem 2.4.4 *Let G be a stable soluble-by-finite group of generic exponent n. Then the exponent of G is finite and divides some power of n.*

Proof: Suppose first that G is soluble. Let us call an element *bad* if its order is infinite or coprime to n. Consider a series $G = G_1 \triangleright G_2 \triangleright \cdots \triangleright G_m = \{1\}$ of definable normal subgroups of G, such that every quotient G_i/G_{i+1} is abelian. If G contains a bad element, then there is a minimal $k < \omega$ such that G_k/G_{k+1} contains a bad element. Clearly we may assume that $G_{k+1} = \{1\}$, so G_k is abelian. By the minimality of k, for $i < k$ the group G_i/G_{i+1} is abelian of finite exponent dividing a power of n.

Now let $G_k \geq M_0 \geq M_1 \geq \cdots \geq M_{k-1}$ be such that M_i is minimal infinitely definable, G_i^0-invariant, and contains a bad element.

Claim. For $k \geq i \geq 0$, the centralizer-connected component G_i^{cc} centralizes M_{i-1}.

Proof of Claim: Inductively, we may assume that G_{i+1}^{cc} centralizes M_i. Thus $C := G_i^0/G_{i+1}^{cc}$ acts on M_i by conjugation. Note that C has finite exponent; it is abelian by Lemma 1.1.7.

Let R be the ring of endomorphisms of M_i generated by C, and consider $\sigma \in R$. Then $\ker(\sigma)$ and $\mathrm{im}(\sigma)$ are still G_i^0-invariant (C is abelian), and one of them has to contain a bad element. Therefore by the minimality of M_i, σ is either 0 or surjective, so R is a division ring. But R is also commutative, hence it has only finitely many elements of order $\exp(C)$. Now C is connected, thus the image of C in R, being finite, must be trivial, and C centralizes M_i. So $C_{G_i}(M_i)$ has finite index in G_i, and G_i^{cc} centralizes M_i. But since G_{i-1} normalizes G_i, it normalizes G_i^{cc} as well, and the centralizer $C_{M_{i-1}}(G_i^{cc})$ is G_{i-1}^0-invariant and must equal M_{i-1} by minimality. $\square$

Therefore a bad element $x \in M_0$ is centralized by the principal generic type p of G. But if y realizes p over x, then xy is also generic and $x^n = x^n \cdot 1 = x^n \cdot y^n = (xy)^n = 1$. This contradicts the badness of x.

Now if G is soluble-by-finite, then it has a relatively definable soluble normal subgroup S of finite index, and S has generic exponent n. Hence S does not contain a bad element. But if $g \in G$ is arbitrary and $s \in S$ is principal generic over g, then gs is generic. So $g^n S = (gS)^n = (gsS)^n = (gs)^n S \in S$, whence $g^n \in S$, and g cannot be bad either. $\square$

This proof is interesting in so far as it does not provide a bound for the exponent of G. Furthermore, by Fact 1.2.16 the group G is nilpotent-by-finite.

Already the existence of one generic type of small order places serious restrictions on the group.

Theorem 2.4.5 *A stable group with a generic involution i is abelian-by-finite, and i inverts a subgroup of finite index.*

Proof: Let x realize the principal generic type over i. Then i and xi have the same type, so xi is an involution and $x^i = x^{-1}$. If y is another principal generic element independent of x and i, then x, y, and xy have the same type over i and are all inverted by conjugation by i. Hence

$$x^i y^i = (xy)^i = (xy)^{-1} = y^{-1}x^{-1} = y^i x^i,$$

so x and y commute. The assertion follows. $\square$

Theorem 2.4.6 *A stable group with a generic element g of order 3 is nilpotent-by-finite.*

Proof: If x realizes the principal generic type over g, then x commutes with x^g and $x^{g^{-1}}$ as in the proof of Theorem 2.4.3, and more generally x^{G^0} commutes with x^{gG^0} and $x^{g^{-1}G^0}$. But $1 = xgxgxg = xx^{g^{-1}}x^g$; hence $(x^{-1})^{G^0}$ is contained in $x^{g^{-1}G^0}x^{gG^0}$. Thus every generic element of G^0 is contained in an abelian normal subgroup; by Theorem 1.1.12 the group G^0 is nilpotent, whence G is nilpotent-by-finite. $\square$

Definition 2.4.2 A (definable) automorphism φ of order n of a group G is *splitting* if the identity

$$(\dagger) \qquad\qquad x x^\varphi \cdots x^{\varphi^{n-1}} = 1$$

holds identically on G. It is called *generically splitting* if the identity $(\dagger)$ holds for generic elements of any type-definable characteristic subgroup of G.

If we consider φ as an element in the semi-direct product $G \rtimes \langle\varphi\rangle$, then $(\dagger)$ amounts to $(x\varphi^{-1})^n = 1$; putting $y = x^{-1}$, we get $(y\varphi)^n = 1$ and φ^{-1} is (generically) splitting as well. Furthermore, for any generic element g of some type-definable characteristic subgroup of G, $g\varphi$ is also (generically) splitting (xg will be a generic element of the type-definable characteristic product of the two subgroups x and g are generic for).

It has been proved by Khukhro that a soluble group with splitting automorphism of prime order p is nilpotent, and the class is bounded in terms of p and the derived length.

Theorem 2.4.7 *If G is a stable soluble group with a generically splitting automorphism of prime order p, then G is nilpotent-by-finite.*

Proof: We shall work in a $|T|^+$-saturated model. Note that if the connected component G^0 is nilpotent, then G itself is nilpotent-by-finite: G^0 will be contained in a definable nilpotent group, which must have finite index in G. Replacing G by G^0, we may assume that it is connected, and also that it is saturated. By induction on the derived length of G, we may further assume that $dc(G')$ is nilpotent (this is a characteristic type-definable subgroup of G of smaller derived length, which allows φ as generically splitting automorphism). We shall prove inductively that all the iterated centres of $dc(G')$ are contained in some iterated centre of G. This is clearly true for $Z_0(dc(G'))$. So assume $Z_i(dc(G')) \leq Z_k(G)$ for some $k < \omega$, and consider the action of G on $Z := Z_{i+1}(dc(G'))/Z_k(G)$.

First we notice that φ also acts on $G/C_G(Z)$, and this group is abelian. It follows that (†) holds not only generically, but identically for all elements of the group ring $\mathbb{Z}[G/C_G(Z)]$.

Put $N := C_Z^\omega(G)$. If Z were not contained in N, we could find a minimal type-definable G-invariant subgroup A of Z not contained in N. We want to consider the ring R of endomorphisms of A generated by G. If r is an endomorphism in R, then $\mathrm{im}(r)$ and $\ker(r)$ are type-definable G-invariant subgroups; if they are not the whole of A, they are contained in N. Therefore the ring $\bar{R}$ of endomorphisms of $A/(A \cap N)$ induced by R has no zero-divisors, and any non-zero element is an automorphism.

Now Z is characteristic in G, so (†) holds generically and because of commutativity even identically on Z. In particular, $\sum_{j=0}^{p-1} \varphi^j a = 0$ for all elements a of A. So there are a minimal $k < p$ and endomorphisms r_i of R which are not all zero in $\bar{R}$, such that an equation

$$\text{(‡)} \qquad\qquad \sum_{j=0}^{k} \varphi^j r_j a = 0$$

holds identically on A modulo N. (By saturation, it must already hold modulo some iterated centre $Z_i(G)$; we replace G by the quotient $G/Z_i(G)$.) Clearly, r_0 and r_k are non-zero in $\bar{R}$.

But for any element a in A and m in G we also have that $m^{\varphi^k} a \in A$ and we get

$$0 = \sum_{j=0}^{k} \varphi^j r_j m^{\varphi^k} a.$$

On the other hand, multiplying (‡) on the left by m yields

$$0 = \sum_{j=0}^{k} m \varphi^j r_j a = \sum_{j=0}^{k} \varphi^j m^{\varphi^j} r_j a.$$

Subtracting these two equations gives

$$\sum_{j=0}^{k-1} \varphi^j r_j(m^{\varphi^k} - m^{\varphi^j})a = 0$$

for all $a \in A$. By the minimality of k, all coefficients $r_j(m^{\varphi^k} - m^{\varphi^j})$ must be zero in $\bar{R}$, for $0 \le j \le k-1$. But $\bar{R}$ does not contain zero-divisors and $r_0 \ne 0$ in $\bar{R}$, whence $m^{\varphi^k} - m = 0$ in $\bar{R}$. As k is coprime to the order p of φ, $m^{\varphi} = m$ as elements of $\bar{R}$.

The equation (†) applied to $G/C_G(A/N)$ therefore yields $g^p = 1$ in $\bar{R}$ for all $g \in G$. So $G/C_G(A/N)$ is mapped to the finitely many p-th roots of unity of $\bar{R}$; if $g_1, \ldots, g_p$ are representatives of the pre-images of those roots, then for every $g \in G$ the image $(g - g_i)A$ must be contained in N for some $i = i(g) \le p$, and by saturation it must be contained in $C_Z^j(G)$ for some $j = j(g) < \omega$. Again by saturation, there is a finite bound j_0 on $j(g)$ for $g \in G$, and $G/C_G(A/C_A^{j_0}(G))$ is a definable group of order at most p. By connectivity of G, it is trivial and G stabilizes A modulo $C_A^{j_0}(G)$. This means $A \le C_A^{j_0+1}(G)$, contradicting the choice of A.

Hence $Z_{i+1}(\mathrm{dc}(G')) \le Z_\omega(G)$, and by saturation there is some k' with $Z_{i+1}(\mathrm{dc}(G')) \le Z_{k'}(G)$. This finishes the induction, and hence the proof. $\square$

Corollary 2.4.8 *A stable soluble group with a generic element of prime order is nilpotent-by-finite.*

Proof: Let H be a characteristic type-definable subgroup of G^0. Then H is type-definable over $\emptyset$. If y is a generic element of order p and h is an independent generic element of H, then conjugation by y is an automorphism of order p of H, and $hh^y \cdots h^{y^{p-1}} = (hy^{-1})^p = 1$, as $\mathrm{tp}(hy^{-1}) = \mathrm{tp}(y^{-1})$ and y^{-1} has order p as well. So the assumptions of Theorem 2.4.7 are satisfied. $\square$

We now give an example to show that this result cannot be generalized to non-prime exponents, even for groups of finite Morley rank:

Example 2.4.1 Let K be an algebraically closed field and $G = (K^+ \oplus K^+) \rtimes K^\times$, with multiplication $(x,y,z)(a,b,c) = (xc + a, yc^{-1} + b, zc)$. Then $\varphi : (x,y,z) \mapsto (-y, x, z^{-1})$ is an automorphism of order 4, and for any $g \in G$, $gg^\varphi g^{\varphi^2} g^{\varphi^3} = 1$. But G is connected, centreless and has Morley rank 3.

In the generic context, Theorems 1.4.3 and 1.5.7 have the following generic corollaries:

Corollary 2.4.9 *A stable generically left Engel (right Engel) group is nilpotent.*

Proof: By compactness the Engel elements must be uniformly Engel. Let C be the set of all conjugates of generic elements. Then C is an invariant subset of G and generates a nilpotent group by Theorem 1.4.3, which must be the whole of G (as every element is the product of two generic elements). $\square$

Corollary 2.4.10 *A stable group such that any two conjugates of generic elements generate a finite p-group must be a nilpotent p-group of finite exponent.*

Proof: By compactness the orders of the groups generated by any two conjugates of generic elements must be bounded. By Theorem 1.5.7 the conjugates of generic elements generate a nilpotent p-group, which again must be the whole group, as it contains all generic elements. Again by compactness, the exponent must be finite. $\square$

Finally we should like to give an alternative proof of the existence of the Fitting subgroup for a stable group. Although the result is weaker than Theorem 1.2.11, it illustrates the use of generic techniques.

Proposition 2.4.11 *Let G be a stable group. Then the Fitting subgroup of G is nilpotent.*

Proof: $F := F(G)$ is locally nilpotent and hence soluble by Lemma 1.2.5. So we may use induction on the derived length of F and consider a definable G-invariant soluble supergroup H of F'. Then F' is contained in the Fitting subgroup of H, and inductively we may assume that F' is nilpotent. We may envelop F' in a definable G-invariant nilpotent subgroup, which we again call H.

Suppose that F were not nilpotent. We shall prove by induction on i that $Z_i(H)$ is contained in some iterated centre of F. This is clearly true for $Z_0(H)$. So suppose $Z_i(H) \leq Z_k(F)$ and consider the action of F on $Z := Z_{i+1}(H)/Z_k(F)$. As G' is contained in H, it centralizes Z, so the ring R of endomorphisms of Z generated by G is commutative.

Suppose there were some sequence $(f_i : i < \omega)$ of elements of F such that the image $\prod_{i<j}(f_i - 1)Z$ is not contained in $Z_s(F)$ for any $j, s < \omega$. The sequence of groups $\bigcap_{i \geq j}(f_i - 1)Z$ is uniformly definable and increasing with j, hence stationary from some j_0 onwards by the ascending chain condition. So for all $j < \omega$ there is some finite set J of indices bigger than j such that $\bigcap_{i \in J}(f_i - 1)Z = \bigcap_{i \geq j_0}(f_i - 1)Z$. In particular we get

(†) $$\prod_{i \in J}(f_i - 1)Z \leq (f_{j_0} - 1)Z.$$

On the other hand, $\langle Z, f_{j_0} \rangle$ is contained in a normal nilpotent subgroup of G (since f_{j_0} is contained in some normal nilpotent N, and then ZN is again normal and nilpotent). Therefore there is $n < \omega$ with $(f_{j_0} - 1)^n Z = 0$. But there are n disjoint sets $J_m \subset \omega$ $(m < n)$ satisfying (†), and $\prod_{i \in \bigcup J_m} (f_i - 1) Z \leq (f_{j_0} - 1)^n Z = 0$, contradicting our choice of the f_i.

Hence there is a finite sequence $f_1, \ldots, f_n$ of elements of F such that $A := \prod_{i \leq n} (f_i - 1) Z$ is not contained in $Z_s(F)$ for any $s < \omega$, but for any f in F the image $(f - 1)A$ is contained there for some s (depending on f). Note that since R is commutative, A must be G-invariant.

If B is a definable subgroup of Z which is not contained in $Z_s(F)$ for any $s < \omega$, then there is a generic element g of $\mathrm{dc}(F)$ such that $(g - 1)B$ is not contained in $Z_s(F)$ for any s either. For if not, then by compactness there must be some $s < \omega$ such that for all generic g the image $(g-1)B$ is contained in $Z_s(F)$, and $B \leq Z_{s+1}(F)$.

Therefore, taking for B first A, then $(g_0 - 1)A$, $(g_1 - 1)(g_0 - 1)A$, etc., we can inductively find a sequence $g_0, g_1, g_2, \ldots$ of generic elements of $\mathrm{dc}(F)$ such that the image $\prod_{i<j} (g_i - 1)A$ is not contained in $Z_s(F)$ for any $j, s < \omega$. However, the intersection $S = \bigcap_{i<\omega} (g_i - 1)A$ equals a finite subintersection $\bigcap_{i<j_0} (g_i - 1)A$ and thus contains $\prod_{i<j_0} (g_i - 1)A$. Therefore it is not contained in $Z_s(F)$ either, and there is a generic $g_{j_0} \in \mathrm{dc}(F)$ with $(g_{j_0} - 1)A \cap S = S$.

On the other hand, for $f \in F$ the image $(f - 1)A$ is contained in the hypercentre, whence $(f - 1)A \cap S < S$ for all $f \in F$. By Theorem 2.1.19 this inequality also holds for all generic elements of $\mathrm{dc}(F)$, a contradiction. Hence F is nilpotent. $\square$

2.5   Historical and Bibliographical Remarks

Theorem 2.1.19 was proved by Newelski in [82] for substable groups; he also suggested the generalization to semi-groups. Precursors are Corollary 2.1.18, due to Poizat [92], who uses stabilizers, and Theorem 2.1.17, which Hrushovski proved in his thesis [51] together with Theorem 2.1.16. Our definition of generic types follows [121] and [113]; the approach via generic formulæ was first developed by Cherlin and Shelah [37]. The alternative route of section 2.2 via the stratified fundamental order comes (for groups) from Poizat [92].

Generic properties were analysed by Poizat [96]; Theorems 2.4.4 and 2.4.6 come from [115], Corollary 2.4.8 from [117]. Splitting automorphisms are also considered in a more group-theoretic context by Khukhro [63]. The original proof of the nilpotency of the Fitting subgroup for stable groups is in [122].

Chapter 3

Groups & Grandeur

In this chapter model theory will play a more prominent rôle. In the noughth section we shall lay the model-theoretic foundations for the subsequent chapters and introduce the notions of "foreign" and "internal" due to Hrushovski. This allows us to generalize the concept of "small" (e.g. finite) and "large" (e.g. generic) sets: a definable subset is "small" if the generic type is foreign to it, and "large" if the generic type is internal or analysable in it. It will turn out in section 1 that these notions are particularly well-behaved in the context of groups: if the generic type of a group is not foreign to some set X, then there is a definable X-internal quotient of the group. This means that quite often it is sufficient to consider only definable subsets which define groups. Furthermore there is some form of compactness: if a group is internal in a *class* of definable sets, then it is internal in a sub*set* of them, of bounded cardinality.

In section 2, we shall consider various components of a stable group. In particular we shall define the Φ-component, which enjoys very strong connectivity properties. It is the intersection of all relatively definable subgroups not only of finite, but of "small" index (defined in terms of analysability of the whole group using the quotient), and it generalizes both the characteristic normal subgroup of monomial U-rank given by Berline and Lascar, and Hrushovski's p-connected component (where p is a regular type in which the group is internal).

Similarly to what happens in the theory of finite groups, involutions and their centralizers play a special rôle in the study of stable groups, as we could already see in the section on Sylow theory of chapter 1. So centralizers of involutions will be investigated in section 3. Unfortunately, however, involutions need not exist: there is no analogue of the Feit-Thompson Theorem stating that groups of odd exponent (replacing "odd order") are soluble. (This is related to the problem in the section on Engel conditions of chapter 1, where very bad groups prevented us from proving the equivalence of bounded

142

right Engelness and ω-hypercentrality.)

In section 4 we shall analyse the Φ-component of a group without abelian normal subgroups. We shall see that it is the direct sum of finitely many $\mathfrak{R}$-connected normal subgroups, i.e. it decomposes into finitely many factors with particularly good analysability properties.

Section 5 sets up a general rank machinery, generalizing Lascar rank, which allows us to measure the size of a type or a set in relation to a family P of types, excluding all influence not connected with P. This will then be used in section 6 and 7 to study ranked groups and fields, and in section 8 to study bad groups, i.e. a particular kind of – hypothetical – counter-example to the Cherlin-Zil'ber Conjecture.

In this chapter, all groups $G, H, \ldots$ are assumed to be type-definable in a stable theory, unless otherwise indicated.

3.0 Foreignness and Internality

In chapter 1 we have already encountered theorems deducing a property of the group (commutativity, say) from finiteness of some set (the set of commutators in this case). The following definition allows us to consider a new collection of small sets:

Definition 3.0.1 A type p is *foreign* to some class Σ of partial types if over every set A of parameters containing the domain of p every realization a of a non-forking extension of p to A is independent of all realizations of any type in Σ which is based on A. Conversely, we shall say that Σ is *co-foreign* to p.

Suppose X is a type-definable set and p is foreign to X. Then clearly p is foreign to every Cartesian power of X, and if X is mapped definably onto some set Y, then p is also foreign to Y. More generally, if every element of Y is algebraic over finitely many elements of X, then p must be foreign to Y, as algebraic closure cannot induce forking.

A particularly useful instance of foreignness is connected with the notion of a *regular* type. These will feature more prominently in the penultimate section and in the next chapter.

Definition 3.0.2 A type p is *regular* if it is stationary and foreign to all its forking extensions.

At this point we should elaborate shortly on the distinction between *classes* and *sets*. A class is just an arbitrary collection, whereas a set is a class which is small enough that all parameters occurring can be captured in some model. In particular, due to our convention about models being elementary substructures of the monster model of smaller cardinality, a class is a set iff it has smaller cardinality than $\mathfrak{C}$. (If we take a monster "model" $\mathfrak{C}$ whose domain is a proper class, then set means set and class means class in the usual set-theoretic sense.) On a more practical note: we may quite freely adjoin parameters for a set to the language, but we cannot do this for a proper class of things. For instance, Σ in the definition above may well contain a type σ based on new parameters A (and indeed will do so if Σ is a proper class); the definition says that when we extend p non-forkingly to A, it may not fork with a realization of σ. Even more: after extending p non-forkingly to A, we may take realizations of *forking* extensions of σ, which must be independent of realizations of non-forking extensions of p.

Every type is foreign to every finite set, so finite sets are always "small". However, we may often replace *finite* by *co-foreign to the (principal) generic type of the group* in the statement of a theorem and it will remain true. (As the generic types are all translates of one another, it does not matter which one we take; we shall usually refer to "the generic type" being foreign to something.) The following lemma illustrates this principle; it generalizes Proposition 1.1.3 from chapter 1.

Lemma 3.0.1 *Let G be a stable group with a type-definable G-invariant subset A. If $\mathrm{gen}(G)$ is foreign to A, then A is centralized by the centralizer-connected component G^{cc}.*

Most often this will be applied to the case when A is a normal subgroup of G.

Proof: If $\mathrm{gen}(G)$ is foreign to A, then a generic element g is independent of $(a, a^g : a \in A)$, and hence of $gC_G(a)$. But this can only happen if the index of $C_G(a)$ in G is finite, and a is centralized by G^{cc}. $\square$

Corollary 3.0.2 *If the generic type of a stable group is foreign to all sets $[g, G]$, then G is abelian-by-finite.*

Proof: In this case the generic type is also foreign to g^G, so any group element is centralized by G^{cc}. In particular G^{cc} is abelian. $\square$

Corollary 3.0.3 *Suppose σ is a definable automorphism of the stable group G such that there is a σ-invariant subgroup N co-foreign to $\mathrm{gen}(G)$. If σ stabilizes the coset space G/N pointwise, then σ stabilizes G^0.*

Proof: Let g realize $\mathrm{gen}(G)$. Then $\sigma(g) = gn$ for some $n \in N$; as g cannot fork with n, for any other principal generic element g' we also have $\sigma(g') = g'n$. Then gg' is again principal generic, so

$$gg'n = \sigma(gg') = \sigma(g)\sigma(g') = gng'n.$$

Hence $g = gn$, and $n = 1$. Therefore $C_G(\sigma)$ contains G^0. $\square$

We shall now define the opposite notion to *foreign*:

Definition 3.0.3 Let π be a partial type over A and Σ a class of partial types.

- π is Σ-*internal* if there is some superset B of parameters containing A such that any realization of π is definable over B and finitely many realizations of types over B from Σ. It is *almost Σ-internal* if there is such a B such that any realization of π is algebraic over B and finitely many realizations of types over B from Σ.

- π is Σ-*analysable* if for any realization a of π there is a sequence $(a_i : i \leq \alpha)$ of elements in $\mathrm{acl}(Aa)$, such that $a \in \mathrm{acl}(Aa_\alpha)$ and $\mathrm{tp}(a_i/A \cup \{a_j : j < i\})$ is Σ-internal for all $i \leq \alpha$.

Remark 3.0.1 The set B of parameters in the definition of internality must contain the parameters of all those types in Σ which are used to witness definability (or algebraicity) of the realizations of π. In particular, if π is Σ-internal for a class of partial types, then there is a subset Σ_0 of Σ such that π is Σ_0-internal. This is not true for analysability: although only a bounded number of types from Σ are used for the first step a_0 of the analysis, already the second step may involve a proper class, as we have to consider all the types $\mathrm{tp}(a_1/a_0)$, which of course vary with the realization a_0. This matter of adjoining parameters may get very complicated; it is conceivable that for sufficiently nasty Σ we need a completely different analysis for every realization of p. In order to avoid such things, we shall usually assume that the class Σ in question is $\emptyset$-invariant.

We should remark at this point that internality is obviously transitive (just compose functions). This also holds for analysability, and foreignness is preserved under analysability:

Lemma 3.0.4 *Let Σ be an $\emptyset$-invariant class of partial types, A a set of parameters, π a partial type over A, and p a complete type over A.*

1. *p is Σ-internal iff there are some $B \supseteq A$ and a non-forking extension p' of p to B such that every realization of p' is definable over finitely many realizations of types over B from Σ.*

2. *If the partial types in Σ are over A, then p is Σ_0-internal for some finite subset Σ_0 of Σ iff it is Σ-internal.*

3. *Suppose $B \supseteq A$. Then π is Σ-analysable iff π is Σ-analysable over B.*

Now suppose that π is Σ-analysable.

4. *If p is π-internal, then it is Σ-analysable.*

5. *If Σ' is an $\emptyset$-invariant class of partial types such that every type in Σ is Σ'-analysable, then π is Σ'-analysable.*

6. *If p is foreign to Σ, then p is foreign to π.*

Proof: In 1.–3. we clearly need only prove the implication from right to left.

1. Let $\mathfrak{M}$ be a $|B|^+$-saturated model containing B, and consider any realization a of p. If a' is a realization of p' independent of $\mathfrak{M}$ over B, then there are partial types $\pi_1, \ldots, \pi_k$ over B from Σ and a B-definable function f such that $a' = f(b_1, \ldots, b_k)$ for some elements $b_i \models \pi_i$ (for $1 \leq i \leq k$). Let $\pi(x_1, \ldots, x_k, B)$ be a finite part of $\pi_1(x_1) \cup \cdots \cup \pi_k(x_k)$. Then

$$\psi_\pi(x, Y) := \exists \bar{x} \, \pi(\bar{x}, Y) \wedge x = f(\bar{x}, Y)$$

(where the formulæ mention only a finite bit of B and hence only a finite subtuple of Y) is contained in the bound $\beta(p')$, and hence also in the bound of $\mathrm{tp}(a/\mathfrak{M}) =: q$. So $\mathfrak{M} \models \exists Y \, d_q x \, \psi_\pi(x, Y)$. By saturation, we can find $B' \subset \mathfrak{M}$ such that $\psi_\pi(x, B') \in q$ for all choices of finite fragments π. But this gives rise to $\emptyset$-conjugates π_i' of π_i (for $1 \leq i \leq k$) and a B'-definable function f', such that a' is definable via f over realizations of $\pi_1', \ldots, \pi_k'$. These partial types however are contained in Σ by $\emptyset$-invariance. Therefore p is Σ-internal.

2. Let B be a superset of A such that some realization a' of p with $a' \underset{A}{\downarrow} B$ is definable over B and finitely many realizations of types $\pi_1, \ldots, \pi_k$ in Σ. Choosing $\mathfrak{M}$ as above, we see that every realization of p is definable over $\mathfrak{M}$ and A-conjugates of $\pi_1, \ldots, \pi_k$. Since Σ is over A, its types are invariant under A-conjugation, and we may put $\Sigma_0 = \{\pi_1, \ldots, \pi_k\}$.

3. Consider a realization a of π, which we may choose independent of B over A by $\emptyset$-invariance of Σ. Take a sequence $(a_i : i \leq \alpha)$ in $\mathrm{acl}(aB)$ such that $\mathrm{tp}(a_i : B \cup \{a_j : j < i\})$ is Σ-internal for all $i \leq \alpha$, and let a_i' be the canonical base of $q := \mathrm{stp}(B \cup \{a_j : j \leq i\}/aA)$. Then $a_i' \in \mathrm{acl}(aA)$. We claim that the sequence $(a_i' : i \leq \alpha)$ is a Σ-analysis of a over A. (Strictly speaking, a_i' is an infinite tuple and should be enumerated itself at this point in the analysis.) Clearly, as a is algebraic over Ba_α, it must be algebraic over Aa_α', since $Ba_\alpha \underset{a_\alpha'}{\downarrow} Aa$.

It remains to check that $\mathrm{tp}(a_i'/Aa_j' : j < i)$ is Σ-internal, for every $i \leq \alpha$. Now a_i' is definable over a Morley sequence $(B^s \cup \{a_j^s : j \leq i\} : s < \omega)$ of q by Corollary 0.2.32. Then $X := (B^s \cup \{a_j^s : j < i\} : s < \omega)$ is a Morley sequence in $\mathrm{stp}(B \cup \{a_j : j < i\}/aA)$ which does not fork over $A \cup \{a_j' : j < i\}$. Therefore a, and hence a_i', is independent of X over $A \cup \{a_j' : j < i\}$, and Σ-internality of $\mathrm{tp}(a_i/B \cup \{a_j : j < i\})$ implies

Σ-internality of $\text{tp}(a_i^s/X)$ by $\emptyset$-invariance of Σ, for all $s < \omega$. Therefore $\text{tp}(a_i'/X)$ is Σ-internal, as is $\text{tp}(a_i'/A \cup \{a_j' : j < i\})$.

4. Let $B \supseteq A$ be a set of parameters such that every realization a of p is definable over A and finitely many realizations $b^1, \dots, b^k$ of π. By part 2. we may add B to the language and assume $B = \emptyset$. Suppose $(b_i^s : i \leq \alpha)$ is a Σ-analysis of b^s, for $1 \leq s \leq k$. Let a_i be the canonical base of $q := \text{stp}(b_j^1, \dots, b_j^k : j \leq i/a)$. Then again $a_i \in \text{acl}(a)$ for all $i \leq \alpha$, and we claim that the sequence $(a_i : i \leq \alpha)$ is a Σ-analysis of a of length α. As a is definable over $b^1, \dots, b^k$, and $b^1, \dots, b^k$ is algebraic over $b_\alpha^1, \dots, b_\alpha^k$, it follows that a must be algebraic over a_α.

It remains to check that $\text{tp}(a_i/a_j : j < i)$ is Σ-internal, for every $i \leq \alpha$. Now a_i is definable over a Morley sequence $((b_j^1, \dots, b_j^k : j \leq i)_s : s < \omega)$ in q. But then $X := ((b_j^1, \dots, b_j^k : j < i)_s : s < \omega)$ is a Morley sequence in $\text{stp}(b_j^1, \dots, b_j^k : j < i/a)$, and this does not fork over $\{a_j : j < i\}$. Therefore a, and hence a_i, is independent of X over $\{a_j : j < i\}$. As $\text{tp}((b_i^1, \dots, b_i^k)_s : s < \omega/X)$ has Σ-internal co-ordinates by $\emptyset$-invariance of Σ, transitivity of internality implies that $\text{tp}(a_i/X)$, and hence also $\text{tp}(a_i/a_j : j < i)$, is Σ-internal.

5. Clearly we may assume that π is a stationary complete type over $\emptyset$. We prove the assertion by induction on the length of the Σ-analysis $(a_i : i \leq \alpha)$ of some realization a of π.

By the inductive hypothesis, $\text{tp}(a_i : i < \alpha)$ is Σ'-analysable. Now there are $\sigma_1, \dots, \sigma_k$ in Σ such that $\text{tp}(a_\alpha/a_i : i < \alpha)$ is $\{\sigma_1, \dots, \sigma_k\}$-internal. We apply part 3. to $\sigma(x_1, \dots, x_k) = \sigma_1(x_1) \cup \cdots \cup \sigma_k(x_k)$. Clearly $\text{tp}(a_\alpha/a_i : i < \alpha)$ is σ-internal, and σ is Σ'-analysable just by concatenating the analyses for the different σ_i. It follows that $\text{tp}(a_\alpha/a_i : i < \alpha)$ is Σ'-analysable, whence $\text{tp}(a)$ is Σ'-analysable.

6. Again we assume that π is complete and stationary, and use induction on the length of an analysis $(a_i : i \leq \alpha)$ of a realization a of π. By the inductive assumption, over any set $B \supseteq A$ a realization c of a non-forking extension of p to B does not fork with $\{a_i : i < \alpha\}$. So c realizes a non-forking extension of p to $B \cup \{a_i : i < \alpha\}$, and does not fork with any realizations $\bar{b}$ of types in Σ. So over any (independent) superset of parameters it does not fork with anything in the algebraic closure of those realizations $\bar{b}$, in particular it is independent of a_α, and thus of a. $\square$

We should note that part 3. did not use $\emptyset$-invariance of Σ.

Lemma 3.0.5 *Let A be a G-invariant subset of G. Then $G/C_G(A)$ is A-internal.*

Proof: The centralizer of A is the centralizer of finitely many elements $a_1, \ldots, a_k$ of A. Clearly, any coset $gC_G(A)$ is then definable over the tuple $(a_1, a_1^g, \ldots, a_k, a_k^g)$. But as A is G-invariant, these are all elements of A. $\square$

In general, the connection between foreignness and internality is given by the following:

Proposition 3.0.6 *Suppose* $\mathrm{tp}(a/A)$ *is not foreign to some type* p *over* A. *Then there is* $a_0 \in \mathrm{acl}(Aa) - \mathrm{acl}(A)$ *such that* $\mathrm{tp}(a_0/A)$ *is* p-*internal.*

Proof: As $\mathrm{tp}(a/A)$ is not foreign to p, there are a superset B of A independent of a over A, and a realization b of p such that $a \not\!\downarrow_B b$. Consider a Morley sequence $(B_i b_i : i < \omega)$ in $\mathrm{stp}(Bb/Aa)$, and an element $a_0 \in \mathrm{Cb}(Bb/Aa)$. Clearly $a_0 \in \mathrm{acl}(Aa)$; as $Bb \not\!\downarrow_A a$, we may choose $a_0 \notin \mathrm{acl}(A)$. Now a_0 is definable over a finite segment $(B_i b_i : i < n)$ of the Morley sequence. However, $B \downarrow_A a$, so $(B_i : i < \omega)$ is a Morley sequence in $\mathrm{stp}(B/A)$ independent of a over A, and hence $a \downarrow_A \{B_i : i < n\}$. Furthermore, every b_i realizes p. It follows that $\mathrm{tp}(a_0/A)$ is p-internal. $\square$

We have already seen that a complete type internal in some set Σ of partial types over its domain is internal in a finite subset of those types. This no longer holds for partial types (e.g. for any A the formula $x = x$ is internal in the set of all 1-types over A, but not necessarily in a finite subset of them) or if Σ is not over A. However, for a class of formulæ it does remain true:

Lemma 3.0.7 *Suppose* π *is* Σ-*internal where* Σ *is a class of formulæ. Then there are a finite fragment* φ *of* π, *a finite subset* Σ_0 *of* Σ *and a definable function* f, *such that* φ *is* Σ_0-*internal via* $f : \left(\bigcup \Sigma_0\right)^n \to \varphi$.

Proof: Let A be a set such that every realization a of π is definable over A and realizations of Σ. By compactness there must be finitely many finite subsets Σ_i of Σ and A-definable functions f_i such that every realization a of π lies in the image $f_i((\bigcup \Sigma_i)^n)$ for some i, and we can assemble these functions in a single function f (of greater arity). Now the image of f must contain a finite fragment φ of π. $\square$

Remark 3.0.2 Even if Σ consists of a single partial type, we may not be able in general to reduce to finitely many functions (and hence to a single one): if a group G is definable but G^0 is not, put $\Sigma = \mathrm{gen}(G)$, $\pi = G$, and let $(a_i : i \in I)$ be a system of representatives for G/G^0. Then we need the functions $f_i : \mathrm{gen}(G)^2 \to G$ given by $f_i(x, y) = a_i xy$ in order to witness internality. However, if G is connected and $\mathrm{gen}(G)$ is Σ-internal, then there are some A, a finite subset Σ_0 of types over A in Σ, and an A-definable function f such that every realization of $\mathrm{gen}(G)|A$ is the image of some realizations of types in Σ_0 under f. But every element of G is the product of two realizations of $\mathrm{gen}(G)|A$, so the whole of G is the image of realizations of Σ_0 under a single function.

Lemma 3.0.8 *If Σ is a set of formulæ and π is a partial type analysable in Σ, then there are a finite fragment φ of π and a finite subset Σ_0 of Σ, such that φ is Σ_0-analysable in finitely many steps.*

We should note that this lemma fails if Σ is a proper class. As we shall see later, however, in the case of groups the difficulties caused by proper classes disappear.

Proof: By adding parameters to the language (and going to a suitable elementary extension), we may assume that all formulæ in Σ are parameter-free. By compactness, we may further assume that π is a complete stationary type: if every type $p \in [\pi]$ over $\mathrm{acl}(\emptyset)$ is covered by a Σ_p-analysable formula φ_p for some finite Σ_p, then finitely many of these formulæ will cover π, and we take the union of the corresponding Σ_p as our Σ_0.

We shall use transfinite induction and assume that the assertion is true for all types which are analysable by analyses of length less than α, and suppose that a realization a of π has an analysis $(a_i : i \leq \alpha)$ of length α. Since $\mathrm{tp}(a_\alpha/\{a_i : i < \alpha\})$ is Σ-internal, there are some finite fragment φ', a finite subset Σ_1 of Σ and a definable function f such that φ' is Σ_1-internal via f (using some parameters A). But φ' mentions only finitely many of the $\{a_i : i < \alpha\}$, say $\bar{a}$. By the inductive hypothesis there are a finite fragment φ'' of $\mathrm{tp}(\bar{a})$ and a finite subset Σ_2 of Σ such that φ'' is Σ_2-analysable in finitely many steps. Finally, there is a formula $\psi(a, a_\alpha, \bar{a})$ which says that a is algebraic over a_α and $a_\alpha, \bar{a}$ are algebraic over a. Putting this together, we get a finite fragment φ in $\mathrm{tp}(a)$ such that for all realizations a' of φ there are a'_α and $\bar{a}'$ such that $\psi(a', a'_\alpha, \bar{a}')$ holds, $\mathrm{tp}(a'_\alpha/\bar{a}')$ is Σ_1-internal via f, and $\mathrm{tp}(\bar{a}')$ is Σ_2-analysable, namely

$$\exists x', \bar{x} \left[\psi(x, x', \bar{x}) \wedge \varphi''(\bar{x}) \wedge \varphi'(x, \bar{x}) \wedge \exists Z \, \forall y \left[\varphi'(y, \bar{x}) \rightarrow \exists \bar{y} \in \Sigma_1 \; y = f(\bar{y}, Z) \right] \right].$$

We put $\Sigma_0 = \Sigma_1 \cup \Sigma_2$ and are done. $\square$

Note that this proof also works if Σ is a definable class of formulæ (all formulæ of the form $\varphi(x, \bar{a})$ with $\models \vartheta(\bar{a})$, say), a condition considerably weaker than analysability in the formula $\exists \bar{y} \, \vartheta(\bar{y}) \wedge \varphi(x, \bar{y})$. This case may occur if for a realization a of π and a Σ-analysis $(a_i : i \leq \alpha)$ the type σ_i from Σ such that $\mathrm{tp}(a_i/\{a_j : j < i\})$ is σ_i-internal depends not only on i, but also on the choice of a. In general there will then not be a finite fragment of π which is Σ-analysable. However, we still have

Lemma 3.0.9 *Let Σ be a class of formulæ and p a Σ-analysable type. Then p is Σ-analysable in finitely many steps.*

Proof: Again we shall use induction on the length of an analysis. If a is a realization of p and $(a_i)_{i\leq\alpha}$ a Σ-analysis of a, then as above there is a finite $\bar{a}$ in $(a_i)_{i<\alpha}$ such that $\mathrm{tp}(a_\alpha/\bar{a})$ is Σ-internal. By the inductive assumption $\mathrm{tp}(\bar{a})$ is Σ-analysable in finitely many steps, and so is $\mathrm{tp}(a) = p$. $\square$

3.1 Analysability and Groups

The fundamental result connecting model-theoretic analysability and group-theoretic structure is a theorem due to Hrushovski, which shows the existence of π-internal quotient groups under certain circumstances.

Definition 3.1.1 Two types p and q are *orthogonal*, written $p \perp q$, if over any set A containing the bases for p and q any realization a of a non-forking extension of p to A is independent of any realization b of a non-forking extension of q to A.

So p is foreign to q iff p is orthogonal to all extensions of q.

Theorem 3.1.1 *Let G be a group such that* $\mathrm{gen}(G)$ *is not foreign to some partial type π. Then there is a relatively definable normal subgroup H of infinite index such that G/H is π-internal. If* $\mathrm{gen}(G)$ *is non-orthogonal to a type q, then we can find a q-internal quotient such that* $\mathrm{gen}(G/H)$ *is non-orthogonal to q.*

Note that the converse is obvious: if H is a subgroup of G, then $\mathrm{gen}(G)$ is foreign to G/H iff $H \geq G^0$ (as already remarked in the proof of Lemma 3.0.1).

Proof: We may assume that G is $|T|^+$-saturated. Let $p = \mathrm{gen}(G)$ and consider dependent realizations a of p and b of π. Let B be a base for $\mathrm{tp}(a, b/G)$ of cardinality $|T|$, and put $K := \{g \in G : \mathrm{tp}(ga, b/B) = \mathrm{tp}(a, b/B)\}$. Then K is a subgroup of G type-definable over B; we claim that it has infinite index.

Indeed, if $g \in G$ is a principal generic element over B, then b, a and g must be independent over B, so g is generic over B, b, a. Therefore ga is generic over B, b, whereas a is not generic. Hence $\mathrm{tp}(ga, b/B) \neq \mathrm{tp}(a, b/B)$, and g does not lie in K. So K must have infinite index; in particular there is a relatively B-definable supergroup K_0 of K of infinite index in G.

Now consider a Morley sequence $(a_i, b_i)_{i<\omega} \subset G$ in $\mathrm{stp}(a, b/B)$. If h is an element in G realizing the principal generic type over $B \cup \{a_i, b_i : i < \omega\}$, then both $\{a_i : i < \omega\}$ and $\{ha_i : i < \omega\}$ realize a Morley sequence in $p|B, h$. Hence there are elements b'_i of G such that $\mathrm{stp}(a_i, b'_i : i < \omega/B, h) = \mathrm{stp}(ha_i, b_i : i < \omega/B, h)$. But the latter is also a Morley sequence in $\mathrm{stp}(ha, b/B, h)$, so this type is definable over $\{a_i, b'_i : i < \omega\}$.

Since $\mathrm{tp}(ha, b/G)$ does not fork over B, h, and for any $h' \in G$ we have $\mathrm{tp}(ha, b/G) = \mathrm{tp}(h'a, b/G)$ iff $\mathrm{tp}(h^{-1}h'a, b/G) = \mathrm{tp}(a, b/G)$ iff $h^{-1}h' \in K$ iff $hK = h'K$, the coset hK must also be type-definable over $\{b'_i : i < \omega\}$, using parameters B and $\{a_i : i < \omega\}$. So over these parameters hK_0 is $\{b'_i : i < \omega\}$-definable, and only finitely many of the b'_i are needed (hK_0 is a finitary object). Thus $\mathrm{gen}(G^0/K_0)$ is π-internal.

But then for any g in G the type $\mathrm{gen}(G^0/K_0^g)$ is π-internal as well. The intersection $H := \bigcap_{g \in G} K_0^g$ of all these conjugates is a finite subintersection $\bigcap_{i=1}^n K_0^{g_i}$, and the map from $\prod_{i=1}^n (G/K_0^{g_i})$ to G/H is definable and surjective. Therefore $\mathrm{gen}(G^0/H)$ is π-internal. But any two generic types are translates of one another, and any group element is a product of two generic types. So the whole of G/H is π-internal.

Finally, if $\mathrm{gen}(G)$ is non-orthogonal to q, then we may take $\pi = q$, and $\mathrm{tp}(b/B)$ a non-forking extension of q. It is then obvious that $\mathrm{gen}(G/K)$, and hence $\mathrm{gen}(G/H)$, is non-orthogonal to q. $\square$

Note that we may even intersect H with its image under finitely many definable automorphisms of G, and we may therefore obtain a subgroup which is not only normal, but invariant under any type-definable group of automorphisms, or more generally under any group consisting of automorphisms σ (not necessarily definable ones) which map H to a uniformly (in σ) definable subgroup H^σ such that G/H^σ is still π-internal (e.g. if σ fixes π).

Corollary 3.1.2 *If the generic type p of a definably simple stable group G or of a stable division ring D is not foreign to some partial type π, then G (resp. D) is π-internal.*

Proof: This is obvious for G from Theorem 3.1.1. In the case of a division ring, Theorem 3.1.1 yields a relatively definable additive subgroup H of D such that D/H is π-internal. But then D/dH is also π-internal for every non-zero d in D, and as $I = \bigcap_{d \in D} dH$ is a finite subintersection, D/I is π-internal as well. However, I is an ideal of infinite index, hence trivial. $\square$

As, for a generic element g, the coset gH is generic for the quotient G/H and definable in g, this is the first step towards an analysis of g. This process can be iterated and enables us to give a particularly well-behaved criterion for analysability in groups:

Lemma 3.1.3 *Let Σ be a class of partial types. A stable group G is Σ-analysable iff there is a descending sequence $G = G_0 > G_1 > \cdots > G_\alpha = \{1\}$ of type-definable normal subgroups, such that $G_\lambda = \bigcap_{i < \lambda} G_i$ for limit ordinals λ, and every quotient G_i/G_{i+1} is Σ-internal.*

In fact, we may assume that G_{i+1} is relatively definable in G_i for all $i < \alpha$.

Proof: As every group element is a product of two generic elements, G is Σ-internal (-analysable) iff gen(G) is.

($\Rightarrow$) Let $G = G_0 > G_1 > \cdots > G_\beta$ be a maximal sequence of type-definable normal subgroups of G such that $G_\lambda = \bigcap_{i<\lambda} G_i$ for limit ordinal λ, and G_i/G_{i+1} is Σ-internal for every $i < \beta$. Such a β must exist by the $|T|^+$cc; by Theorem 3.1.1 and maximality, gen(G_β) is foreign to Σ.

Consider a realization a of gen(G) with Σ-analysis $(a_i : i < \alpha)$. Suppose there were some (minimal) i_0 such that a_{i_0} is not algebraic over aG_β. As gen(G_β) is a translate of tp(a/aG_β), the assumption implies that tp(a/aG_β), and hence also tp(a_{i_0}/aG_β), is foreign to Σ. On the other hand, tp($a_{i_0}/\{a_i : i < i_0\}$) is Σ-internal and $\{a_i : i < i_0\}$ is algebraic in aG_β, so tp(a_{i_0}/aG_β) must be Σ-internal. Together this means that a_{i_0} is algebraic in aG_β, contradicting our assumptions. Therefore a is algebraic over aG_β, so G_β must be finite and we can put $G_{\beta+1} = \{1\}$.

($\Leftarrow$) Suppose $(G_i : i < \alpha)$ is such a sequence and a realizes gen(G). Then $a_i := aG_i$ is a Σ-analysis of a. $\square$

We should note that we can choose the G_i in such a way that they are definable over the parameters needed for the definition of G and Σ. As we have allowed Σ to be a class, we get the validity of Lemma 3.0.8 for classes of formulæ:

Corollary 3.1.4 *Let G be a type-definable stable group analysable in a class Σ of types. Then there is a subset Σ_0 of Σ of cardinality at most $|T|$ such that G is Σ_0-analysable. If Σ is a class of formulæ, we can even choose Σ_0 to be finite.*

Proof: Choose a sequence $(G_i : i \leq \alpha)$ as in Lemma 3.1.3. Then every quotient is by definition analysable in finitely many elements of Σ; as α has cardinality at most $|T|$, we find the required Σ_0.

But now we may apply Lemma 3.0.8 to the set Σ_0 and obtain a finite subset in which G, and hence a definable supergroup, is still analysable. $\square$

The following lemma allows transfinite chain arguments:

Lemma 3.1.5 *Suppose H is normal in G. If a complete type q is G-internal via some function $f(x_1, \ldots, x_n)$, but foreign to G/H, then it is H-internal via some translate $f(g_1 x_1, \ldots, g_n x_n)$ of f, for certain $g_1, \ldots, g_n$ in G such that $f(g_1, \ldots, g_n)$ realizes q.*

Proof: Suppose a realizes q and $\bar{h}$ is in G such that $a = f(\bar{h})$. As q is foreign to G/H, we see that a must be independent of $h_1 H, \ldots, h_n H$. If we choose $\bar{g}$ realizing tp($\bar{h}/\bar{h}H$) in G such that a is independent of $\bar{g}$ over $\bar{h}H$, then a is independent of $\bar{g}$. But $g_i^{-1} h_i$ lies in H for $1 \leq i \leq n$; this shows the lemma. $\square$

So if q is G internal, but foreign to G/H_i for a descending chain $(H_i : i \leq \alpha)$ of subgroups, then q is $(\bigcap_{i \leq \alpha} H_i)$-internal via a translate of the original function witnessing internality.

Finally, we shall investigate the relationship between analysability and commutator subgroups.

Lemma 3.1.6 *Let G be a type-definable and H an arbitrary subgroup of some stable group, put $K := \mathrm{dc}[G^0, H]$, and suppose $\mathrm{gen}(G)$ is foreign to some partial type π. Then $\mathrm{gen}(K)$ is foreign to π.*

Proof: Suppose there is some relatively definable $G^0 H$-invariant subgroup N in K such that K/N is π-internal. For h in H and independent principal generic elements g and g' of G, we get modulo N

$$[g, h] = [g', h] = [g'g, h] = [g', h]^g [g, h] = 1,$$

as this is an element of K/N and hence independent of g and g'. Thus $[G^0, H]$ is contained in N and so is K (it is connected by Lemma 1.1.7). The result now follows from Theorem 3.1.1. $\square$

Corollary 3.1.7 *If $\mathrm{gen}(G)$ is foreign to $\mathrm{dc}(G')$, then G is abelian-by-finite.*

Proof: By Lemma 3.1.6 the generic type of $\mathrm{dc}[G^0, G]$ is foreign to $\mathrm{dc}(G')$, but this implies that $[G^0, G]$ is finite. Hence G^0 is abelian. $\square$

Lemma 3.1.8 *Suppose G is nilpotent. Then G is $Z(G)$-analysable.*

Proof: We show inductively that $Z_i(B)$ is $Z(B)$-analysable. As $B = Z_n(B)$ for some $n < \omega$, this proves the lemma. Clearly, the assertion is trivial for $i = 1$. So suppose $Z_i(B)$ is $Z(B)$-analysable, and consider an element $b \in Z_{i+1}(B)$. There are $b_1, \ldots, b_k \in B$ such that $Z(B) = C_B(b_1, \ldots, b_k)$, and $[b, b_i] \in Z_i(B)$ for all $i = 1, \ldots, k$. Therefore these commutators have $Z(B)$-analysable types by the inductive hypothesis. But b is determined by $[b, b_1], \ldots, [b, b_k]$ modulo $C_B(b_1) \cap \cdots \cap C_B(b_k) = Z(B)$, so $\mathrm{tp}(b)$ is $Z(B)$-analysable as well. $\square$

3.2 Components

In this section we shall define various components of a group G. A particular rôle will be played by normal subgroups N whose generic type is foreign to the quotient G/N.

Lemma 3.2.1 *Let H be a subgroup of G such that the generic type of H is foreign to G/H. Then H^0 is definably characteristic in G (and, in particular, it is normal).*

Proof: Let σ be a definable automorphism of G. As the generic type of H is foreign to G/H, and since $G/H^\sigma = G^\sigma/H^\sigma$ is definably isomorphic to G/H, the generic type of H is also foreign to G/H^σ, and hence to $G/(H \cap H^\sigma)$ (as $g(H \cap H^\sigma) \in \mathrm{dcl}\{gH, gH^\sigma\}$). But this can only happen if $H \cap H^\sigma$ has finite index in H, i.e. $H^\sigma \geq H^0$. $\square$

Lemma 3.2.2 *Let G be a definable stable group and $(H_i : i < \alpha)$ a descending sequence of type-definable normal subgroups such that* $\mathrm{gen}(H_i)$ *is foreign to G/H_i. Put $H := \bigcap_{i<\alpha} H_i$. Then the generic type of H is foreign to G/H.*

Proof: Suppose not. Then by Theorem 3.1.1 there is a relatively definable normal subgroup N of infinite index in H such that H/N is (G/H)-internal. We may assume that H is connected. Then there is a definable function φ from $\prod_n G$ to some superset of H/N which only depends on the class of a tuple $\bar{g}$ modulo H. So there are definable supergroups H_i and H_j of H such that φ depends only on the class of $\bar{g}$ modulo H_i and maps $\prod_n G$ surjectively to H_j/N. By compactness there is $\alpha_0 < \alpha$ such that H_i and H_j both contain H_{α_0}, and φ induces a function from $\prod_n(G/H_{\alpha_0})$ to H_{α_0}/N, contradicting that $\mathrm{gen}(H_{\alpha_0})$ is foreign to G/H_{α_0}. $\square$

Note that this lemma seems to need definability of G.

Definition 3.2.1 Let G be a group and Σ a class of partial types. The Σ-*connected component* G^Σ of G is the intersection of all relatively definable subgroups H such that the quotient G/H is Σ-analysable. G is Σ-*connected* if $G = G^\Sigma$.

Note that if G/H is Σ-analysable, so is G/H^g for any $g \in G$, and therefore also $G/\bigcap_{g \in G} H^g$. Thus it is sufficient to intersect only *normal* relatively definable subgroups with Σ-analysable quotient. By Theorem 3.1.1 the generic type of G^Σ must be foreign to Σ, as otherwise the analysis could be extended; as finite quotients are internal in any set, G^Σ is connected. Furthermore G/G^Σ (i.e. every quotient in the directed system) is Σ-analysable. In particular the generic type of G^Σ is foreign to G/G^Σ, and G^Σ is definably characteristic. If Σ' is a second class of partial types, then the $(\Sigma \cup \Sigma')$-connected component is contained in $G^\Sigma \cap G^{\Sigma'}$.

Lemma 3.2.3 *If H is a subgroup of G, then $(G/H)^\Sigma = G^\Sigma H/H$.*

Proof: The generic type of $G^\Sigma H/H$ is foreign to Σ, hence $G^\Sigma H/H \leq (G/H)^\Sigma$. On the other hand, if G/K is Σ-analysable for some relatively definable normal subgroup K of G, then G/KH is Σ-analysable, and the reverse inequality follows. $\square$

Note that this makes sense even if H is not normal in G.

Definition 3.2.2 Let $\mathfrak{X}$ be a class of groups, and G a stable group.

- G is $\mathfrak{R}$-*connected modulo* $\mathfrak{X}$, if G is not in $\mathfrak{X}$, but every type-definable normal subgroup N whose generic type $\mathrm{gen}(N)$ is foreign to G/N lies in $\mathfrak{X}$.

- A minimal type-definable normal subgroup N of G not in $\mathfrak{X}$ whose generic type is foreign to G/N is called an $\mathfrak{R}$-*component modulo* $\mathfrak{X}$.

If $\mathfrak{X}$ is closed under finite extensions, then an $\mathfrak{R}$-connected group must be connected modulo $\mathfrak{X}$. Clearly, an $\mathfrak{R}$-connected group is its own $\mathfrak{R}$-component. Examples for such $\mathfrak{X}$ are the classes of groups which are finite, soluble-by-finite, nilpotent-by-finite, central in G^0, or ω-central in G^0. We shall usually omit "modulo finite groups" and just talk of $\mathfrak{R}$-connectivity and the $\mathfrak{R}$-component. An $\mathfrak{R}$-component modulo $\mathfrak{X}$ is invariant under definable automorphisms of G by Lemma 3.2.1. If $\mathfrak{X}$ contains the trivial group and the intersection of a chain of type-definable groups not in $\mathfrak{X}$ is not in $\mathfrak{X}$ either, then by Lemma 3.2.2 a definable stable group has an $\mathfrak{R}$-component modulo $\mathfrak{X}$. Note that all of the above $\mathfrak{X}$ possess this closure property. However, even if the intersection of two groups not in $\mathfrak{X}$ is again not in $\mathfrak{X}$, the $\mathfrak{R}$-component modulo $\mathfrak{X}$ need not be unique, as it may happen that $\mathrm{gen}(N_i)$ is foreign to G/N_i for $i = 0, 1$, but $\mathrm{gen}(N_0 \cap N_1)$ is not foreign to $G/(N_0 \cap N_1)$. Furthermore, an $\mathfrak{R}$-component modulo $\mathfrak{X}$ need not be $\mathfrak{R}$-connected modulo $\mathfrak{X}$; e.g. G might have subgroups $G > G_1 > G_2$ such that $\mathrm{gen}(G_1)$ is foreign to G/G_1 and $\mathrm{gen}(G_2)$ is foreign to G_1/G_2, but $\mathrm{gen}(G_2)$ need not be foreign to G/G_2. Finally, as in the case of ordinary connectivity, the relativization of a property defining $\mathfrak{X}$ to the $\mathfrak{R}$-component modulo $\mathfrak{X}$ may well result in a different class $\mathfrak{X}'$ of groups.

Note that all of the above $\mathfrak{X}$ possess the closure properties mentioned.

Definition 3.2.3 A type p is $\mathfrak{R}$-*primary* if p is analysable in any partial type it is not foreign to. We call p strongly $\mathfrak{R}$-*primary* if p is internal in any partial type it is not foreign to.

By Corollary 3.1.2, the generic type of a definably simple group or a division ring is strongly $\mathfrak{R}$-primary.

Lemma 3.2.4 *A group G is $\mathfrak{R}$-connected if and only if it is connected and its generic type is $\mathfrak{R}$-primary.*

Proof: Suppose the generic type of G is $\mathfrak{R}$-primary, but G is not $\mathfrak{R}$-connected, as witnessed by some infinite subgroup H. If G is connected, $\mathrm{gen}(G)$ is not foreign to G/H and must be (G/H)-analysable by $\mathfrak{R}$-primarity. Hence G, and also H, are (G/H)-analysable; in particular $\mathrm{gen}(H)$ cannot be foreign to G/H, a contradiction.

On the other hand, suppose that G is $\mathfrak{R}$-connected and its generic type is not foreign to some partial type π. Let H be the π-connected component of G. By $\mathfrak{R}$-connectivity, H is trivial, and G is π-analysable. $\square$

Note that if p is a regular type, the same lemma holds with $\mathfrak{R}$ replaced by p, with Hrushovski's notions of p-connectivity and p-primarity.

Lemma 3.2.5 *If G_1 is a connected $\mathfrak{R}$-component modulo $\mathfrak{X}$ of G and G_2 is a connected proper subgroup such that the generic type of G_2 is foreign to G_1/G_2, then $\mathrm{dc}[G_1, G_2]$ is contained in an infinitely definable group in $\mathfrak{X}$.*

Proof: By Lemma 3.1.6 the generic type of $\mathrm{dc}[G_1, G_2]$ is foreign to G/G_1 and G_1/G_2, hence to G/G_2. Let G_3 be the (G/G_2)-connected component of G (we consider G/G_2 as a directed system of quotients, and hence as a set of partial types). Then G_3 is a proper subgroup of G_1 whose generic type is foreign to G/G_3, so it must lie in $\mathfrak{X}$. However, $\mathrm{dc}[G_1, G_2]$ is contained in G_3, as its generic type is foreign to G/G_2, hence to G/G_3, and therefore to $\mathrm{dc}[G_1, G_2]/G_3$. $\square$

Corollary 3.2.6 *An $\mathfrak{R}$-component modulo soluble-by-finite groups is $\mathfrak{R}$-connected modulo soluble-by-finite groups. If G_1 is an $\mathfrak{R}$-component of G modulo groups ω-central in G^0, it is $\mathfrak{R}$-connected modulo groups ω-central in G_1^0.*

Proof: Let G_1 be an $\mathfrak{R}$-component modulo soluble-by-finite groups, and suppose there is a normal proper subgroup G_2 of G_1 whose generic type is foreign to G_1/G_2. Then $[G_1, G_2]$ is soluble (it is connected by Lemma 1.1.7), and so is G_2.

If G_1 is an $\mathfrak{R}$-component modulo groups ω-central in G^0 of G and G_2 is a proper normal subgroup whose generic type is foreign to G_1/G_2, then $\mathrm{dc}[G_1, G_2]$ is ω-central in G^0. By compactness, it is contained in some finitely iterated centralizer $C_G^i(G^0)$, and G_2 is $(i+1)$-central in G_1^0. $\square$

We shall now define a kind of strongly connected component, which is the intersection not only of all subgroups of finite index, but of all subgroups with small quotient.

Definition 3.2.4 Let G be a stable group.

- A formula $\varphi(x)$ is *Frattini for G* if for all sets Σ of formulæ such that G is $(\Sigma \cup \{\varphi\})$-analysable, G is already Σ-analysable. We denote the class of Frattini formulæ for G by $\Phi(G)$.

- G is *Φ-connected* if it is $\Phi(G)$-connected, i.e. $\mathrm{gen}(G)$ is foreign to all Frattini formulæ for G.

- The Φ-*component* G^{Φ} of G is the $\Phi(G)$-connected component of G.

A Frattini formula is, in a way, small as compared to G.

Remark 3.2.1 A group of monomial U-rank ω^{α} or a p-connected group (for some regular type p) is Φ-connected. The converse need not hold (even in a superstable group): if A_1 and A_2 are two groups with mutually foreign generic types and $U(A_1) = \omega^{\alpha_1} \cdot n_1$ and $U(A_2) = \omega^{\alpha_2} \cdot n_2$, then $A_1 \oplus A_2$ is Φ-connected, but has U-rank $\omega^{\alpha_1} \cdot n_1 \oplus \omega^{\alpha_2} \cdot n_2$, which need not be monomial. However, in a rank-free context, we cannot expect to be able to distinguish between the "sizes" of A_1 and A_2.

Lemma 3.2.7 G^{Φ} *is non-trivial, connected, normal, and its generic type is foreign to all Frattini formulæ. In particular* $\mathrm{gen}(G^{\Phi})$ *is foreign to* G/G^{Φ}. *Furthermore,* $\Phi(G^{\Phi}) = \Phi(G)$. *Hence* $(G^{\Phi})^{\Phi} = G^{\Phi}$.

Proof: Connectivity is obvious, as any finite formula is Frattini. Non-triviality follows immediately from Corollary 3.1.4 (finiteness of group analyses in formulæ), foreignness from Theorem 3.1.1, and normality from Lemma 3.2.1. Lastly, if φ is a Frattini formula for G^{Φ} and G is $(\Sigma \cup \{\varphi\})$-analysable, then G^{Φ} is $(\Sigma \cup \{\varphi\})$-analysable and hence Σ-analysable. Therefore G is $(\Sigma \cup \Phi(G))$-analysable and hence Σ-analysable by Corollary 3.1.4. Thus φ is a Frattini formula for G.

Conversely, if $\varphi \in \Phi(G)$ and G^{Φ} is $(\Sigma \cup \{\varphi\})$-analysable, it follows that G is $(\Sigma \cup \{\varphi\} \cup \Phi(G))$-analysable and hence Σ-analysable. Therefore φ is Frattini for G^{Φ}. $\square$

If G is definable, G^{Φ} is the intersection of all definable groups H such that G/H is Frattini. (To motivate the terminology, remember that for a finite group G the Frattini subgroup $\Phi(G)$ is the intersection of all maximal proper subgroups, and $g \in \Phi(G)$ iff g can be omitted from any generating set for G.) Note that G is X-analysable for any definable superset X of G^{Φ}.

Lemma 3.2.8 *If G is definable and $Z(G^{\Phi}) = \{1\}$, then G is G^{Φ}-analysable. More generally, if G^{Φ} is inter-analysable with some definable set (in particular if T has the ωdcc^0), then G is G^{Φ}-analysable.*

Proof: Let φ be a formula inter-analysable with G^{Φ}. Then G is $(\{\varphi\} \cup \Phi(G))$-analysable. By Corollary 3.1.4, we only need finitely many formulæ, and we can leave out those in $\Phi(G)$. Hence G is φ-analysable and thus G^{Φ}-analysable.

Now suppose $Z(G^{\Phi}) = \{1\}$. Then $G/C(G^{\Phi})$ is definable and contains definably a copy of G^{Φ}, as $C_G(G^{\Phi})$ intersects G^{Φ} trivially. So G^{Φ} is $(G/C_G(G^{\Phi}))$-internal. On the other hand, the quotient $G/C_G(G^{\Phi})$ is G^{Φ}-internal by Lemma 3.0.5, and we finish by the first paragraph.

Finally, in a theory with the ωdcc^0 there is a definable group F with $F^0 = G^\Phi$, and then F is clearly inter-analysable with G^Φ. $\square$

The lemma after the next definition shows a typical manipulation using Frattini formulæ.

Definition 3.2.5 Let M be a group of automorphisms of some group A. We say that A is M-*minimal* if it is infinite and has no proper type-definable M-invariant infinite subgroup.

Note that if A is M-invariant, so is A^0. Hence an M-minimal group must be connected.

Lemma 3.2.9 *Let G be definable and H a connected subgroup of G whose generic type is foreign to all types which are co-foreign to* $\mathrm{gen}(G^\Phi)$. *Suppose N is a normal subgroup of H, and M is a group of automorphisms of H/N. If A/N is an M-minimal normal subgroup of H/N, then either H centralizes A/N, or the generic type of A/N is not foreign to $H/C_H(A/N)$.*

By Lemma 3.1.6 any definable closure of a commutator subgroup involving G^Φ will satisfy the hypotheses on H.

Proof: Suppose neither conclusion holds. As $C_H(A/N)$ is properly contained in H, $\mathrm{gen}(H)$ is not foreign to $H/C_H(A/N)$. Hence $\mathrm{gen}(G^\Phi)$ is not foreign to this quotient, so by Theorem 3.1.1 there is a definable normal subgroup K of G which does not contain G^Φ, such that G^Φ/K is $(H/C_H(A/N))$-internal. We claim that G/K is a Frattini formula, contradicting $G^\Phi \not\le K$.

So consider a set Σ of formulæ such that G is analysable in $\Sigma \cup \{G/K\}$. Then in particular A/N is analysable in that set; as the generic type of A/N is foreign to $H/C_H(A/N)$ and thus also to G^Φ/K, it cannot be foreign to $\Sigma \cup \Phi(G)$. By Theorem 3.1.1 and M-minimality, it must actually be almost $(\Sigma \cup \Phi(G))$-internal. But $H/C_H(A/N)$ is (A/N)-internal, and so is G^Φ/K. Therefore G is $(\Sigma \cup \Phi(G))$-analysable, hence Σ-analysable. $\square$

We shall now look at quotient and commutator subgroups.

Lemma 3.2.10 *Suppose φ is a formula, H is a normal subgroup of G, and Σ and Σ' are two sets of formulæ.*

1. *If φ is Frattini for H and G/H, then φ is Frattini for G.*

2. *If G/H is $(\Sigma \cup \Sigma')$-analysable, then $[G^{\Sigma'}, G^\Sigma]$ is contained in H.*

Proof:

1. Suppose G is analysable in $\Sigma \cup \{\varphi\}$. Then both H and G/H are analysable in $\Sigma \cup \{\varphi\}$, and hence already Σ-analysable. Therefore G is Σ-analysable, implying that φ is Frattini for G.

2. As $G^{\Sigma'}$ is foreign to Σ' and G^Σ is foreign to Σ, by Lemma 3.1.6 the generic type of $dc[G^{\Sigma'}, G^\Sigma]$ is foreign to $\Sigma \cup \Sigma'$. But G/H is analysable in $\Sigma \cup \Sigma'$. Hence H must contain $dc[G^{\Sigma'}, G^\Sigma]^0 = dc[G^{\Sigma'}, G^\Sigma]$. $\square$

Lemma 3.2.11 *Suppose G and H are normal in some stable group, and φ is Frattini for $dc[G, H]$. If Σ is a set of formulæ such that G is analysable in $\Sigma \cup \{\varphi\}$, then $G/C_G(H)$ is Σ-analysable. Indeed, $[G^\Sigma, H] = 1$.*

Proof: Note that $dc[G, H] \leq G \cap H$. As φ is Frattini for $dc[G, H]$, we know that $dc[G, H]$ is Σ-analysable, and hence the generic type of G^Σ is foreign to $dc[G, H]$. By Lemma 3.1.6 the generic type of $dc[G^\Sigma, H]$ is foreign to $dc[G, H]$, but this can only be if it is trivial. $\square$

These lemmas suggest, for non-soluble groups, a generalization of Φ-connectivity.

Definition 3.2.6 Let $\Phi_s(G)$ be the set of all formulæ φ with the property that for all sets Σ of formulæ, if the $(\Sigma \cup \{\varphi\})$-connected component of G is soluble, then the Σ-connected component of G is soluble.

Remark 3.2.2 $G^{\Phi_s(G)}$ is a definably characteristic non-soluble subgroup of G whose generic type is foreign to $G/G^{\Phi_s(G)}$.

Lemma 3.2.12 *Let H be a relatively definable normal subgroup of G.*

1. *If G is $\Phi_s(G)$-connected, then $\Phi_s(G)$ is contained in $\Phi_s(G/H)$.*

2. *If G is $\Phi_s(G/H)$-connected, then $\Phi_s(G/H)$ is contained in $\Phi_s(G)$.*

Proof:

1. Suppose φ lies in $\Phi_s(G)$ and the $(\Sigma \cup \{\varphi\})$-connected component of G/H is contained in some soluble group S/H. Then by Lemma 3.2.10 the commutator $[G^\Sigma, G^\varphi]$ is contained in S, but $G = G^\varphi$. Therefore $(G/H)^\Sigma = G^\Sigma H/H$ is soluble and $\varphi \in \Phi_s(G/H)$.

2. Suppose φ lies in $\Phi_s(G/H)$ and the $(\Sigma \cup \{\varphi\})$-connected component of G is contained in some soluble subgroup S. Then again by Lemma 3.2.10 the commutator $[G^\Sigma, G^\varphi]$ is contained in S, but as $G^\varphi/H = G/H$, the quotient G^Σ/H is soluble. $\square$

3.3   Involutions

In this section, we shall continue our investigation of involutions and 2-elements in stable groups from section 1.5 of chapter 1.

Theorem 3.3.1 *Let G be a group of finite exponent containing some involution i. Then either $\mathrm{gen}(G)$ is not foreign to the centralizer of i, or G has an abelian subgroup of finite index on which conjugation by i acts as inversion.*

Proof: Suppose $p = \mathrm{gen}(G)$ is foreign to $C_G(i)$. Let g be a principal generic element over i and consider $c = [g, i] = i^g i$, an element of finite order $2^n m$, with m odd. We shall use induction on n.
Case 1: $n = 0$.
As g is not centralized by i, we must have $m > 0$. Put $i_k(g) := [g, i]^{2^{k-1}} i$. As m is odd, there is some $k > 1$ such that $2^k \equiv 1 \mod(m)$ (say $k = \varphi(m)$, where φ is the Euler function), so $i_{k+1}(g) = [g, i] i = i^g$. On the other hand it is easily verified that $i_{k+1}(g) = i^{i_k(g)}$. Hence $j := i_k(g)$ lies in $C_G(i)g$. But g is generic over the elements of $C_G(i)$, so j must be a generic involution. By Theorem 2.4.5 there is an abelian subgroup A of finite index inverted by j. Now $j = i_k = i^{i_{k-1}}$, so i inverts the abelian group $A^{i_{k-1}}$ of finite index.
Case 2: $n > 0$.
Put $k := [g, i]^{2^{n-1}m}$. Then k is an involution centralized by both i and i^g. Hence k and $k^{g^{-1}}$ are independent of g – they both belong to $C_G(i)$ – and $gC_G(k)$ is independent of g, whence $C_G(k)$ must have finite index in G. The assumptions of the inductive hypothesis hold for the definable group $C_G(k)/\{1, k\}$, so there is an abelian group $\bar{A}$ of finite index inverted by i. By Corollary 1.1.4 the pre-image of $\bar{A}$ has an abelian subgroup A of finite index, and for some principal generic element a of A the conjugate a^i is one of $\{a^{-1}, a^{-1}k\}$. If b is a second principal generic element, then $(ab)^i = a^i b^i = a^{-1} b^{-1} = (ab)^{-1}$ (as $k^2 = 1$), but ab is principal generic as well. Therefore an abelian subgroup of finite index is inverted by i. $\square$

Theorem 3.3.2 *Let G be a group of finite exponent and i a 2-element. If the generic type p of G is foreign to $C_G(i)$, then G has an abelian subgroup A inverted by i^{2^k} for some $k < \omega$, such that p is not foreign to A.*

Proof: Suppose p is foreign to $C_G(i)$, and let j be a maximal 2^k-th power of i (for some $k < \omega$) such that p is foreign to $C_G(j)$. Then p is not foreign to $C_G(j^2)$. We consider the $C_G(j)$-connected component A of $C_G(j^2)$. Clearly $C_G(j)$ and hence A are invariant under conjugation by i, and p is foreign to the quotient $C_G(j^2)/A$. So p cannot be foreign to A. On the other hand the generic type of A is foreign to $C_G(j)$ and *a fortiori* to $C_A(j)$, whence by Theorem 3.3.1, applied to $A \rtimes (\langle j \rangle / \langle j^2 \rangle)$, an abelian subgroup of finite index of A is inverted by j. $\square$

If G has infinite exponent, we still get a result for an involution operating on an abelian group with few fixed points:

Theorem 3.3.3 *Let i be an involution operating on the abelian stable group A. If $\mathrm{gen}(A)$ is foreign to $C_A(i)$, then i inverts a subgroup of finite index.*

Proof: We look at the homomorphism $\varphi : a \mapsto a^i a$ from A to to $C_A(i)$. As the image of a principal generic element is independent of the pre-image, this map must be zero on the connected component, as for two independent principal generic elements a and b we have $\varphi(a) = \varphi(b) = \varphi(ab) = \varphi(a)\varphi(b) = 1$. So i inverts principal generic elements, whence a subgroup of finite index. $\square$

This theorem is trivial for finite $C_A(i)$ and does not use stability: obviously the index of $\varphi^{-1}(0)$ in A must be finite. Finally we want to remark that in certain cases (for non-abelian groups) smallness of $C_G(i)$ may imply the genericity of i^g for generic g; we may then apply Theorem 2.4.5. This is possible e.g. if $C_G(i)$ is finite and G has property $\mathfrak{R}$ from chapter 5, or if $\mathrm{gen}(G)$ is regular.

3.4 Groups without an Abelian Normal Subgroup

In this section we shall analyse the structure of a stable group without an abelian normal subgroup. First we shall consider a slightly more general situation:

Proposition 3.4.1 *Let G be a type-definable stable group such that $\mathrm{gen}(G)$ is foreign to all relatively definable normal abelian subgroups. Then $G^{cc}/Z(G^{cc})$ has no normal abelian subgroup.*

Proof: Suppose A is a relatively definable normal subgroup of G^{cc} such that $A/Z(G^{cc})$ is abelian. Then A and its finitely many G-conjugates generate a relatively definable nilpotent G-invariant subgroup N of G^{cc}. Now $Z(N)$ is normal in G, so by assumption $\mathrm{gen}(G)$ is foreign to $Z(N)$. But N is $Z(N)$-analysable by Lemma 3.1.8, so $\mathrm{gen}(G)$ is foreign to N, and N is centralized by G^{cc} by Lemma 3.0.1. $\square$

For the rest of this section G will be a stable type-definable group without an abelian normal subgroup.

Lemma 3.4.2 *A normal subgroup N of G has no abelian normal subgroup either.*

Proof: Let A be a non-trivial abelian normal subgroup of N. By Theorem 1.1.12 the G-conjugates of A generate a nilpotent normal subgroup, whose centre would be an abelian normal subgroup of G, a contradiction. $\square$

Lemma 3.4.3 G *has no locally soluble normal subgroup.*

Proof: Otherwise, by Proposition 1.3.7 there would be an abelian normal subgroup. $\square$

Corollary 3.4.4 *An $\Re$-component of G is $\Re$-connected.*

Proof: As any infinite normal subgroup is not soluble-by-finite, an $\Re$-component is the same as an $\Re$-component modulo soluble-by-finite groups. But modulo soluble-by-finite groups, an $\Re$-component is $\Re$-connected by Corollary 3.2.6. $\square$

Lemma 3.4.5 *Let H_1 and H_2 be two distinct $\Re$-components of G, with generic types p_1 and p_2, respectively. Then p_1 is foreign to H_2, and H_1 intersects H_2 trivially. In particular, H_1 and H_2 commute.*

Proof: Put $I := H_1 \cap H_2$. By $\Re$-connectivity, I is a proper subgroup of both H_1 and H_2, and H_2 is (H_2/I)-analysable. So if p_1 were not foreign to H_2, it would be analysable in H_2 and hence in H_2/I by $\Re$-primarity. But then it could not be foreign to G/H_1 as H_2/I definably embeds into there, a contradiction.

Hence p_1 is foreign to H_2. By Lemma 3.1.6, the generic type of $dc[H_1, H_2]$ is foreign to H_2. But $dc[H_1, H_2]$ is contained in H_2, so it must be finite, and trivial by connectivity. Hence H_1 and H_2 commute, so I is abelian, hence trivial. $\square$

Theorem 3.4.6 *Let G be a definable stable group, N a definable normal subgroup, and suppose G/N has no abelian normal subgroups. Then G^{Φ}/N decomposes as a direct product of finitely many $\Re$-connected normal subgroups.*

Proof: By Lemma 3.4.5 any two distinct $\Re$-components of G/N have pairwise foreign generic type and trivial intersection. They are non-abelian, so Corollary 1.0.6 gives us a finite bound n on their number. Let them be $H_1, H_2, \ldots, H_n$ with generic types $p_1, p_2, \ldots, p_n$, and set $H = \prod_{i=1}^n H_i$, with generic type p. This product is direct, as H_j is centralized by H_i for $i \neq j$, but $C_{G/N}(H_i)$ intersects H_i trivially. Hence p is inter-algebraic with $p_1 \otimes \cdots \otimes p_n$, and foreign to G/HN.

Now if the p-connected component G^p/N were infinite, it would contain an $\Re$-component H_i. But $G/C_G(H)$ is p-internal, so H_i must be contained in $C_{G/N}(H)$ and be abelian, a contradiction. Therefore G/N is p-analysable.

Suppose p were not foreign to some Frattini formula φ. Then we could find some i such that p_i is not foreign to φ. But p_i is $\Re$-primary, so p_i is φ-analysable. Therefore $G/C_G(H_i)$ is φ-analysable, hence a Frattini formula.

On the other hand every p_j is inter-internal with $G/C_G(H_j)$. It follows that G is analysable in the set $\{G/C_G(H_j) : 1 \leq j \leq n\}$, and none of the formulæ can be omitted, as their generic types are pairwise foreign. So $G/C_G(H_i)$ cannot be Frattini after all, a contradiction. Hence p is foreign to $\Phi(G)$ and H must be contained in G^Φ/N.

We now want to show that $G^\Phi/N = H$. So let H_1 be a definable super-group of H. Since p is foreign to G/HN, it must be foreign to G/H_1N. We claim that G/H_1 is Frattini. So let Σ be a set of formulæ such that G is analysable in $\Sigma \cup \{G/H_1\}$. Then all the p_i are $(\Sigma \cup \{G/H_1\})$-analysable; as they are foreign to G/H_1, they cannot be foreign to Σ. By $\Re$-primarity, they are Σ-analysable. Hence p, and also G, is Σ-analysable. But this shows that $H_1 \geq G^\Phi/N$. $\square$

So $\mathrm{gen}(H_i)$ is $\Re$-primary for all i. In fact, it is even strongly $\Re$-primary:

Proposition 3.4.7 *Suppose G is a type-definable $\Re$-connected stable group without a normal abelian subgroup. Then* $\mathrm{gen}(G)$ *is strongly $\Re$-primary.*

Proof: By Lemma 1.1.17 the socle $S(G)$ of G is type-definable, and a direct product of definably simple normal subgroups N_i. Now if $\mathrm{gen}(G)$ is not foreign to some type p, then G, and hence N_i, is p-analysable for all i. So $\mathrm{gen}(N_i)$ cannot be foreign to p; as a definably simple group is strongly $\Re$-primary, N_i is p-internal for all i, and so is $S(G)$. But $S(G)$ has trivial centralizer in G; as $G/C_G(S(G))$ is $S(G)$-internal, it follows that G is p-internal. $\square$

3.5 Localized Lascar Rank

In this section, we shall adapt the theory of Lascar rank to a local context, i.e. given an $\emptyset$-invariant family P of types, we shall restrict the rank to measure only forking with types in P. If P is the family of all types, then this rank will equal Lascar rank; in any case it gives a more precise measure for the size of a type (of rank less than ∞, of course) in relation to P than mere P-analysability and foreignness to P.

Throughout this section, P will denote an $\emptyset$-invariant class of stationary types (as usual over parameters in $\mathfrak{C}$) closed under non-forking extensions. As we may (almost) always add parameters to the language, this includes the case where P is reduced to a single stationary type over some set A of parameters (and its non-forking extensions): we just add A to the language.

Definition 3.5.1 Let q be a type. The U_P-*rank of q*, written $U_P(q)$, is the smallest function from the class of all types (over subsets of $\mathfrak{C}$) to On^+ satisfying

$U_P(q) \geq \alpha + 1$ iff there are an extension $q' \supseteq q$ over some set A, a type $p \in P$ over A, and realizations $a \models q'$ and $b \models p$ with $a \underset{A}{\cancel{\smile}} b$ and $U_P(a/Ab) \geq \alpha$.

Lemma 3.5.1 *If q' is a non-forking extension of p, then $U_P(q') = U_P(q)$.*

Proof: We use induction on α to show that if $U_P(q) \geq \alpha$ then $U_P(q') \geq \alpha$, the other inequality being obvious. This is clear for $\alpha = 0$ or limits. So suppose the assertion holds for rank α, and assume $q \in S(A_0)$ has a non-forking extension $q' \in S(A_1)$, and that there are $A \supseteq A_0$, $a \models q$, and $\mathrm{tp}(b/A) \in P$, with $a \underset{A}{\cancel{\smile}} b$ and $U_P(a/Ab) \geq \alpha$. By the $\emptyset$-invariance of P we may assume that $a \models q'$ and $bA \underset{aA_0}{\cancel{\smile}} A_1$. Hence $A_1 \underset{A_0}{\cancel{\smile}} abA$, and $\mathrm{tp}(a/AA_1b)$ is a non-forking extension of $\mathrm{tp}(a/Ab)$. By the inductive hypothesis $U_P(a/AA_1b) \geq \alpha$; as also $\mathrm{tp}(b/AA_1)$ is a non-forking extension of $\mathrm{tp}(b/A)$ and hence in P, we get $b \underset{AA_1}{\cancel{\smile}} a$ and therefore $U_P(a/A_1) \geq \alpha + 1$. $\square$

Clearly, $U_P(q) = 0$ iff q is co-foreign to P. However, q may well be foreign to P and still have non-zero U_P-rank.

If P is the family of all types, then U_P-rank is just Lascar rank; in fact this is also the case when P is merely the class of all types of U-rank one. U_P-rank satisfies the Lascar inequalities:

Theorem 3.5.2 $U_P(a/bA) + U_P(b/A) \leq U_P(ab/A) \leq U_P(a/bA) \oplus U_P(b/A)$.

Remark 3.5.1 In this inequality, $+$ denotes the usual non-commutative ordinal sum. In order to define the commutative ordinal sum $\oplus$, note first that every ordinal α can be written as a finite sum $\sum_{i=1}^{k} \omega^{\alpha_i} \cdot n_i$ for ordinals $\alpha_1 > \cdots > \alpha_k$ and natural numbers $n_1, \ldots, n_k$, and this sum is unique if we require all summands to be non-zero. (This is called the *Cantor normal form*.) If $\beta = \sum_{i=1}^{k} \omega^{\alpha_i} \cdot m_i$, then $\alpha \oplus \beta = \sum_{i=1}^{k} \omega^{\alpha_i} \cdots (n_i + m_i)$. Clearly, if α and β are finite, then $\alpha + \beta = \alpha \oplus \beta$, so in the case of finite rank the Lascar inequality becomes an equality.

Proof of Theorem: We prove inductively on α that $U_P(b/A) \geq \alpha$ implies $U_P(ab/A) \geq U_P(a/bA) + \alpha$. For $\alpha = 0$ this amounts to $U_P(ab/A) \geq U_P(a/bA)$, which is obvious. The limit case being trivial, we assume the inequality to hold for α and $U_P(b/A) \geq \alpha + 1$. So there are $B \supseteq A$ and some $\mathrm{tp}(c/B) \in P$ with $b \underset{B}{\cancel{\smile}} c$ and $U_P(b/Bc) \geq \alpha$. We may choose $Bc \underset{Ab}{\cancel{\smile}} a$, whence $U_P(a/bA) = U_P(a/bBc)$. By the inductive hypothesis, $U_P(ab/Bc) \geq U_P(a/bBc) + \alpha = U_P(a/bA) + \alpha$. On the other hand, clearly $c \underset{B}{\cancel{\smile}} ab$, so $U_P(ab/A) \geq U_P(ab/Bc) + 1 \geq U_P(a/bA) + \alpha + 1$.

For the second inequality we prove inductively on α that $U_P(ab/A) \geq \alpha$ implies $U_P(a/bA) \oplus U_P(b/A) \geq \alpha$. Again this is clear for $\alpha = 0$ and limit ordinals. So suppose $U_P(ab/A) \geq \alpha + 1$, i.e. there are $B \supseteq A$ and $\mathrm{tp}(c/B) \in P$

with $ab \underset{B}{\not\downarrow} c$ and $U_P(ab/Bc) \geq \alpha$. By the inductive hypothesis, $U_P(a/bBc) \oplus U_P(b/Bc) \geq \alpha$. But either $b \underset{B}{\not\downarrow} c$, whence $U_P(b/A) \geq U_P(b/Bc) + 1$ and $U_P(a/bA) \oplus U_P(b/A) \geq \alpha + 1$, or $b \underset{B}{\downarrow} c$. In that case $\mathrm{tp}(c/Bb) \in P$, so $U_P(a/bA) \geq U_P(a/bBc) + 1$ and again $U_P(a/bA) \oplus U_P(b/A) \geq \alpha + 1$. $\square$

Theorem 3.5.3 *If a and b are independent over A, then $U_P(ab/A) = U_P(a/A) \oplus U_P(b/A)$.*

Proof: If, say, $U_P(a/A) \geq \alpha + 1$, then there are a superset B of A and a realization c of some type in P (over B) such that $a \underset{B}{\not\downarrow} c$ and $U_P(a/Bc) \geq \alpha$. We may choose $Bc \underset{Aa}{\downarrow} b$, whence $Bca \underset{A}{\downarrow} b$. Therefore $U_P(b/Bc) = U_P(b/A)$, $a \underset{Bc}{\downarrow} b$, and $c \underset{B}{\not\downarrow} ab$. By the inductive assumption

$$U_P(ab/A) \geq U_P(ab/Bc) + 1 \geq (U_P(a/Bc) \oplus U_P(b/Bc)) + 1 \geq (\alpha + 1) \oplus U_P(b/A).$$

The zero and limit cases being trivial, the proposition now follows by induction and symmetry. $\square$

Theorem 3.5.4 *If $U_P(a/Ab) < \infty$ and $U_P(a/A) \geq U_P(a/Ab) \oplus \alpha$, then $U_P(b/A) \geq U_P(b/Aa) + \alpha$.*

Proof: By induction on $U_P(a/Ab) \oplus \alpha$, the cases $\alpha = 0$ and α a limit ordinal being trivial. So suppose $U_P(a/A) \geq U_P(a/Ab) \oplus \alpha + 1$. There is some $B \supseteq A$ and $\mathrm{tp}(c/B) \in P$ such that $a \underset{B}{\not\downarrow} c$ and $U_P(a/Bc) \geq U_P(a/Ab) \oplus \alpha$. We may choose $Bc \underset{Aa}{\downarrow} b$, so $U_P(b/Bca) = U_P(b/Aa)$. Now if $b \underset{B}{\not\downarrow} c$, then since $U_P(a/Bc) \geq U_P(a/Bcb) \oplus \alpha$, we get by the inductive hypothesis

$$U_P(b/A) \geq U_P(b/Bc) + 1 \geq U_P(b/Bca) + \alpha + 1 = U_P(b/Aa) + \alpha + 1.$$

Otherwise $U_P(a/Bb) \geq U_P(a/Bbc) + 1$, so

$$U_P(a/Bc) \geq U_P(a/Ab) \oplus \alpha \geq U_P(a/Bb) \oplus \alpha \geq U_P(a/Bbc) \oplus (\alpha + 1),$$

whence again by the inductive hypothesis

$$U_P(b/A) \geq U_P(b/Bc) \geq U_P(b/Bca) + \alpha + 1 = U_P(b/Aa) + \alpha + 1. \square$$

Corollary 3.5.5 *Suppose $U_P(a/Ab) < \infty$ and $U_P(a/A) \geq U_P(a/Ab) + \omega^\alpha \cdot n$. Then $U_P(b/A) \geq U_P(b/Aa) + \omega^\alpha \cdot n$.*

Proof: For $\alpha = 0$ ordinal and symmetric sums coincide. For $\alpha > 0$ and $\beta < \omega^\alpha \cdot n$ we have $U_P(a/Ab) + \omega^\alpha \cdot n > U_P(a/Ab) \oplus \beta$. The corollary now follows by continuity from Theorem 3.5.4. $\square$

Lemma 3.5.6 *Suppose $U_P(a/A) < \infty$ and $B \supseteq A$. Then $U_P(a/A) = U_P(a/B)$ iff $U_P(c/A) = 0$ for every $c \in \mathrm{Cb}(a/B)$.*

Proof: Put $C = \mathrm{Cb}(a/B)$ and suppose $U_P(c/A) = 0$ for every $c \in C$. Then $U_P(a/B) = U_P(a/AC) = U_P(a/A)$ by Theorem 3.5.2. Conversely, for any $c \in C$ consider a Morley sequence $(a_i)_{i<\omega}$ in $\mathrm{stp}(a/B)$ such that c is algebraic over A and $a_1,\dots,a_n$ for some $n < \omega$. Then

$$U_P(a_1\dots a_nc/A) \geq U_P(a_1\dots a_n/Ac) + U_P(c/A)$$
$$\geq U_P(a_1\dots a_n/AC) + U_P(c/A)$$
$$= U_P(a_1/AC) \oplus \cdots \oplus U_P(a_n/AC) + U_P(c/A).$$

On the other hand,

$$U_P(a_1\dots a_nc/A) \leq U_P(c/Aa_1\dots a_n) \oplus U_P(a_1\dots a_n/A)$$
$$\leq 0 + U_P(a_1/a_2\dots a_nA) \oplus \cdots \oplus U_P(a_n/A)$$
$$\leq U_P(a_1/A) \oplus \cdots \oplus U_P(a_n/A).$$

As $U_P(a_i/AC) = U_P(a/AC) = U_P(a/B) = U_P(a/A) = U_P(a_i/A)$, we get $U_P(c/A) = 0$ for all $c \in C$. $\square$

Corollary 3.5.7 *Suppose $U_P(p) < \infty$ and p is foreign to all types of zero U_P-rank. Then any forking extension of p has smaller U_P-rank.*

Proof: Suppose $p = \mathrm{tp}(a/A)$ and $B \supseteq A$. If $U_P(a/A) = U_P(a/B)$, then $U_P(\mathrm{Cb}(a/B)/A) = 0$ by Lemma 3.5.6. Hence $a \underset{A}{\downarrow} \mathrm{Cb}(a/B)$, i.e. $a \underset{A}{\downarrow} B$. $\square$

Remark 3.5.2 Given any type $\mathrm{tp}(a/A)$, we may consider $A_0 = \{a_0 \in \mathrm{acl}(aA) : U_P(a_0/A) = 0\}$, a superset of A. We claim that $\mathrm{tp}(a/A_0)$ is foreign to all types of U_P-rank zero, and $U_P(a/A) = U_P(a/A_0)$.

Proof: The second assertion follows from Corollary 3.5.5. Suppose $\mathrm{tp}(a/A_0)$ is not foreign to some type p of U_P-rank zero. By Proposition 3.0.6 there is some p-internal $a_0 \in \mathrm{acl}(aA_0) - \mathrm{acl}(A_0)$. But $U_P(a_0/A) = 0$ by p-internality, so $a_0 \in A_0$, a contradiction. $\square$

Proposition 3.5.8 *Let $U_P(a/A) = \omega^\alpha \cdot n + \beta$ with $\beta < \omega^\alpha$. Then there is $a' \in \mathrm{acl}(Aa)$ such that $U_P(a/a'A) = \omega^\alpha \cdot n$ and $U_P(a'/A) = \beta$.*

Proof: Let B be a set such that $U_P(a/BA) = \omega^\alpha \cdot n$. We may assume that B is a subset of $\mathrm{Cb}(a/B)$; as the rank is ordinal-valued, it can only go down finitely often and we take B to be finite. As B is algebraic over a Morley sequence in $\mathrm{stp}(a/B)$, it follows from Corollary 3.5.5 that $U_P(B/A) < \omega^\alpha$, as the coefficient of ω^α can never be affected by any forking. Now consider $a' \in \mathrm{Cb}(B/Aa)$ such that $U_P(B/Aa') = U_P(B/Aa)$. So $U_P(a/Aa') = U_P(a/ABa')$. But a' is algebraic over A in a Morley sequence in $\mathrm{stp}(B/Aa)$ and must have U_P-rank less than ω^α, whence it cannot affect the coefficient of ω^α either: $\omega^\alpha \cdot n = U_P(a/AB) = U_P(a/ABa') = U_P(a/Aa')$.

Now $U_P(a/A) = U_P(aa'/A)$, so Theorem 3.5.2 yields $U_P(a'/A) = \beta$. $\square$

Note that we can now repeat the process and get elements a_i algebraic over Aa such that $U_P(a/Aa_i)$ is the sum of the first i terms in the Cantor normal form of $U_P(a/A)$, and $U_P(a_i/A)$ equals the remainder.

Proposition 3.5.9 *Let $U_P(p) = \omega^\alpha$ and suppose p is stationary and foreign to all types of zero U_P-rank. Then p is regular.*

Proof: By Corollary 3.5.7 a forking extension of p has smaller U_P-rank. But then Corollary 3.5.5 implies that p is foreign to any forking extension. $\square$

Proposition 3.5.10 *Suppose $\mathrm{tp}(a/A)$ is foreign to all types of zero U_P-rank, and $U_P(a/A) = \beta + \omega^\alpha \cdot n$, with $n > 0$ and $\beta \geq \omega^{\alpha+1}$ or $\beta = 0$. Then there are a superset B of A independent of a over A, and some regular type $\mathrm{tp}(b/B)$ of U_P-rank ω^α and foreign to all types of U_P-rank zero, such that $a \not\perp_B b$.*

Proof: Let b be such that $U_P(a/Ab) = \beta + \omega^\alpha \cdot (n-1)$; we may assume that b is finite and contained in $\mathrm{Cb}(a/Ab)$. By Corollary 3.5.5 the rank $U_P(b/A) \geq \omega^\alpha$, and there is some extension B of A such that $U_P(b/B) = \omega^\alpha$ and $\mathrm{tp}(b/B)$ is stationary and foreign to all types of U_P-rank zero. We may assume that $B \downarrow_{Ab} a$, then $U_P(a/Bb) = U_P(a/Ab) = \beta + \omega^\alpha \cdot (n-1)$. As b is contained in $\mathrm{Cb}(a/Ab) = \mathrm{Cb}(a/Bb)$, but is not algebraic over B, a must fork with b over B; this must affect the U_P-rank. But now from Corollary 3.5.5 we get $U_P(a/B) \geq \beta + \omega^\alpha \cdot n = U_P(a/A)$, and equality holds. So $a \downarrow_A B$ by Corollary 3.5.7. $\square$

Notice that the mere fact that U_P-rank is ordinal gives rise to the existence of regular types which may well be foreign to P.

Proposition 3.5.11 *Let P_α be the family of all types of U_P-rank ω^α. Suppose $U_{P_\alpha}(p) = \sum_{i=1}^k \omega^{\alpha_i} n_i$, with $\alpha_1 > \alpha_2 > \cdots > \alpha_k$. Then*

$$\sum_{i=1}^k \omega^{\alpha+\alpha_i} n_i \leq U_P(p) < \sum_{i=1}^k \omega^{\alpha+\alpha_i} n_i + \omega^\alpha.$$

Proof: Note first that $U_{P_\alpha}(p) = 0$ iff $U_P(p) < \omega^\alpha$, so we may assume $n_i > 0$ for all i. We show the first inequality by induction on $U_{P_\alpha}(p)$. It follows for $\alpha_k = 0$ from the inductive hypothesis by Corollary 3.5.5; for α_k a limit ordinal it is clear from the continuity of the rank. So suppose that $\alpha_k = \beta + 1$. Then

$$U_{P_\alpha}(p) \geq \sum_{i=1}^{k-1} \omega^{\alpha_i} n_i + \omega^{\alpha_k}(n_k - 1) + \omega^\beta m$$

for all $m < \omega$, whence by the inductive hypothesis

$$U_P(p) \geq \sum_{i=1}^{k-1} \omega^{\alpha+\alpha_i} n_i + \omega^{\alpha+\alpha_k}(n_k - 1) + \omega^{\alpha+\beta} m$$

and therefore $U_P(p) \geq \sum_{i=1}^{k} \omega^{\alpha+\alpha_i} n_i$.

For the second inequality, note that there are unique $l < \omega$, ordinals β_i and positive integers m_i for $1 \leq i \leq l$, and $\beta' < \omega^\alpha$ such that $U_P(p) = \sum_{i=1}^{l} \omega^{\alpha+\beta_i} m_i + \beta'$. We prove by induction on $U_P(p)$ that $U_{P_\alpha}(p) \geq \sum_{i=1}^{l} \omega^{\beta_i} m_i$. The assertion is trivial for $l = 0$, i.e. $U_P(p) < \omega^\alpha$. If $\beta_l = 0$, there is an extension p' of p with $U_P(p') = \sum_{i=1}^{l} \omega^{\alpha+\beta_i} m_i$ and foreign to all types of U_P-rank zero; by Proposition 3.5.10 it can be made to fork with a type of U_P-rank ω^α, and the assertion now follows from the inductive hypothesis by Corollary 3.5.5. The limit case again being trivial, we assume that $\beta_l = \beta+1$. Then

$$U_P(p) \geq \sum_{i=1}^{l-1} \omega^{\alpha+\beta_i} m_i + \omega^{\alpha+\beta_l}(m_l - 1) + \omega^{\alpha+\beta} m$$

for all $m < \omega$, whence by the inductive hypothesis

$$U_{P_\alpha}(p) \geq \sum_{i=1}^{l-1} \omega^{\beta_i} m_i + \omega^{\beta_l}(m_l - 1) + \omega^{\beta} m,$$

for all $m < \omega$. This finishes the induction.

Putting the two inequalities together now yields $k = l$, and $\alpha_i = \beta_i$ and $n_i = m_i$ for all $1 \leq i \leq k$. The result follows. $\square$

In particular, $U_P = U_{P_0}$, and if $U_P(p) = \omega^\alpha n + \beta$ for some $\beta < \omega^\alpha$, then $U_{P_\alpha}(p) = n$. This will reduce many problems to the case of finite U_P-rank. The families P_α defined above are examples of particularly useful families of types:

Definition 3.5.2 A family P of types is *regular* if any type $p \in P$ is stationary and foreign to all forking extensions of types in P.

Note that any type in a regular family P is regular, of U_P-rank one. Furthermore, for regular P every P-internal type has finite U_P-rank, bounded by the number of realizations of types in P needed to witness internality. This also holds for any P-analysable type if the analysis is in finitely many steps (by Theorem 3.5.2), but not in general.

3.6   Ranked Groups

In this section, P is again an $\emptyset$-invariant family of stationary types. We shall study groups which have ordinal U_P-rank. This forces them to be rather well-behaved: they have big abelian subgroups (of comparable U_P-rank), strong connected components (the P-connected component), and good definability properties. In particular, we shall prove an indecomposability theorem (originally due to Zil'ber) and derive the result that for a P-connected group of ordinal U_P-rank the commutator subgroups and the elements of the descending central series are all definable.

If G is a type-definable group, we define $U_P(G) = U_P(\text{gen}(G))$. As U_P-rank is preserved under definable bijections, the U_P-ranks of all generic types are equal (they are translates of one another), and the U_P-rank of any type in G is bounded by the U_P-rank of any generic type (as they are all translates of some extension of a generic type). Similarly, we can define the U_P-rank of a coset space to be the U_p-rank of any generic type for it. We then obtain a version of the Lascar inequalities for groups:

Proposition 3.6.1 *Let H be a subgroup of G. Then $U_P(H) + U_P(G/H) \leq U_P(G) \leq U_P(H) \oplus U_P(G/H)$.*

Proof: For a generic element $g \in G$, the coset gH is generic for G/H, and $\text{tp}(g/gH)$ is generic for the coset gH. But gH is a translate of the subgroup H and must have the same U_P-rank. By Theorem 3.5.2 we have $U_P(g/gH) + U_P(gH) \leq U_P(g, gH) \leq U_P(g/gH) \oplus U_P(gH)$; as gH is definable over g, we get $U_P(g, gH) = U_P(g)$ and the result follows. $\square$

Note that gH denotes both the canonical parameter for the coset gH (as defined at the end of section 2.2) and the coset itself.

Definition 3.6.1 A group G is *P-connected* if it has no subgroup H with $U_P(G/H) = 0$. The *P-connected component* G^P is the intersection of all subgroups H with $U_P(G/H) = 0$.

Clearly, $U_P(G/H) = 0$ iff G/H is co-foreign to P. So if Σ is the class of all types co-foreign to P, then the P-connected component is just the Σ-connected component. In particular, it is definably characteristic, and $(G^P)^P = G^P$. If P is the family of all types, then $G^P = G^0$; if P is empty, then $G^P = \{1\}$.

Remark 3.6.1 One should not confuse the P-connected component G^P, where P is a class of types, with the Σ-connected component G^Σ from Definition 3.2.1, where Σ is a class of partial types (and in particular closed under extensions); in a way one is the opposite of the other. For an $\emptyset$-invariant family P of stationary types one might define $\Sigma(P)$ to be the class of partial types co-foreign to all types in P, and for an $\emptyset$-invariant family Σ one could put $P(\Sigma)$ to be the class of types foreign to Σ (or even, in reminiscence of group theory, one might use a (P, P')-terminology). Then it is easy to see that $G^P = G^{\Sigma(P)}$ and $G^\Sigma = G^{P(\Sigma)}$.

If G is p-simple in the sense of Hrushovski [51], then our $\{p\}$-connected component is just his p-connected component.

Lemma 3.6.2 $U_P(G^P) = U_P(G)$, and G is P-connected iff $\text{gen}(G)$ is foreign to all types of U_P-rank zero. If G is P-connected and $U_P(G) < \infty$, then every forking extension of a generic type has smaller U_P-rank, and a subset X of G is generic iff $U_P(X) = U_P(G)$.

Proof: As $U_P(G/G^P) = 0$, the equality follows from Proposition 3.6.1; the second assertion is obtained from Theorem 3.1.1 together with the fact that a group internal in types of U_P-rank zero has U_P-rank zero. The third one follows from Corollary 3.5.7, and immediately implies the fourth statement. □

If $U_P(G)$ is ordinal, then we cannot have an infinite descending chain of subgroups of ever smaller U_P-rank of G, and every type-definable subgroup is contained in a relatively definable subgroup of the same U_P-rank.

Lemma 3.6.3 Suppose G is P-connected, and H is the definable hull of a commutator involving G. Then H is P-connected.

Proof: H is connected by Lemma 1.1.7, and $\text{gen}(H)$ is foreign to all types to which $\text{gen}(G)$ is foreign, in particular to all types of U_P-rank zero. Hence H is P-connected by Lemma 3.6.2. □

Definition 3.6.2 A type p is P-semi-regular if it is non-algebraic, P-internal, and foreign to every type q which is co-foreign to all types in P.

This generalizes the notion of a p-semi-regular type, where p is a regular type.

Proposition 3.6.4 A connected group G is P-internal and P-connected iff $\text{gen}(G)$ is P-semi-regular.

Proof: Immediate from Lemma 3.6.2. (Recall that q is co-foreign to all types in P iff $U_P(q) = 0$.) □

Theorem 3.6.5 Suppose $U_P(G) = \sum_{i=1}^{k} \omega^{\alpha_i} n_i$, with $\alpha_1 > \alpha_2 > \cdots > \alpha_k$ and $0 < n_i < \omega$. Then for every j with $1 \leq j \leq k$ there is a unique P-connected subgroup H_j of G with $U_P(H_j) = \sum_{i=1}^{j} \omega^{\alpha_i} n_i =: \beta_j$.

Proof: Let H_j be a type-definable subgroup of minimal U_P-rank greater than or equal to β_j. This exists as the rank cannot decrease infinitely often, and we may assume (after replacing H_j by H_j^P) that H_j is P-connected. If $U_P(H_j) > \beta_j$, then by Proposition 3.5.8 its generic type is not foreign to some type of U_P-rank less than ω^{α_j}, and by Theorem 3.1.1 there is a relatively definable normal subgroup N with $U_P(H_j/N) < \omega^{\alpha_j}$. Hence $U_P(N) \geq \beta_j$ by Corollary 3.5.5, contradicting the minimality of H_j.

Therefore $U_P(H_j) = \beta_j$, and $U_P(G/H_j) < \omega^{\alpha_j}$ by Proposition 3.6.1. So $\mathrm{gen}(H_j)$ is foreign to G/H_j by Corollary 3.5.5; since H_j is connected, it is definably characteristic in G by Lemma 3.2.1.

If H'_j is another P-connected subgroup of rank β_j, then $U_P(G/(H_j \cap H'_j)) \leq U_P(G/H_j) \oplus U_P(G/H'_j) < \omega^{\alpha_j}$, so $U_P(H_j \cap H'_j) = \beta_j$ as well. Now $H_j = H_j \cap H'_j = H'_j$ by P-connectedness. $\square$

Remark 3.6.2 We thus get a series $G \trianglerighteq G_1 \triangleright \cdots \triangleright G_k \triangleright G_{k+1} = \{1\}$ such that $G_1 = G^P$, every G_i is P-connected, and $U_P(G_i/G_{i+1}) = \omega^{\alpha_i} n_i$ is monomial for all i. In fact, this series can be further refined: by Proposition 3.5.10 the generic type of every P-connected group is non-orthogonal to a regular type. If p is a regular type of smallest U_P-rank such that $\mathrm{gen}(G^P)$ is non-orthogonal to p, then there is a p-internal quotient G^P/N by Theorem 3.1.1, and as $U_P(p)$ was minimal, $\mathrm{gen}(G^P)$ and hence $\mathrm{gen}(G^P/N)$ are foreign to all forking extensions of p, and $\mathrm{gen}(G^P/N)$ is p-semi-regular. Now $U_P(N) < U_P(G)$, so inductively we see that there is a sequence $G = G_0 \triangleright G_1 \triangleright \cdots \triangleright G_m = \{1\}$ such that G_i^P/G_{i+1} is p_i-semi-regular for some regular type p_i, for all $i < m$.

Note that Theorem 3.6.5 also implies that a skew field of ordinal U_P-rank has rank $\omega^{\alpha} n$ for some α and $n < \omega$, as there are no non-trivial proper definably characteristic additive subgroups (i.e. ideals).

Remark 3.6.3 It had been conjectured that a superstable field must have Lascar rank ω^{α} for some ordinal α (and it is easy to construct such fields for arbitrary α). However, recently the amalgamation method of Hrushovski mentioned after Example 0.3.4 has been adapted by Poizat in order to construct a field of Morley rank $\omega \cdot 2$.

Proposition 3.6.6 *Suppose $U_P(G) \geq \omega^{\alpha}$. Then G has an abelian subgroup H with $U_P(H) \geq \omega^{\alpha}$.*

Proof: Let H be a type-definable subgroup of minimal U_P-rank greater than or equal to ω^{α}; after replacing it by H^P, we may assume that H is P-connected. Then $U_P(H) = \omega^{\alpha} n$ for some $n < \omega$ by Theorem 3.6.5. Now if $g \in H$ is non-central, we get $U_P(g^H) = U_P(H/C_H(g))$; as $U_P(C_H(g)) < \omega^{\alpha}$, this yields $U_P(g^H) = U_P(H)$ and g^H is generic. As H is connected, $H/Z(H)$ has a single non-trivial conjugacy class, contradicting Theorem 1.0.3. $\square$

Therefore a group of ordinal U_P-rank contains a big abelian subgroup of comparable U_P-rank.

Corollary 3.6.7 *Let G be a connected group with regular generic type. Then G is abelian, and any definable endomorphism of G is zero or surjective. In particular, any definable involutive endomorphism is the identity or the inversion $x \mapsto -x$.*

Note that by Proposition 3.5.9 a P-connected group of U_P-rank ω^α has a regular generic type.

Proof: As every type is a translate of an extension of $p = \text{gen}(G)$, every non-generic type has U_p-rank zero, and $U_p(G) = 1$. However, by Proposition 3.6.6 there is an abelian subgroup A of G with $U_p(A) \geq 1$. This can only happen if A is generic, and equals G by connectivity. If r is a definable endomorphism of G, then $U_p(\text{im}(r)) = U_p(G/\text{ker}(r))$; the Lascar inequalities, Proposition 3.6.1, imply that either $U_p(\text{im}(r)) = 1$ or $U_p(\text{ker}(r)) = 1$, so r is surjective or zero. It follows that the ring of definable endomorphisms of G has no zero-divisors; in particular the equality $r^2 = 1$ factors uniquely as $(r-1)(r+1) = 0$. $\square$

Proposition 3.6.8 *Suppose G is P-connected of U_P-rank $\sum_{i=1}^{k} \omega^{\alpha_i} n_i$ (with $k \geq 1$, $0 < n_i < \omega$ for all $1 \leq i \leq k$, and $\alpha_1 > \cdots > \alpha_k$), and σ is a definable endomorphism of G with $U_P(\text{ker}(\sigma)) < \omega^{\alpha_k}$. Then σ is surjective.*

In particular an endomorphism with finite kernel is surjective for any P-connected group of ordinal U_P-rank.

Proof: $U_P(\text{im}(\sigma)) = U_P(G/\text{ker}(\sigma))$. It follows that

$$U_P(\text{ker}(\sigma)) + U_P(\text{im}(\sigma)) \leq U_P(G) \leq U_P(\text{ker}(\sigma)) \oplus U_P(\text{im}(\sigma));$$

the proposition follows by the P-connectivity of G. $\square$

We can now connect the P-connected component with the Φ-component.

Proposition 3.6.9 *Let G be a definable stable group of U_P-rank $\omega^\alpha n + \beta$, with $0 < n < \omega$ and $\beta < \omega^\alpha$. Suppose H is a definable subgroup such that G/H is Frattini. Then $U_P(G/H) < \omega^\alpha$. In particular, G^Φ contains the unique subgroup of U_P-rank $\omega^\alpha n$.*

Proof: Suppose $U_P(G/H) \geq \omega^\alpha$. We clearly may assume that H is normal in G. We can then find a maximal sequence $G = G_0 \triangleright G_1 \triangleright \cdots \triangleright G_k = \{1\}$ of definable normal subgroups of G with $U_P(G_i) \geq U_P(G_{i+1}) + \omega^\alpha$ for $i < k$, and such that $H = G_{i_0}$ for some $1 \leq i_0 \leq k$. Note that $k \leq n$. Let Σ be the set of formulæ of U_P-rank less than ω^α, and Σ' the set of quotients G_i/G_{i+1} which are not analysable in $\Sigma \cup \{G/H\}$. Then G is analysable in $\Sigma \cup \Sigma' \cup \{G/H\}$. We claim that G is not analysable in $\Sigma \cup \Sigma'$, contradicting the fact that G/H is a Frattini formula.

Suppose that G is analysable in $\Sigma \cup \Sigma'$, and consider $p = \text{gen}(G^\Sigma/H)$, which is non-trivial, since $U_P(G/H) \geq \omega^\alpha$. As p is foreign to Σ, there is some $G_i/G_{i+1} \in \Sigma'$ such that p is not foreign to G_i/G_{i+1}. Put $p_i = \text{gen}(G_i^\Sigma/G_{i+1})$, again a non-trivial type, since $U_P(G_i/G_{i+1}) \geq \omega^\alpha$. We claim that p is non-orthogonal to p_i. Note first that both types are foreign to all types of U_P-rank less than ω^α, as otherwise by Theorem 3.1.1 there would be a relatively definable infinite quotient G^Σ/N (resp. G_i^Σ/N) of U_P-rank less than ω^α, and then G/N (resp. G_i/N) would be in Σ, contradicting Σ-connectivity of G^Σ (resp. G_i^Σ). Now p_i is non-orthogonal to some regular type q with $U_P(q) = \omega^\beta$ for some ordinal β by Proposition 3.5.10, and clearly $\beta = \alpha$. By Theorem 3.1.1 there is a definable normal subgroup N with $G_i > N \geq G_{i+1}$ and $N \not\geq G_i^\Sigma$, such that G_i^Σ/N is q-internal. Now p_i is foreign to types of U_P-rank less than ω^α, so $U_P(G_i^\Sigma/N) \geq \omega^\alpha$ and hence $U_P(G_i) \geq U_P(N) + \omega^\alpha$; by the maximality of the sequence $U_P(N) < U_P(G_{i+1}) + \omega^\alpha$, whence $U_P(N/G_{i+1}) < \omega^\alpha$ and $N/G_{i+1} \in \Sigma$. So G_i/G_i^Σ and N/G_{i+1} are both Σ-analysable, and G_i/G_{i+1} is analysable in $\Sigma \cup \{q\}$. Therefore p cannot be foreign to $\Sigma \cup \{q\}$; since it is foreign to Σ, it is not foreign to q. As forking extensions of q have U_P-rank less than ω^α, p is foreign to them, and therefore must be non-orthogonal to q. But then there are a set A of parameters and realizations $a \models p|A$, $b \models q|A$ and $c \models p_i|A$ such that $a \not\!\downarrow_A b$ and $b \not\!\downarrow_A c$. As $U_P(b/Ac) < \omega^\alpha$ we get $a \downarrow_{Ac} b$, whence $a \not\!\downarrow_A c$ and $p \not\perp p_i$. Hence by Theorem 3.1.1 there is a relatively definable p-internal infinite quotient G_i^Σ/M with $G_i > M \geq G_{i+1}$, and as above we see that $M/G_{i+1} \in \Sigma$. It follows that G_i/G_{i+1} is analysable in $\Sigma \cup \{p\}$ and hence in $\Sigma \cup \{G/H\}$, contradicting the definition of Σ'. $\square$

The requirement that G is definable can be omitted if G is a type-definable subgroup of a definable group of ordinal U_P-rank, as it is then contained in a definable group of the same rank. However, in general the condition seems necessary, as it may happen that no definable set at all has zero U_P-rank.

Definition 3.6.3 Let G be a group of U_P-rank $\omega^\alpha n + \beta$, with $0 < n < \omega$ and $\beta < \omega^\alpha$. A type-definable subset X is called α_P-*indecomposable* if whenever H is a subgroup of G with $U_P(X/H) < \omega^\alpha$ (i.e. $U_P(\{xH : x \in X\}) < \omega^\alpha$), then X/H consists of a single coset. If X is an α_P-indecomposable subgroup of G, it is called α_P-*connected*.

Remark 3.6.4 Suppose X is α_P-indecomposable and H is a type-definable subgroup of G, say $H = \bigcap_{i \in I} H_i$ for some relatively definable subgroups H_i of G. Then either X lies in a unique coset of H_i for all $i \in I$ and hence in a unique coset of H, or there is some $i \in I$ such that $U_P(\{xH_i : x \in X\}) \geq \omega^\alpha$. But then also $U_P(\{xH : x \in X\}) \geq \omega^\alpha$.

Note that a subgroup H of G is α_P-connected iff $U_P(H/K) \geq \omega^\alpha$ for any proper relatively definable subgroup $K \leq H$ iff H is P_α-connected. In particular, α_P-connectivity implies P-connectivity, and 0_P-connectivity equals P-connectivity.

Remark 3.6.5 It can be shown, by an easy induction on the rank, that in a group of finite Morley rank every definable subset decomposes uniquely into a finite disjoint union of indecomposable subsets, its *irreducible components*.

(Note that here $\alpha = 0$ and P is the family of all types; in this case both are dropped from the notation.)

Lemma 3.6.10 Let $U(G) = \omega^\alpha n + \beta$ with $0 < n < \omega$ and $\beta < \omega^\alpha$, and suppose G has a definable group S of automorphisms and X is a type-definable S-invariant subset of G. In order to determine the α_P-indecomposability of X it is sufficient to consider only relatively definable S-invariant subgroups of G.

Proof: Suppose H is a relatively definable subgroup of G and $U_P(X/H) < \omega^\alpha$. The intersection $N := \bigcap_{s \in S} H^s$ is a finite subintersection $\bigcap_{s \in S_0} H^s$, and for every $s \in S_0$ we have $U_P(X/H^s) = U_P(X/H)$ by s-invariance of X. As any coset xN is determined by the cosets $(xH^s : s \in S_0)$, we have

$$U_P(X/N) \le \bigoplus_{s \in S_0} U_P(X/H^s) = \bigoplus_{s \in S_0} U_P(X/H) < \omega^\alpha,$$

and either X is contained in a single coset of N (and therefore also of H), or N witnesses that X is not α_P-indecomposable. $\square$

Theorem 3.6.11 INDECOMPOSABILITY THEOREM *If* $U_P(G) = \omega^\alpha n + \beta$ *(where* $0 < n < \omega$ *and* $\beta < \omega^\alpha$*) and* $\mathfrak{X} := \{X_i : i \in I\}$ *is a family of* α_P-*indecomposable type-definable subsets of* G *containing the identity element, then* $\mathfrak{X}$ *generates a type-definable* α_P-*connected group* H*. Furthermore,* $H = (X_{i_1} \cdots X_{i_m})^2$ *for some indices* $i_1, \ldots, i_m$ *in* I *with* $m \le n$.

In particular, if all the X_i are definable, so is H.
Proof: As the rank is bounded by $U_P(G)$, there is a set $X := X_{i_1} \cdots X_{i_m}$ such that $\omega^\alpha m \le U_P(X) < \omega^\alpha(m+1)$ and such that $U_P(XX_i) < \omega^\alpha(m+1)$ for all $i \in I$. Let p be a type over G extending the partial type $x \in X$, with $U_P(p) = \omega^\alpha m$, and put $H := \mathrm{stab}_G(p)$. We claim that $X_i \subseteq H$ for all $i \in I$. So suppose not. Then by α_P-indecomposability we can find $g \in X_i$ such that $U_P(gH) = \omega^\alpha$. Let h realize p independently of g over G.
Claim. $g \underset{G,gH}{\cup} gh$, and $gH \underset{G}{\cancel{\cup}} gh$, if gH is not realized in G.
Proof of Claim: Let g' realize the same type as g over G, gH independently of g, h over G, gH. Then g, g' is independent of h over G and $g^{-1}g' \in H$, whence $\mathrm{tp}(g^{-1}g'h/G, g, g') = \mathrm{tp}(h/G, g, g')$ and $\mathrm{tp}(gh/G, g, g') = \mathrm{tp}(g'h/G, g, g')$. Now $g, h \underset{G,gH}{\cup} g'$ and $g \underset{G,gH}{\cup} h$ imply $g \underset{G,gH}{\cup} g', h$, and therefore $\mathrm{tp}(gh/G, g)$ does not fork over G, gH.

Now suppose that gH is not realized in G, and let g' realize $\mathrm{tp}(g/G)$ independently of g, h over G. Then $gH \ne g'H$; as $h \underset{G}{\cup} g, g'$ we get $\mathrm{tp}(gh/G, g, g') \ne \mathrm{tp}(g'h/G, g, g')$. Note that $\mathrm{tp}(g, h/G) = \mathrm{tp}(g', h/G)$. But $gh \underset{G,g}{\cup} g'$, and by the first part $\mathrm{tp}(gh/G, g)$ does not fork over G, gH. If tp$(gh/G, g, g')$ did not fork over G, then by symmetry tp$(g'h/G, g, g')$ would not fork over G either, and both would constitute the unique heir of tp(gh, G), a contradiction. Therefore tp$(gh/G, gH)$ must fork over G. $\square$

Now $U_P(gh/G, g) = U_P(h/G, g) = U_P(h/G) = \omega^\alpha m$; as $gh \underset{G}{\not\smile} gH$, Corollary 3.5.5 implies that $U_P(gh/G) \geq \omega^\alpha(m + 1)$, contradicting the choice of m. Therefore $X \subset H$. Proposition 2.1.24 implies that p is generic for H; if N is a subgroup of H with $U_P(H/N) < \omega^\alpha$, then $X_i \subseteq N$ by α_P-indecomposability. It follows that H is α_P-connected, and $H = X^2$ since X is a generic subset. $\square$

Corollary 3.6.12 *A definably simple non-abelian type-definable group G of ordinal non-zero U_P-rank is simple and definable.*

Proof: Note first that by Theorem 3.6.5 the group must have monomial U_P-rank $\omega^\alpha n$; furthermore G is P-connected. Consider an element $g \in G$. Then g^G is G-invariant; by Lemma 3.6.10 and definable simplicity, g^G is α_P-indecomposable, and generates a definable normal subgroup of G containing g. By definable simplicity again, either g is central (and then $g = 1$, since $Z(G)$ is definable normal), or g^G generates G. It follows that G must be simple, and definable. $\square$

Corollary 3.6.13 *Let $U_P(G) = \omega^\alpha n + \beta$ with $0 < n < \omega$ and $\beta < \omega^\alpha$. Suppose H is an α_P-connected subgroup of G. Then $\langle [X, H] \rangle$ is type-definable and α_P-connected for any subset X of G. In particular, H^i and $H^{(i)}$ are type-definable and α_P-connected for all $i < \omega$, and both series must become stationary after finitely many steps. Furthermore, the hypercentre of H is relatively definable and equals some finitely iterated centre.*

The intersection $\bigcap_{i<\omega} H^i$ is called the *hypocentre* of H.

Proof: Let $x \in X$. We claim that x^H is α_P-indecomposable. By Lemma 3.6.10 it is sufficient to consider an H-invariant relatively definable subgroup K of G. But x^H/K is in definable bijection with $H/C_H(x/K)$; by α_P-connectivity this is trivial or has rank at least ω^α. It follows that x^H is α_P-indecomposable; clearly the left translate $x^{-1}x^H$ is then α_P-indecomposable as well. So the family $\{x^{-1}x^H : x \in X\}$ generates an α_P-connected type-definable subgroup $\langle [X, H] \rangle$ by Theorem 3.6.11. The assertion about the derived and descending central series follow immediately.

As all the subgroups in both series are P-connected, by Lemma 3.6.2 either $H^i = H^{i+1}$ (resp. $H^{(i)} = H^{(i+1)}$) or $U_P(H^i) > U_P(H^{i+1})$ (resp. $U_P(H^{(i)}) > U_P(H^{(i+1)})$). It follows that both series must become stationary after finitely many steps.

For the last assertion, consider $i < \omega$ such that $U_P(Z_i(H)) + \omega^\alpha$ is maximal. Now gen(H) is foreign to all types of U_P-rank less than ω^α, in particular to

$Z_j(H)/Z_i(H)$ for all $i \leq j < \omega$. So $Z_j(H)/Z_i(H)$ is central in $H/Z_i(H)$ by Lemma 3.0.1, and $Z_\omega(H) = Z_\infty(H) = Z_{i+1}(H)$. $\square$

Corollary 3.6.14 *Let G be P-connected of ordinal U_P-rank. Then the soluble radical $R_s(G)$ is relatively definable.*

Proof: Let $U_P(G) = \sum_{i<k} \omega^{\alpha_i} n_i$ for some $\alpha_0 > \cdots > \alpha_{k-1}$ and natural numbers $(n_i : i < k)$. By Theorem 3.6.5 there is a sequence of P-connected characteristic subgroups $G = G_k > \cdots > G_0 = \{1\}$ such that $U_P(G_j) = \sum_{i<j} \omega^{\alpha_i} n_i$. We claim that there are relatively definable normal soluble subgroups S_j for all $j \leq k$ such that $R_s(G) \cap G_j \leq S_j$ and $R_s(G) \leq C_G(G_j/S_j)$. This is clearly true for $j = 0$. So suppose it holds for j. Since $R_s(G) \cap G_j \leq S_j$, Lascar's inequalities imply that for every normal soluble subgroup $S \geq S_j$ we have $U_P(S/S_j) \leq \sum_{i=j}^{k-1} \omega^{\alpha_i} n_i$. Let S be normal, soluble, and such that $U_P(S/S_i) + \omega^{\alpha_j}$ is maximal possible. Because $G_j \leq C_G(S/S_j)$ and $G_{j+1}/C_G(S/S_j)$ is P_{α_j}-connected (since G_{j+1} is), any normal soluble supergroup $\bar{S} \geq S$ must be centralized by G_{j+1} modulo S by Lemma 3.0.1, as $U_P(\bar{S}/S) < \omega^{\alpha_j}$. Therefore we may take any relatively definable soluble supergroup of $Z(G_{j+1}/S)$ as S_{j+1}.

But then $S_k = R_s(G)$. $\square$

Note that this implies that $R_s(G)$ is soluble for any P-connected group of ordinal U_P-rank.

Corollary 3.6.15 *Let G be P-connected of ordinal U_P-rank. Then there is a series $G = G_0 \geq G_1 \geq \cdots \geq G_n = \{1\}$ of relatively definable normal subgroups, such that every quotient G_i/G_{i+1} (for $i < n$) is either abelian or simple.*

Proof: We use induction on $U_P(G)$. If it is zero, then G is trivial by P-connectivity. So assume $U_P(G) = \omega^\alpha n + \beta > 0$ for some $\beta < \omega^\alpha$, and suppose the corollary holds for all groups of smaller rank. By Corollary 3.6.14 we may assume that G has no normal abelian subgroup; by Theorem 3.6.5 it has a characteristic P-connected subgroup H of rank $\omega^\alpha n$. As H cannot have normal abelian subgroups either, a minimal normal subgroup M of H must be definably simple by Lemma 1.1.17, and hence definable and simple by Corollary 3.6.12. Furthermore M has only finitely many G-conjugates; by connectivity of G it is G-invariant. We may hence divide out by M; as $U_P(G/M) < U_P(G)$ we finish, using the inductive hypothesis. $\square$

3.7   Fields

We have already noted after Remark 3.6.2 that a (skew) field of ordinal U_P-rank must have monomial rank. We shall now analyse the structure of such fields further.

Lemma 3.7.1 *Let K be a type-definable stable field, or a type-definable stable group without relatively definable normal subgroups. If K is infinite and P-analysable, then the following are equivalent:*

1. $\mathrm{gen}(K)$ *is P-semi-regular.*

2. K *is P-connected.*

3. $U_P(K) > 0$.

If $U_P(K)$ is finite and non-zero, then K is P-semi-regular.

Proof: Clearly P-semi-regularity of $\mathrm{gen}(K)$ implies P-connectivity of K and $U_P(K) > 0$. Since K is infinite, P-connectivity implies $U_P(K) > 0$. For the last implication from (3) to (1), we use strong $\Re$-primarity of $\mathrm{gen}(K)$ (Corollary 3.1.2). Firstly, if $\mathrm{gen}(K)$ is not foreign to some type q of U_P-rank zero, it is q-internal and has U_P-rank zero itself. But K is P-analysable; since K is infinite, there must be some $p \in P$ such that $\mathrm{gen}(K)$ is not foreign to p. Again by strong $\Re$-primarity, K is p-internal. Hence (3) implies (1).

For the second assertion, it suffices to show that K is P-analysable. So let $\mathrm{gen}(K)$ be realized by a. By definition and finiteness of rank there are some A and a realization b of a type $q \in P$ over A such that $a \mathop{\smash{\not\!\raise1pt\hbox{$\smile$}}}_A b$ and $U_P(a/A) = U_P(a)$. Since K is P-connected, $a \mathop{\smash{\raise1pt\hbox{$\smile$}}} A$ by Corollary 3.5.7. It follows that $\mathrm{gen}(K)$ is not foreign to P, and hence it is P-internal by strong $\Re$-primarity. $\square$

For the rest of this section, let K be a ring without zero-divisors, of U_P-rank $\omega^\alpha n > 0$.

Theorem 3.7.2 *K is an algebraically closed commutative field.*

Proof: Firstly, K is a skew field by Lemma 1.0.1.
Claim. If K is commutative, then K is algebraically closed.
Proof of Claim: First note that any finite extension of K is interpretable in K as a finite-dimensional vector space; it must therefore have ordinal non-zero U_P-rank as well, and is P-connected by Lemma 3.7.1. Then consider the endomorphisms $x \mapsto x^n$ on $K^\times$ (for all $n < \omega$) and $x \mapsto x^p - x$ on K^+ (where p is the characteristic of K, if it is not zero). Both maps have finite kernel, and are surjective by Proposition 3.6.8. Therefore K and all of its finite extensions have n-th roots for all $n < \omega$ and p-th pseudo-roots (solutions to the equation $x^p - x = a$ in characteristic $p \neq 0$).

Now suppose that K is not algebraically closed. As K must be perfect, it has a normal finite extension L. Consider an element $\sigma \in \mathrm{Gal}(L/K)$ of prime order p, and put $K' = L^\sigma$, the fixed field of σ. Then K' is a finite extension of K and has a normal extension L of prime degree p; we choose K' and L such that p is minimal possible. So the p-th roots of unity must be contained in K' (they would yield an extension of smaller degree), and L is obtained from K' by adjoining a p-th root (for $\operatorname{char}(K) \neq p$) or a p-th pseudo-root (in characteristic p). Both alternatives contradict the first paragraph. $\square$

Claim. K has a commutative subfield of rank at least ω^α, and degree at most n in K.

Proof of Claim: By Proposition 3.6.6 there is an abelian subgroup M of $K^\times$ of rank at least ω^α, and $M \subseteq Z(C_K(M)) =: F$. However, F is clearly closed under addition and hence a commutative subfield of K of rank at least ω^α. Now if the index of F in K were greater than n, we could interpret an $(n+1)$-dimensional vector space over F in K, and then $U_P(K) \geq \bigoplus_{n+1} U_P(F) \geq \omega^\alpha(n+1)$, a contradiction. $\square$

Lemma 3.7.3 *If K has a commutative subfield F with $[K : F] = k < \omega$, then $[K : Z(K)] \leq k^2$.*

Proof: This is pure algebra. Suppose that $k_1, \ldots, k_m$ in K are linearly independent over $Z(K)$.
Claim. There are no $d_1, \ldots, d_m \in K$, not all zero, such that $\sum_{i=1}^m k_i x d_i = 0$ for all $x \in K$.
Proof of Claim: Suppose otherwise, and let I be minimal non-empty such that there are non-zero $d_i \in K$ for $i \in I$ with $\sum_{i \in I} k_i x d_i = 0$ for all $x \in K$. Replacing x by $x d_{i_0}^{-1}$ for some fixed $i_0 \in I$, we may assume that $d_{i_0} = 1$.
    Now suppose $d \in C_K(d_i) - C_K(d_j)$ for some $i, j \in I$. Then

$$0 = \sum_{i \in I} k_i x d_i = \sum_{i \in I} k_i (x d^{-1}) d_i = \sum_{i \in I} k_i x d_i^d,$$

and subtraction yields $\sum_{i \in I} k_i x (d_i - d_i^d) = 0$ for all $x \in K$. However, this sum has fewer terms and is non-trivial, contradicting the minimality of I. Hence $C_K(d_i) = C_K(d_{i_0}) = K$ for all $i \in I$, and $d_i \in Z(K)$ for all $i \in I$, contradicting the linear independence of $(k_i : i \in I)$ over $Z(K)$. $\square$

It follows that left multiplication by k_i, for $1 \leq i \leq n$, yields right automorphisms of the right F-vector-space D, which are linearly independent. But the dimension of the space of linear transformations of a k-dimensional vector space is k^2, whence $m \leq k^2$. This proves the lemma. $\square$

It follows that every element of K generates a commutative extension of $Z(K)$ of finite degree. But $Z(K)$ must be infinite and algebraically closed, and has no finite extensions. This finishes the proof of Theorem 3.7.2. $\square$

Corollary 3.7.4 *If F is a skew field or not algebraically closed, then $\operatorname{gen}(F)$ is orthogonal to all regular types.*

Proof: Suppose gen(F) is not orthogonal to the regular type p. Then by strong $\Re$-primarity F is p-internal, and hence has finite U_P-rank; as gen(F) is non-orthogonal to p, we get $U_p(F) > 0$. The result now follows from Theorem 3.7.2. $\square$

Proposition 3.7.5 *Every subring R of K is a subfield with $U_P(R) < \omega^\alpha$.*

Proof: R is a field by Lemma 1.0.1; if $U_P(R) \geq \omega^\alpha$, then $[K : R] \leq n$ by the same rank considerations as above, contradicting the fact that R is algebraically closed by Theorem 3.7.2. $\square$

Theorem 3.7.6 *Let σ be a definable automorphism of K. Then the fixed field K^σ is not algebraically closed, $U_P(K^\sigma) = 0$, and σ has infinite order. There is no infinite family of uniformly definable automorphisms of K. In particular, every automorphism of K is $\mathrm{acl}(\emptyset)$-definable.*

Proof: Let F be the fixed field of σ. Then $U_P(F) < \omega^\alpha$ by Proposition 3.7.5. But $\rho(x) := \sigma(x)/x$ is an endomorphism of $K^\times$ with kernel $F^\times$, which is surjective by Proposition 3.6.8. Suppose first that $\mathrm{char}(K) \neq 2$. We then find some $k \in K^\times$ with $\rho(k) = -1$. So $\sigma(k) = -k$ and $\sigma(k^2) = k^2$, whence $k \notin F$, but $k^2 \in F$. In $\mathrm{char}(K) = 2$, we consider a third root of unity, ζ, and k with $\rho(k) = \zeta$. Then $\sigma(k) = \zeta k$, and $\sigma(k^3) = k^3$, whence $k \notin F$, but $k^3 \in F$. Therefore F is not algebraically closed. By Theorem 3.7.2 it must have U_P-rank zero.

For the second assertion, suppose that σ has finite order q, say. Replacing σ by some finite power, we may assume that q is prime. Let again F be the fixed field of σ, so $U_P(F) = 0$. Now consider $\tau = \sum_{i=0}^{q-1} \sigma^i$, an endomorphism of K^+. The image of τ lies in F, so gen(K) is foreign to it by P-semi-regularity. Hence for two independent generic elements k and k' we get

$$\tau(k) = \tau(k') = \tau(k + k') = \tau(k) + \tau(k') = 0.$$

Hence τ is generically, and therefore identically, zero. Since $\sigma(1) = 1$, we get $0 = \sum_{i=0}^{q-1} 1 = q$, and $q = \mathrm{char}(K)$. Furthermore $q \neq 2$, as otherwise $\tau = 0$ implies $\sigma = \mathrm{id}_K$. Considering again $x \mapsto \sigma(x)/x$, we find $k \in K$ with $\sigma(k) = -k$. Then $\sigma^i(k) = (-1)^i k$, and $\tau(k) = k \neq 0$, a contradiction.

For the last assertion, suppose that Σ is an infinite family of uniformly definable automorphisms of K. Then the family of fixed fields of $\sigma^{-1}\sigma'$, for $\sigma, \sigma' \in \Sigma$, is uniformly definable. By the ucc there is a maximal such fixed field F such that there is an infinite subfamily $\Sigma' \subseteq \Sigma$ of automorphisms which agree on F.

Claim. F is algebraically closed.

Proof of Claim: If F' is a finite extension of F, then an infinite subfamily of Σ' must agree on F', contradicting the maximality of F. $\square$

But now $F = K^\tau$ for some $\tau = \sigma^{-1}\sigma'$, and F cannot be algebraically closed, by the first part. $\square$

In particular, a superstable field of characteristic zero has no definable automorphism, as the prime field is contained in the fixed field and is infinite. In non-zero characteristic, any automorphism of a superstable field is determined by its restriction to the algebraic closure k of the prime field. Hence the group of definable automorphisms of K embeds into the group of automorphisms of k, which is $\bar{\mathbb{Z}}$, the pro-finite completion of $\mathbb{Z}$. In particular, it is commutative.

Proposition 3.7.7 *Let M be a multiplicative subgroup of K with $U_P(M) \geq \omega^\alpha$. Then M generates K additively.*

Proof: We claim that M is additively indecomposable. By Lemma 3.6.10 it is sufficient to test indecomposability with an M-invariant subgroup A of K^+. But for such A we may consider $R := \{k \in K : kA \leq A\}$, a subring of K containing M. Then $U_P(R) \geq U_P(M) \geq \omega^\alpha$, so $R = K$ by Proposition 3.7.5, and $A = K$ or $A = \{0\}$. The result now follows from Theorem 3.6.11. $\square$

Remark 3.7.1 Suppose K has no relatively definable subfields. Then it has no non-trivial proper relatively definable additive subgroups either, and any additive endomorphism $\sigma : K^n \to K^m$ is linear.

The condition holds in particular for a field of finite Lascar rank in characteristic zero.

Proof: Let A be an additive subgroup of K^+. Then $R := \{k \in K : kA \leq A\}$ is a relatively definable subring, and hence subfield, of K, and must equal K by assumption. So $A = K$ or $A = \{0\}$. For the second assertion, put $R := \{k \in K : \forall x \in K^n \, \sigma(kx) = k\sigma(x)\}$ (scalar multiplication of vectors). Then again R must be a subfield of K and therefore equal K, so σ is linear. $\square$

3.8   Bad Groups

In this section, we shall look at a minimal counter-example to the Cherlin-Zil'ber conjecture. Recall that the conjecture states that a simple group of finite Morley rank should be isomorphic to a linear group over a definable algebraically closed field. In particular, there ought to be a definable field. It has been shown by Nesin (and we shall see in chapter 5) that a soluble non-nilpotent group of finite Morley rank interprets a field. So if there is no interpretable field, then every soluble subgroup must be nilpotent.

Definition 3.8.1 A *bad group* is a non-nilpotent connected group such that every connected soluble subgroup is nilpotent.

If the above holds for a group G when we replace "connected" by "P-connected", we call G a *P-bad group*.

Theorem 3.8.1 *Let G_0 be a type-definable P-bad group of finite U_P-rank. Then G_0 has a type-definable simple P-bad section G, such that in addition G is the disjoint union of the conjugates of some nilpotent P-connected subgroup B. For any $b \in B$ the centralizer $C_G(b)$ is contained in B; furthermore $N_G(B)/B$ has U_P-rank zero and contains no elements of finite order. In particular, G has no involutions. Finally, B is definable relative to G.*

Proof: Clearly, we may assume that G_0 is $|T|^+$-saturated. Let G_1 be a type-definable P-connected non-nilpotent subgroup of G_0 of minimal U_P-rank, and F the Fitting subgroup of G_1, which is relatively definable and nilpotent by Theorem 1.2.11. Put $Z_1 := Z(G_1/F)$ and $G := G_1/Z_1$.
Claim 1. G is a simple P-connected non-soluble group such that any proper P-connected type-definable subgroup is nilpotent.
Proof of Claim: First note that G is again P-connected, as it is the quotient of a P-connected group. Furthermore, Z_1 is soluble and G_1 is not soluble, so G is not soluble either. If H/Z_1 is a proper type-definable P-connected subgroup of G, then H is a proper subgroup of G_1, and the P-connected component H^P has smaller U_P-rank and must be nilpotent. Furthermore

$$U_P(H^P Z_1/Z_1) = U_P(H^P Z_1) - U_P(Z_1) = U_P(H) - U_P(Z_1) = U_P(H/Z_1),$$

so $H^P Z_1/Z_1 = H/Z_1$ by P-connectivity. Thus any proper type-definable P-connected subgroup of G is nilpotent.
    Let H/Z_1 be a proper type-definable normal subgroup of G. Then H^P is a proper normal P-connected subgroup of G_1, and hence nilpotent and contained in F. Hence $U_P(H/F) = 0$, and $\text{gen}(G_1)$ is foreign to H/F. So H/F is central in G_1/F by Lemma 3.0.1, and H is contained in Z_1. Therefore G is definably simple, and simple by Corollary 3.6.12. □

We shall call a maximal type-definable P-connected proper non-trivial subgroup B of G a *Borel* subgroup. By the finiteness of U_P-rank, every P-connected proper non-trivial subgroup of G is contained in a Borel subgroup. On the other hand, by Proposition 3.6.6 there is an abelian subgroup A of G with $U_P(A) > 0$, and clearly $A < G$, since G is not soluble. As A^P is P-connected and non-trivial, Borel subgroups exist. Note that they are nilpotent by claim 1, and the elements of their descending central series are type-definable and P-connected by Corollary 3.6.13.
Claim 2. The Borel subgroups intersect in subgroups of U_P-rank zero.

Proof of Claim: Suppose B_1 and B_2 are two distinct Borel subgroups which intersect in a subgroup I of maximal non-zero U_P-rank. We note first that $N_{B_1}(I)/I$ has U_P-rank at least 1: if B_1^i is the first element of the descending central series of B_1 which is contained in I, then $B_1^i \neq B_1$ since the Borels are distinct. Now B_1^{i-1} is contained in $N_{B_1}(I)$, and $I \cap B_1^{i-1}$ is a proper subgroup of B_1^{i-1}. By the P-connectivity of B_1^{i-1} we get $U_P(B_1^{i-1}) > U_P(I \cap B_1^{i-1})$, whence

$$U_P(N_{B_1}(I)/I) \geq U_P(B_1^{i-1}/(B_1^{i-1} \cap I)) \geq 1.$$

Therefore $U_P(N_{B_1}(I)^P) = U_P(N_{B_1}(I)) > U_P(I)$, and similarly $U_P(I) < U_P(N_{B_2}(I)^P)$. Now if N is the P-connected component of $N_G(I)$, it normalizes I and cannot be the whole of G, which is simple. So there must be a Borel B_3 of G containing N. But then B_3 contains $N_{B_1}(I)^P$ and $N_{B_2}(I)^P$, so B_3 intersects B_1 and B_2 in subgroups of greater U_P-rank than $U_P(I)$. By the maximality of the rank of the intersection, we get $B_1 = B_3 = B_2$, a contradiction. □

Claim 3. The Borel subgroups intersect trivially.
Proof of Claim: Suppose $1 \neq b \in B_1 \cap B_2$. Then $C_G(b)$ contains both $Z(B_1)$ and $Z(B_2)$; as B_1 is $Z(B_1)$-analysable of non-zero U_P-rank by Lemma 3.1.8, also $U_P(Z(B_1)) > 0$, whence $U_P(C_G(b)) > 0$. So the P-connected component $C_G(b)^P$ is non-trivial, centralizes b, and cannot be the whole of G; it must therefore be contained in a Borel subgroup B_3. But then B_3 intersects B at least in $Z(B_1)^P$ and B_2 at least in $Z(B_2)^P$, i.e. the intersections have non-zero U_P-rank. By claim 2 we get $B_1 = B_3 = B_2$. □

Claim 4. The Borel subgroups are all conjugate.
Proof of Claim: Let B be a Borel subgroup. If $U_P(N_G(B)) > U_P(B)$, then $N_G(B)^P$ would be a P-connected proper (since G is simple) subgroup of G of greater U_P-rank than B, a contradiction. Hence $U_P(N_G(B)) = U_P(B)$. It follows that

$$U_P(G/N_G(B)) = U_P(G) - U_P(N_G(B)) = U_P(G) - U_P(B) = U_P(G/B).$$

But now the union X of all G-conjugates of $B - \{1\}$ is a disjoint union of rank $U_P(G/N_G(B))$ of disjoint sets of rank $U_P(B)$, and hence has rank $U_P(G/N_G(B)) + U_P(B) = U_P(G/B) + U_P(B) = U_P(G)$. So X is generic. Since G is connected, there cannot be two disjoint generic subsets; as any two Borel subgroups intersect trivially, they must all be conjugate in G. □

Claim 5. If $U_P(C_B(n)) = 0$ for some $n \in N_G(B)$, then the map $\sigma : x \mapsto [n, x]$ is surjective on B.
Proof of Claim: Clearly σ acts on B^i for all $i < \omega$, and also on the quotients B^i/B^j for $0 \leq i \leq j \leq k$. Trivially σ acts surjectively on B^i if i is at least the nilpotency class of B (i.e. $B^i = \{1\}$). Suppose σ is surjective on B^j for all

$j \geq i$, and consider the action on B^{i-1}/B^i. If $[n, z] \in B^i$ for some $z \in B^{i-1}$, then there is $z' \in B^i$ with $[n, z'] = [n, z]$, that is $n^{z'} = n^z$, and $z'z^{-1} \in C_B(n)$. Therefore $C_{B^{i-1}/B^i}(n) \leq C_B(n)B^i/B^i$ and the kernel of the action of σ on B^{i-1}/B^i has U_P-rank bounded by $U_P(C_B(n)) = 0$. So σ induces a surjective homomorphism of B^{i-1}/B^i by Proposition 3.6.8.

We now assume that σ is surjective on B^{i-1}/B^j for some $j \geq i$, and show that it is surjective on B^{i-1}/B^{j+1}. Indeed, if $z \in B^{i-1}$, surjectivity of σ on B^{i-1}/B^j yields $z_1 \in B^{i-1}$ and $z^1 \in B^j$ such that $z = [n, z_1]z^1$. By surjectivity of σ on B^j we find $z_2 \in B^j$ such that $z^1 = [n, z_2]$. Since B^j is central in B^{i-1}/B^{j+1}, we get (modulo B^{j+1})

$$[n, z_2 z_1] = [n, z_1][n, z_2]^{z_1} = [n, z_1][n, z_2] = z,$$

so σ is surjective on B^{i-1}/B^{j+1}. It follows inductively that σ is surjective on B^{i-1}, and finally on B. $\square$

Note that if a normalizes B with $U_P(C_B(a)) = 0$, and $b \in B$, then there is $b' \in B$ such that $ab = a[a, b'] = a^{b'}$, so a and ab are conjugate under B (as are a and ba by symmetry).

Claim 6. The Borel subgroups cover the group.

Proof of Claim: If $g \in G$ has a centralizer of U_P-rank zero, then $U_P(g^G) = U_P(G/C_G(g)) = U_P(G)$ and the conjugacy class of g is generic. But then it must intersect X, so g lies in some conjugate of B and centralizes its centre, which has non-zero U_P-rank, a contradiction. Hence every non-trivial element g determines a unique Borel $B(g)$, namely the Borel subgroup which contains $C_G(g)^P$, and clearly g normalizes $B(g)$.

If $b \in B(g)$ and $U_P(C_{B(g)}(gb)) = 0$, then gb and $gbb^{-1} = g$ are conjugate by some element of $B(g)$, contradicting $U_P(C_{B(g)}(g)) \geq 1$. Hence $C_{B(g)}(gb)^P$ is non-trivial. But it is contained in $C_G(gb)^P \leq B(gb)$ and in $B(g)$; so $B(gb) = B(g)$ by claim 3. Let Y denote the set of $g \in G$ such that $g \notin B(g)$, and suppose for a contradiction that Y is non-empty. If $g \in Y$ and $b \in B(g)$, then clearly $gb \notin B(g)$ and we have just seen that $B(gb) = B(g)$, so the whole coset $gB(g)$ is contained in Y, as are all its G-conjugates. But for $h \notin N_G(B(g))$, the sets $gB(g)$ and $(gB(g))^h$ are disjoint, as an element in the first set has $B(g)$ as its Borel subgroup, whereas an element in the second one has $B(g)^h$ as its Borel subgroup. Hence Y contains the disjoint union of rank $U_P(G/N_G(B(g)))$ of sets of rank $U_P(gB(g))$, whence

$$U_P(Y) \geq U_P(G/N_G(B)) + U_P(gB) = U_P(G/B) + U_P(B) = U_P(G)$$

and Y is generic. But clearly any element h of a Borel B centralizes $Z(B)$ and $U_P(Z(B)) > 0$, so $B(h) = B$ and Y must be disjoint from X, contradicting connectivity of G. Therefore Y is empty, and $g \in B(g)$ for all $g \in G$. $\square$

Note that if $g \in C_G(B)$, then $B \leq C_G(g)$, so $B = B(g)$ and $g \in B$. Thus $C_G(B) \leq B$. Furthermore, if $b \notin B$, then $C_G(b)^P \cap B \leq B(b) \cap B = \{1\}$, so

$$U_P(C_B(b)) = U_P(C_B(b)/C_G(b)^P) \leq U_P(C_G(b)) - U_P(C_G(b)^P) = 0.$$

Claim 7. G contains no involutions.
Proof of Claim: Suppose G contains an involution. Then B contains an involution i and a distinct conjugate of B contains an involution j. Both i and j invert the product ij and hence normalize $B(ij)$. It follows that $N_G(B) - B$ contains an involution, say k. Note that $k \in B(k)$ implies $B \neq B(k)$. Setting $B_0 = \langle B, k \rangle = B \cup Bk$, it follows that

$$U_P(k^B) = U_P(B/C_B(k)) = U_P(B) - U_P(C_B(k)) = U_P(B_0) - 0 = U_P(B_0),$$

so k^B is generic in B_0 and there is a generic involution s in B_0 conjugate to k. Then B is abelian and inverted by s by Theorem 2.4.5. As k is conjugate to s under B, it must also invert B; indeed any involution in $N_G(B) - B$ inverts B; they are thus all congruent modulo $C_G(B) = B$ and form a single coset kB.

If i and j are two involutions in B, they are both centralized by k and normalize (and invert) $B(k)$. Therefore ij centralizes $B(k)$ and must lie in $B(k) \cap B = \{1\}$: every Borel subgroup contains a unique involution.

We now obtain a projective plane, where the points and lines are both given by the set of involutions, and a point i lies on a line j iff i and j are distinct and commute. This means that the line j consists of the involutions in $C_G(j) - B(j)$, which form a single coset of $B(j)$. Now

- Given two distinct lines i and j, they intersect in at most $|B(i) \cap B(j)| = 1$ point. But if k is the involution in $B(ij)$, then i and j invert $B(ij)$ and commute with k, so k lies on $i \cap j$, and any two lines intersect in a unique point. By symmetry any two distinct points lie on a unique line.

- Given three lines i, j and k, there must be a point not on $i \cup j \cup k$: any fourth line k' (and there are infinitely many involutions) intersects $i \cup j \cup k$ in at most three points, but Borel subgroups, and hence lines, are infinite, and must contain a point outside $i \cup j \cup k$.

But this means that we may apply the Hauptsatz of Bachmann [2], which states that under these circumstances there is a field K of characteristic $\neq 2$, such that G is isomorphic to a subgroup Γ of $SO_3(K)$ preserving a certain symmetric bilinear form F of rank 2 or 3, and the action of G by conjugation on the set of its involutions corresponds to the action of Γ on the projective plane $P(K^3)$. The field K is definable in G (using some parameters), as is the action of G on $P(K^3)$. In particular, this implies that K is P-internal of finite U_P-rank; on the other hand G is interpretable in K, and $U_P(G) > 0$ implies $U_P(K) > 0$. By Lemma 3.7.1 the field K is P-connected of finite U_P-rank, so by Theorems 3.7.2 and 3.7.6 it is algebraically closed without definable automorphisms of finite order. But the projective plane has a polarity (exchanging the point i and the line i) without isotropic vectors (no point lies on its polar line); however, for algebraically closed K we can always solve the equation $F(\bar{x}, \bar{x}) = 0$, so there is always an isotropic point $P(\langle \bar{x} \rangle)$ in $P(K^3)$ with respect to the form F. This contradiction finishes the proof of claim 7. $\square$

Claim 8. If $b \in B$ normalizes B' and $b' \in B'$ normalizes B, then $B = B'$.
Proof of Claim: Suppose $B \neq B'$. Then b' is conjugate under B to $b'b$ and to $b'b^{-1}$ by claim 5; similarly, b is conjugate under B' to $b'b$ and b^{-1} is conjugate under B' to $b'b^{-1}$. Hence there is $g \in G$ with $b^g = b^{-1}$, and clearly $g \in N_G(B)$.

For the same reason (just replacing b^{-1} by b^2 in the above argument) there is $g' \in N_G(B)$ with $b^{g'} = b^2$. Now consider the last iterated centre Z of B such that $b \notin Z$. Then Z is normalized by g and by g', and we may consider the actions of g and of g' on B/Z. Since $(bZ)^2 = (bZ)^{g'} \neq 1Z$, the element bZ is not an involution. Hence $(bZ)^g = (bZ)^{-1} \neq bZ$, and as b is central in B modulo Z we get $g \notin B$.

If g lies in a Borel B'', then $g^2 \neq 1$ also lies in that Borel subgroup and commutes with b. It follows that $b \in N_G(B'') - B''$. So b, bg, and bg^2 are conjugate under B'', and both g and bg, as well as g^2 and bg^2, are conjugate under B. Therefore g and g^2 are conjugate in G by some element h. But $C_G(g^2)$ strictly contains $C_G(g)$ (the former contains b whereas the latter does not), yielding an infinite chain of centralizers $C_G(g) < C_G(g^h) < C_G(g^{h^2}) < \cdots$, which contradicts the ucc. $\square$

In particular, $C_G(b) \leq B$ for any $b \in B$.
Claim 9. $N_G(B)/B$ has no elements of finite order.
Proof of Claim: Let $g \in N_G(B)$ be an element of prime order p modulo B. Then g^p centralizes g, so $g^p \in B \cap B(g) = \{1\}$ and g has order p. Now B must contain a conjugate of g and hence an element of order p. Since B is nilpotent, the subset B_p of elements of p-power order of B forms a characteristic subgroup (see e.g. [107] Theorem 4.3.10) and contains a central element b of order p. Then b and g generate a finite p-group which contains a central element $z \neq 1$. So g and b both lie in $C_G(z)$ and therefore in the Borel subgroup of z, a contradiction. $\square$

Claim 10. B is definable relative to G.
Proof of Claim: Let $B = \bigcap_i B_i$ for some relatively definable groups B_i of the same U_P-rank, and put $\bar{B}_i = \bigcap_{g \in N_G(B)} B_i^g$. Then $\bigcap_i \bar{B}_i = B$ and the

normalizer of each $\bar{B}_i$ contains the normalizer of B. On the other hand, $B = \bar{B}_i^P$ for all i, so B is normalized whenever $\bar{B}_i$ is, and $N_G(B) = N_G(\bar{B}_i)$ is definable relative to G.

Suppose for every i there is some $g \in G - N_G(B)$ such that $\bar{B}_i$ and $\bar{B}_i^g$ intersect non-trivially. By compactness, we find $g \in G - N_G(B)$ such that $B \cap B^g$ is non-trivial, contradicting claim 3. So $\bar{B}_i \cap \bar{B}_i^g = \{1\}$ for some i and all $g \in G - N_G(B)$. Suppose there were some $b \in \bar{B}_i - B$. As the conjugates of B cover G, there is $g \in G - N_G(B)$ with $b \in B^g$, whence $b \in \bar{B}_i \cap \bar{B}_i^g = \{1\}$, contradicting our choice of b. Therefore $B = \bar{B}_i$, and B is definable relative to G. $\square$

This finishes the proof of Theorem 3.8.1. $\square$

Note that if P contains all minimal types (in particular if $U(G)$ is finite), then P-connectedness equals connectedness. But the Borel subgroups in the section G given by Theorem 3.8.1 are relatively definable and connected. It follows that a Borel subgroup B has finite index in its normalizer $N_G(B)$, and by claim 9 the Borel subgroups must be self-normalizing.

Corollary 3.8.2 *If G is a P-bad group of finite U_P-rank, then $U_P(G) \geq 3$. In particular, a P-connected group of U_P-rank 2 is soluble.*

Proof: We may assume that G satisfies the conclusion of Theorem 3.8.1. If $U_P(G) = 1$, then G is abelian by Corollary 3.6.7. Suppose $U_P(G) = 2$, and consider a Borel subgroup B of U_P-rank one. Now $B \cap B^g = \{1\}$ for any $g \notin N_G(B)$, so

$$U_P(BgB) = U_P(B^gB) = U_P(B^g) + U_P(B) = 2 = U_P(G),$$

so BgB and $Bg^{-1}B$ are both generic sets for G by P-connectivity. But any two generic sets intersect, so there are $a, b \in B$ with $ag = g^{-1}b$. But this implies $(ag)^2 \in B$; since $g \notin B$ we have $ag \notin B$, so either $(ag)^2 = 1$, contradicting the non-existence of involutions, or ag normalizes $B((ag)^2) = B$, contradicting the fact that $N_G(B)/B$ is torsion-free.

Now let G be P-connected of rank 2. Then either it is nilpotent, or any proper P-connected subgroup has rank one and is abelian, hence nilpotent. As G cannot be P-bad, it must itself be soluble. $\square$

3.9   Historical and Bibliographical Remarks

Foreignness and internality were defined and used by Hrushovski [51]; we follow the exposition in Poizat [95] (in particular for Theorem 3.1.1).

Sections 2 and 4 come from [116]; the decomposition in section 4 (Theorem 3.4.6) generalizes an earlier decomposition of Baldwin and Pillay [7] of a connected semi-simple group of monomial U-rank $\omega^\alpha \cdot n$ into a finite direct product of α-semi-regular groups. Section 3 on involutions is taken from [115].

Localized Lascar rank was developed in [124], generalizing the properties of the usual Lascar rank (see Poizat [94] for a reference). This set-up unifies the standard finite rank approach with the superstable machinery due to Berline and Lascar [17, 18] via Proposition 3.5.11 connecting U_P-rank and U_{P_α}-rank. The Indecomposability Theorem is originally due to Zil'ber [130] and was generalized by Berline [16]; Corollary 3.6.15 was proven (in the superstable case) by Baudisch [9]; fields were studied first by Macintyre [70] in the finite rank case, and then by Cherlin and Shelah [37] and Hrushovski [53] in the superstable case. Finally, the bad group analysis (again for groups of finite Morley rank) is due independently to Corredor [38] and Borovik and Poizat [21], building on earlier work by Cherlin [34], who also proves Corollary 3.8.2, and Nesin [76]. The local version proved here can be found in [123].

Chapter 4

Groups & Geometry

In this chapter, we shall analyse the dependence relation associated with forking. After some remarks in section 0 about geometries and pre-geometries in general and the geometry of forking in particular, we add additional conditions on the family P of types whose forking geometry we want to study. In sections 1 and 2 this is local modularity; put crudely, a locally modular theory should essentially behave either like a set with no structure (the *trivial* case), or like a module over some ring. In section 1 the general theory of local modularity is developed, and it is then applied to groups in section 2: if the generic type of a group is locally modular, then the group is abelian-by-finite; if the generic type is analysable in a locally modular family P of types, then the P-connected component of the group is nilpotent. In the case where the generic type of a group is locally modular and regular, we shall show that the module structure is actually present: generically the group behaves like a vector space over a certain division ring, namely the ring of *quasi-endomorphisms*. A particular case of local modularity is one-basedness; we shall show that a group is one-based iff it is an abelian structure.

Section 3 introduces an important tool in obtaining a definable group from structural considerations, the *group configuration*. We shall show in particular that a locally modular non-trivial family P of types gives rise to a type-definable P-semi-regular group acting faithfully and transitively on some set X, and we classify the possibilities for that action in the case $U_P(X) = 1$. In particular, every locally modular regular type is either trivial (and so its forking geometry is uninteresting) or equivalent to the generic type of a group (and we may apply the results of the preceding section).

However, even in the non-regular case we can obtain structural results about the ring of quasi-endomorphisms, in particular for a small group. These satisfy a dichotomy: either they have a connected-by-finite definable quotient, or the ring of quasi-endomorphisms is locally finite. This corresponds to the two possibilities we have for a group of Lascar rank one, which may be either connected-by-finite and strongly regular, or weakly minimal with a commutative field of quasi-endomorphisms. On the way, we also prove a chain condition for images of quasi-endomorphisms, which implies that a small group is ABD.

The next two sections introduce a slightly weaker geometric notion, namely CM-triviality. Again we develop the basic notions first in section 5 and then apply them to groups in section 6: under certain conditions a CM-trivial group must be nilpotent-by-finite

Section 7 looks at a notion akin to categoricity, namely *dimensionality*. We shall see that groups of finite Lascar rank are dimensional, and that simple groups of finite Morley rank are uncountably categorical. Finally, the penultimate section introduces the binding group between two types, which measures the interaction between them. It ends with Hrushovski's proof that unidimensional theories are superstable.

4.0 Pre-geometries

A fundamental result in algebra is the classification of algebraic structures (vector spaces, or algebraically closed fields) in terms of the cardinality of a basis, i.e. a maximal independent subset of the structure. For this to work, independence has to satisfy certain properties.

Definition 4.0.1 A *pre-geometry* is a set X together with a closure function $\mathrm{cl}(.)$ from the power set of X to itself, satisfying:

1. Monotonicity: $Y \subseteq Y'$ implies $\mathrm{cl}(Y) \subseteq \mathrm{cl}(Y')$,

2. Finite Character: $\mathrm{cl}(Y) = \bigcup \{\mathrm{cl}(Y_0) : Y_0 \subseteq Y,\ Y_0 \text{ finite}\}$,

3. Transitivity: $Y \subseteq \mathrm{cl}(Y')$ implies $\mathrm{cl}(Y) \subseteq \mathrm{cl}(Y')$,

4. Exchange: $x \in \mathrm{cl}(Y \cup \{y\}) - \mathrm{cl}(Y)$ implies $y \in \mathrm{cl}(Y \cup \{x\})$.

It defines a *dependence relation:* x depends on Y iff $x \in \mathrm{cl}(Y)$. Otherwise x is *independent* of Y.

A set Y is *independent* if y is independent of $Y - \{y\}$ for all $y \in Y$. A *basis* of a subset $Y \subseteq X$ is a maximal independent subset of Y; the *dimension* $\dim(Y)$ is the cardinality of a basis for Y. If Y is contained in $\mathrm{cl}(I)$ for some $I \subseteq Y$, then I *spans* Y.

Proposition 4.0.1 *In a pre-geometry, every independent subset I of some $Y \subseteq X$ can be extended to a basis of Y, and a basis of Y spans Y. Any two bases have the same cardinality, so the dimension is well-defined.*

Proof: We have to show that if $y \in Y$ is independent of I, then $I \cup \{y\}$ is independent. We may then add independent elements to I one after the other; as the union of an ascending chain of independent sets is independent by the finite character of dependence, it follows from transfinite induction that I can be extended to a basis for Y. Furthermore, any basis of Y must span Y.

So assume that $y \in Y$ is independent of the independent set I, let $x \in I$, and put $I' := I - \{x\}$. Suppose $x \in \mathrm{cl}(I'y)$. As $x \notin \mathrm{cl}(I')$, exchange yields $y \in \mathrm{cl}(I'x) = \mathrm{cl}(I)$, contradicting our assumption. Therefore $x \notin \mathrm{cl}(I'y)$ and $I \cup \{y\}$ is independent.

Now let I and J be two bases for Y. Well-order I and consider the set $\mathfrak{F}$ of functions f from initial segments s of I to J such that $f(s) \cup (I - s)$ is independent. If $\mathfrak{K}$ is a chain in $\mathfrak{F}$, then $f := \bigcup \mathfrak{K}$ is again a function from an initial segment of I (the union of the domains of the functions in $\mathfrak{K}$) to J. By the finite character of dependence, $\mathrm{range}(f) \cup (I - \mathrm{domain}(f))$ is independent, whence $f \in \mathfrak{F}$ is an upper bound for the chain $\mathfrak{K}$. By Zorn's Lemma there is a maximal element $f_0 \in \mathfrak{F}$ with domain I_0. Suppose $I_0 \neq I$, and let $x \in I - I_0$ be minimal. As $f_0(I_0) \cup (I - I_0)$ is independent and J spans Y, there must be some $y \in J$ such that $y \notin \mathrm{cl}(f_0(I_0) \cup (I - I_0 - \{x\}))$. Hence $f_0(I_0) \cup \{y\} \cup (I - I_0 - \{x\})$ is independent and we may extend f_0 by mapping x to y, contradicting the maximality of f_0. Therefore $I = I_0$ and $|I| \leq |J|$. By symmetry $|J| \leq |I|$, so the two bases have the same cardinality. $\square$

Conversely, given a dependence relation, we may try to define a closure relation via $y \in \mathrm{cl}(Y)$ iff y depends on Y. Then properties 1.–4. have obvious reformulations in terms of dependence; it is easy to see from Theorem 0.2.19 that forking dependence satisfies all of them except possibly transitivity. In this chapter we shall analyse how the geometric properties of forking dependence are related to algebraic properties.

Let us remark at this point that both in the theory of vector spaces over a field K and in the theory of algebraically closed fields (in the natural language) forking dependence is exactly the customary, linear (resp. algebraic) dependence.

For the rest of this chapter, P will denote an $\emptyset$-invariant family of stationary types closed under non-forking extensions. We shall consider a closure relation induced by P-analysability, and prove a necessary condition for this to form a pre-geometry.

Definition 4.0.2 The *P-closure* $\mathrm{cl}_P(A)$ of a set A is the collection of all elements a such that $\mathrm{tp}(a/A)$ is P-analysable of U_P-rank zero.

We write cl_p instead of $\mathrm{cl}_{\{p\}}$. This slightly deviates from the usual convention for the p-closure with respect to a regular type p in that it requires $\mathrm{tp}(a/A)$

to be p-analysable, but not to be p-simple. The analysability assumption is not really necessary; if we dropped it from the definition, we would in essence get the same results, but work with bigger classes (note that the closure of an element in general is a proper subclass of the monster model, not a subset); p-simplicity on the other hand is not needed as we work with U_P-rank rather than weight. Finally, we should remark that cl_p satisfies all the conditions from Definition 4.0.1 with the possible exception of the exchange property.

Lemma 4.0.2 *The following are equivalent:*

1. $\mathrm{tp}(a/A)$ *is foreign to all P-analysable types of U_P-rank zero.*

2. $a \underset{A}{\downarrow} \mathrm{cl}_P(A)$.

3. $a \underset{A}{\downarrow} \mathrm{acl}(aA) \cap \mathrm{cl}_P(A)$.

If $\mathrm{tp}(a/A)$ is P-analysable, then $a \underset{A}{\downarrow} \mathrm{cl}_P(A)$ iff $\mathrm{tp}(a/A)$ is foreign to all types of U_P-rank zero.

Proof: 1. $\Rightarrow$ 2. $\Rightarrow$ 3. is trivial. So suppose $\mathrm{tp}(a/A)$ is not foreign to some type p of U_P-rank zero, and p or $\mathrm{tp}(a/A)$ is P-analysable. By Proposition 3.0.6 there is some $a_0 \in \mathrm{acl}(aA) - \mathrm{acl}(A)$ such that $\mathrm{tp}(a_0/A)$ is p-internal. Therefore $\mathrm{tp}(a_0/A)$ is P-analysable, of U_P-rank zero. Hence $a_0 \in (\mathrm{acl}(aA) \cap \mathrm{cl}_P(A)) - \mathrm{acl}(A)$, so $a \underset{A}{\not\downarrow} a_0$. This also proves the last assertion. $\square$

Lemma 4.0.3 *Suppose $B \underset{A}{\downarrow} C$. Then $\mathrm{cl}_P(B) \underset{\mathrm{cl}_P(A)}{\downarrow} \mathrm{cl}_P(C)$.*

Proof: Let $A' = \mathrm{acl}(BA) \cap \mathrm{cl}_P(A)$. Then $B \underset{A'}{\downarrow} C$; since $\mathrm{cl}_P(A) = \mathrm{cl}_P(A')$, Lemma 4.0.2 implies that $\mathrm{tp}(B/A')$ is foreign to all P-analysable types of U_P-rank zero.

Suppose $c \in \mathrm{cl}_P(AC)$. Then $\mathrm{tp}(c/A'C)$ is P-analysable of U_P-rank zero; since $B \underset{A'}{\downarrow} C$, $\mathrm{tp}(B/A'C)$ is foreign to $\mathrm{tp}(c/A'C)$. By transitivity, $B \underset{A'}{\downarrow} cC$, whence $B \underset{A'}{\downarrow} \mathrm{cl}_P(AC)$, and $B \underset{\mathrm{cl}_P(A)}{\downarrow} \mathrm{cl}_P(AC)$.

Since $\mathrm{cl}_P(\mathrm{cl}_P(A)) = \mathrm{cl}_P(A)$, Lemma 4.0.2 implies that $\mathrm{tp}(\mathrm{cl}_P(C)/\mathrm{cl}_P(A))$ is foreign to all P-analysable types of U_P-rank zero. It follows as above that $\mathrm{cl}_P(B) \underset{\mathrm{cl}_P(A)}{\downarrow} \mathrm{cl}_P(C)$. $\square$

Lemma 4.0.4 *Suppose $A \subseteq B \cap C$ satisfies $\mathrm{cl}_P(B) \cap \mathrm{cl}_P(C) = \mathrm{cl}_P(A)$. If $D \underset{A}{\downarrow} BC$, then $\mathrm{cl}_P(BD) \cap \mathrm{cl}_P(CD) = \mathrm{cl}_P(AD)$.*

Proof: The assumptions imply $BD \underset{B}{\downarrow} BC$ and $CD \underset{C}{\downarrow} BC$. By Lemma 4.0.3 we have $\mathrm{cl}_P(BD) \underset{\mathrm{cl}_P(B)}{\downarrow} \mathrm{cl}_P(BC)$ and $\mathrm{cl}_P(CD) \underset{\mathrm{cl}_P(C)}{\downarrow} \mathrm{cl}_P(BC)$. Let $e \in \mathrm{cl}_P(CD) \cap \mathrm{cl}_P(BD)$. Then $\mathrm{Cb}(eD/\mathrm{cl}_P(BC)) \subset \mathrm{cl}_P(B) \cap \mathrm{cl}_P(C) = \mathrm{cl}_P(A)$, so $eD \underset{\mathrm{cl}_P(A)}{\downarrow} \mathrm{cl}_P(BC)$, whence $e \underset{\mathrm{cl}_P(A)D}{\downarrow} \mathrm{cl}_P(BC)D$. Therefore $e \underset{\mathrm{cl}_P(AD)}{\downarrow} \mathrm{cl}_P(BCD)$ by Lemma 4.0.3; as $e \in \mathrm{cl}_P(BCD)$ we get $e \in \mathrm{cl}_P(AD)$. $\square$

Lemma 4.0.5 *Suppose* $\mathrm{tp}(a/A)$ *is P-analysable, and* $B \supseteq A$. *Then* $\mathrm{Cb}(a/\mathrm{cl}_P(B)) \subset \mathrm{cl}_P(\mathrm{Cb}(a/B)A)$. *If* $U_P(a/A)$ *is finite, then* $\mathrm{Cb}(a/B) \subset \mathrm{cl}_P(\mathrm{Cb}(a/\mathrm{cl}_P(B))A)$.

Proof: Put $B_0 = \mathrm{Cb}(a/B) \cup A$ and $B_0' = \mathrm{acl}(aB_0) \cap \mathrm{cl}_P(B_0)$. As $\mathrm{tp}(a/A)$ is P-analysable, $\mathrm{tp}(a/B_0')$ is foreign to all types of U_P-rank zero. Since $a \underset{B_0}{\downarrow} B$ implies $a \underset{B_0'}{\downarrow} B$, the type $\mathrm{tp}(a/BB_0')$ must also be foreign to $\mathrm{cl}_P(B)$ over BB_0'. Hence $a \underset{B_0'}{\downarrow} \mathrm{cl}_P(B)$ and

$$\mathrm{Cb}(a/\mathrm{cl}_P(B)) \subseteq \mathrm{acl}(B_0') \subset \mathrm{cl}_P(B_0) = \mathrm{cl}_P(\mathrm{Cb}(a/B)A).$$

For the second assertion, assume that $\mathrm{tp}(a/A)$ is P-analysable of finite U_P-rank. Put $B_0 = \mathrm{Cb}(a/B)$ and $B_0' = \mathrm{Cb}(a/\mathrm{cl}_P(B))$. Then

$$U_P(a/B_0) = U_P(a/B) = U_P(a/\mathrm{cl}_P(B)) = U_P(a/B_0').$$

Now any $b \in B_0$ is algebraic in an initial segment $\bar{a} = (a_1, \dots, a_n)$ of a Morley sequence in $\mathrm{stp}(a/B)$. Hence

$$
\begin{aligned}
U_P(b/B_0') &= U_P(\bar{a}b/B_0') - U_P(\bar{a}/bB_0') \le U_P(\bar{a}/B_0') - U_P(\bar{a}/\mathrm{cl}_P(B)) \\
&\le n \cdot U_P(a/B_0') - n \cdot U_P(a/B) = 0.
\end{aligned}
$$

So $U_P(B_0/B_0') = 0$; as $\mathrm{tp}(B_0/A)$ is P-analysable, we have $B_0 \subset \mathrm{cl}_P(B_0'A)$. $\square$

Proposition 4.0.6 *Suppose no type in P forks over* $\emptyset$, *denote the class of their realizations (in the monster model* $\mathfrak{C}$*) by* X, *and put* $\mathrm{cl}(Y) := \mathrm{cl}_P(Y) \cap X$ *for any set* Y. *Then* P *is regular iff for any non-algebraic* $x \in X$ *and* $Y \subset X$ *we have* $x \underset{\ }{\not\downarrow} Y \Leftrightarrow x \in \mathrm{cl}(Y)$. *In this case,* $\mathrm{cl}(.)$ *induces a pre-geometry on* X.

Proof: First note that we may remove all algebraic types from P, as their realizations do not fork with anything and are in $\mathrm{cl}_P(\emptyset)$. Hence we assume that all types in P are non-algebraic.

Suppose P is regular, Y is a subset of X, and $x \in X$ forks with Y. Then every type in P is foreign to $\mathrm{tp}(x/Y)$, whence $x \in \mathrm{cl}(Y)$. Conversely, if $x \in \mathrm{cl}(Y)$, then x must depend on Y. Hence $x \underset{\ }{\not\downarrow} Y$ iff $x \in \mathrm{cl}(Y)$.

Now suppose that $x \underset{\ }{\not\downarrow} Y$ iff $x \in \mathrm{cl}(Y)$ for all $x \in X$ and $Y \subset X$. Let x realize a type in P with forking extension $\mathrm{tp}(x/A)$ for some A, and suppose $\mathrm{tp}(y/A)$ is in P. We may replace A by a Morley sequence in $\mathrm{stp}(xy/A)$ and consider it to be a subset of X. Now $x \underset{\ }{\downarrow} A$ implies $x \in \mathrm{cl}(A)$, so $U_P(x/A) = 0$ and $x \underset{A}{\downarrow} y$. Therefore any type in P is foreign to any forking extension of a type in P.

Finally, if $x \underset{\ }{\not\downarrow} Y$ is equivalent to $x \in \mathrm{cl}(Y)$, then cl inherits exchange from forking dependence (or forking dependence inherits transitivity from the closure relation), and therefore induces a pre-geometry on X. $\square$

4.1   Local Modularity

In this section we shall study a case when the forking geometry is particularly well-behaved.

Definition 4.1.1 P is *locally modular* if for any A and elements a and b whose types over A are P-internal, a and b are independent over $\mathrm{cl}_P(aA) \cap \mathrm{cl}_P(bA)$.

Remark 4.1.1 If P is the family of all types and P is locally modular, we call the theory *one-based*. Note that in this case P-closure equals algebraic closure, and every type is trivially P-internal. Hence a theory is one-based iff any tuples a and b are independent over $\mathrm{acl}(a) \cap \mathrm{acl}(b)$.

Lemma 4.1.1 *Suppose* $\mathrm{tp}(a/A)$ *and* $\mathrm{tp}(b/A)$ *are P-internal,* $ab \underset{A}{\downarrow} B$*, and* a *and* b *are independent over* $\mathrm{cl}_P(aAB) \cap \mathrm{cl}_P(bAB)$*. Then they are independent over* $\mathrm{cl}_P(aA) \cap \mathrm{cl}_P(bA)$*.*

Proof: Let $I = \mathrm{cl}_P(aA) \cap \mathrm{cl}_P(bA)$, and note that $I = \mathrm{cl}_P(IA)$. Then $\mathrm{cl}_P(abA) \underset{\mathrm{cl}_P(A)}{\downarrow} \mathrm{cl}_P(B)$, whence $\mathrm{cl}_P(aA)\mathrm{cl}_P(bA) \underset{I}{\downarrow} \mathrm{cl}_P(B)I$, and furthermore $\mathrm{cl}_P(aA)\mathrm{cl}_P(bA) \underset{I}{\downarrow} \mathrm{cl}_P(IB)$. So Lemma 4.0.4 implies that $\mathrm{cl}_P(IB)$ equals $\mathrm{cl}_P(aAB) \cap \mathrm{cl}_P(bAB)$. But $ab \underset{I}{\downarrow} \mathrm{cl}_P(IB)$; therefore $a \underset{\mathrm{cl}_P(IB)}{\downarrow} b$ yields $a \underset{I}{\downarrow} b$. $\square$

Corollary 4.1.2 *Local modularity is preserved under naming and forgetting parameters.*

Proof: Preservation under naming parameters is trivial, and that under forgetting parameters follows from Lemma 4.1.1 by conjugating everything to be independent of the parameters. $\square$

Lemma 4.1.3 *Suppose* $\mathrm{tp}(a)$ *and* $\mathrm{tp}(b)$ *are P-internal and P is locally modular. If* $A \subset \mathrm{cl}_P(a)$ *and* $B \subset \mathrm{cl}_P(b)$*, then A and B are independent over* $\mathrm{cl}_P(A) \cap \mathrm{cl}_P(B)$*.*

Proof: By $\emptyset$-invariance of P we may assume $a \underset{A}{\downarrow} B$ and $b \underset{B}{\downarrow} Aa$. Then a and b are independent over $\mathrm{cl}_P(a) \cap \mathrm{cl}_P(b)$ by local modularity, so A and B are independent over $\mathrm{cl}_P(a) \cap \mathrm{cl}_P(b)$ by Lemma 4.0.3. On the other hand, the assumptions imply $\mathrm{cl}_P(a) \underset{\mathrm{cl}_P(A)}{\downarrow} \mathrm{cl}_P(B)$ and $\mathrm{cl}_P(b) \underset{\mathrm{cl}_P(B)}{\downarrow} \mathrm{cl}_P(a)$. Hence

$$\mathrm{cl}_P(a) \cap \mathrm{cl}_P(b) \subseteq \mathrm{cl}_P(a) \cap \mathrm{cl}_P(B) \subseteq \mathrm{cl}_P(A) \cap \mathrm{cl}_P(B),$$

and clearly equality holds. The lemma follows. $\square$

CHAPTER 4. GROUPS & GEOMETRY

Lemma 4.1.4 *Suppose* $a \in \mathrm{cl}_P(b)$ *and* $b \underset{a}{\downarrow} A$. *Then* $\mathrm{cl}_P(\mathrm{Cb}(a/\mathrm{cl}_P(A))) = \mathrm{cl}_P(\mathrm{Cb}(b/\mathrm{cl}_P(A)))$.

Proof: Let $B = \mathrm{Cb}(a/\mathrm{cl}_P(A))$ and $B' = \mathrm{Cb}(b/\mathrm{cl}_P(A))$, so $b \underset{B'}{\downarrow} A$, whence $\mathrm{cl}_P(b) \underset{\mathrm{cl}_P(B')}{\downarrow} \mathrm{cl}_P(A)$ by Lemma 4.0.3. As, trivially, $a \underset{\mathrm{cl}_P(B')\mathrm{cl}_P(b)}{\downarrow} \mathrm{cl}_P(A)$, we get $a \underset{\mathrm{cl}_P(B')}{\downarrow} \mathrm{cl}_P(A)$ and therefore $\mathrm{cl}_P(B) \subseteq \mathrm{cl}_P(B')$. Conversely, $b \underset{a}{\downarrow} A$ implies by Lemma 4.0.3 that $\mathrm{cl}_P(b)$ and $\mathrm{cl}_P(A)$ are independent over $\mathrm{cl}_P(a)$ and hence over $\mathrm{cl}_P(a)\mathrm{cl}_P(B)$; on the other hand $a \underset{B}{\downarrow} \mathrm{cl}_P(A)$ implies $\mathrm{cl}_P(a) \underset{\mathrm{cl}_P(B)}{\downarrow} \mathrm{cl}_P(A)$, whence $b \underset{\mathrm{cl}_P(B)}{\downarrow} \mathrm{cl}_P(A)$, and $\mathrm{cl}_P(B') \subseteq \mathrm{cl}_P(B)$. $\square$

The following two propositions give two quite different characterizations of local modularity.

Proposition 4.1.5 *P is locally modular iff for any a and sets $A \subseteq B$ such that $\mathrm{tp}(a/A)$ is P-internal, we have $\mathrm{Cb}(a/\mathrm{cl}_P(B)) \subseteq \mathrm{cl}_P(aA)$. Furthermore, it is sufficient to check local modularity for tuples a realizing types in P.*

Proof: Let P be locally modular, suppose $\mathrm{tp}(a/A)$ is P-internal and consider some $B \supset A$. Clearly, we may replace B by a Morley sequence in $\mathrm{stp}(a/B)$ and thus assume that $\mathrm{tp}(B/A)$ is P-internal. For any finite $\bar{b} \in \mathrm{cl}_P(B)$ local modularity of P implies that a and $\bar{b}$ are independent over $\mathrm{cl}_P(aA) \cap \mathrm{cl}_P(\bar{b}A)$; Lemma 4.0.3 implies that $\mathrm{cl}_P(aA)$ and $\bar{b}$ are independent over $\mathrm{cl}_P(aA) \cap \mathrm{cl}_P(\bar{b}A)$, and hence also over $\mathrm{cl}_P(aA) \cap \mathrm{cl}_P(B)$. Therefore a and $\mathrm{cl}_P(B)$ are independent over $\mathrm{cl}_P(aA) \cap \mathrm{cl}_P(B)$. Hence $Cb(a/\mathrm{cl}_P(B)) \subset \mathrm{cl}_P(aA)$.

Conversely, suppose that the second condition holds and consider a, b, A such that $\mathrm{tp}(a/A)$ and $\mathrm{tp}(b/A)$ are P-internal. Let $C = \mathrm{Cb}(a/\mathrm{cl}_P(bA))$. Then $C \subset \mathrm{cl}_P(bA)$, and by assumption $C \subset \mathrm{cl}_P(aA)$. Hence $C \subset \mathrm{cl}_P(aA) \cap \mathrm{cl}_P(bA)$; since $a \underset{C}{\downarrow} \mathrm{cl}_P(bA)$, we obtain that a and b are independent over $\mathrm{cl}_P(aA) \cap \mathrm{cl}_P(bA)$.

The above equivalence works just as well if we require a and b to satisfy types in P over A. It remains to prove that it is sufficient to check local modularity for tuples realizing types in P. So suppose $\mathrm{tp}(a/A)$ is P-internal, and $B \supseteq A$ is arbitrary. Then there are D and a tuple $\bar{d}$ of realizations of types in P such that $a \in \mathrm{dcl}(D\bar{d})$ and $a \underset{A}{\downarrow} D$; we may assume that in addition $D\bar{d} \underset{aA}{\downarrow} B$. Then $D \underset{A}{\downarrow} Ba$. By Lemma 4.1.1 it is sufficient to show that $\mathrm{Cb}(a/BD) \subset \mathrm{cl}_P(aAD)$. So we assume AD to be named.

Now $\bar{d} \underset{a}{\downarrow} B$; by Lemma 4.1.4 and our assumption we have

$$\mathrm{cl}_P(\mathrm{Cb}(a/\mathrm{cl}_P(B))) = \mathrm{cl}_P(\mathrm{Cb}(\bar{d}/\mathrm{cl}_P(B))) \subseteq \mathrm{cl}_P(\bar{d}) \cap \mathrm{cl}_P(B).$$

Since $\bar{d} \underset{a}{\downarrow} B$ implies $\mathrm{cl}_P(\bar{d}) \underset{\mathrm{cl}_P(a)}{\downarrow} \mathrm{cl}_P(B)$, we get $\mathrm{cl}_P(\bar{d}) \cap \mathrm{cl}_P(B) = \mathrm{cl}_P(a)$. $\square$

In the case where the theory is one-based, this simplifies nicely:

Corollary 4.1.6 *A theory is one-based iff every type* $\text{tp}(a/A)$ *is based on* $\text{acl}(a)$. *In particular, in a one-based theory every type is based on a finite set.*

Proof: Clearly $a \underset{A}{\cup} \text{acl}(A)$, and $\text{cl}_P(X) = \text{acl}(X)$ for any X. The result now follows from Proposition 4.1.5. $\square$

Proposition 4.1.7 *A regular family P is locally modular iff for all algebraically closed parameter sets A and all $\bar{x}, \bar{y}$ realizing types in P over A we have*
$$\dim(\bar{x}) + \dim(\bar{y}) = \dim(\bar{x}\bar{y}) + \dim(\text{cl}(\bar{x}) \cap \text{cl}(\bar{y}))$$
whenever $\text{cl}_P(\text{cl}(\bar{x}) \cap \text{cl}(\bar{y}), A) = \text{cl}_P(\bar{x}A) \cap \text{cl}_P(\bar{y}A)$, *where* $\text{cl}(\bar{x}) = \{y \in \text{cl}_P(\bar{x}A) : \text{tp}(y/A) \in P\}$.

Remark 4.1.2 Note that for regular P always $\dim(\bar{x}) + \dim(\bar{y}) \geq \dim(\bar{x}\bar{y}) + \dim(\text{cl}(\bar{x}) \cap \text{cl}(\bar{y}))$, as we can see by extending a basis for $\text{cl}(\bar{x}) \cap \text{cl}(\bar{y})$ to a basis for $\text{cl}(\bar{x})$, and to a basis for $\text{cl}(\bar{y})$. Furthermore, $\text{cl}_P(\text{cl}(\bar{x}) \cap \text{cl}(\bar{y}), A)$ may well be a proper subset of $\text{cl}_P(\bar{x}A) \cap \text{cl}_P(\bar{y}A)$. Consider an affine abelian group, i.e. an abelian group A without addition, but instead with an equivalence relation $R(x, y; z, u)$ which holds iff $x + y = z + u$. If A has prime exponent, there is a unique 1-type p, and for independent elements a, b, c and $d = a + b - c$ we have $\dim(a, b) = \dim(c, d) = 2$, $\dim(a, b, c, d) = 3$, and $\text{cl}(a, b) \cap \text{cl}(c, d) = \emptyset$. However, $\text{cl}_p(a, b) \cap \text{cl}_p(c, d)$ contains the class $[a, b]$ modulo R; after fixing a single group element u_0, every equivalence class $[x, y]$ modulo R becomes equi-definable with the unique group element z such that $R(x, y; z, u_0)$ holds.

Proof of Proposition: Let Q be the set of types in P which are based on A. Then Q is regular as well, and if we denote the set of realizations of types in Q in $\mathfrak{C}$ by X, then $U_P(\bar{x}/A) = U_Q(\bar{x}/A)$ for any $\bar{x} \in X$ by regularity of P. Therefore $\text{cl}(.)$ is the closure operator associated with the family Q, and induces a pre-geometry on X, whose independence is forking independence. So if we extend a basis I of $\text{cl}(\bar{x}) \cap \text{cl}(\bar{y})$ to a basis IX of $\text{cl}(\bar{x})$ and to a basis IY of $\text{cl}(\bar{y})$, then for locally modular P we obtain $X \underset{\text{cl}_P(IXA) \cap \text{cl}_P(IYA)}{\cup} Y$. If $\text{cl}_P(\text{cl}(\bar{x}) \cap \text{cl}(\bar{y}), A) = \text{cl}_P(\bar{x}A) \cap \text{cl}_P(\bar{y}A) = \text{cl}_P(IXA) \cap \text{cl}_P(IYA)$, then $X \underset{\text{cl}_P(IA)}{\cup} Y$, so IXY is an independent set and thus a basis of $\text{cl}(\bar{x}\bar{y})$. The dimension equality follows.

Conversely, suppose $\bar{x}$ and $\bar{y}$ are both tuples of realizations of types in P over some set A. Set $I := \text{cl}_P(\bar{x}A) \cap \text{cl}_P(\bar{y}A)$. Consider a realization $\bar{x}'$ of $\text{stp}(\bar{x}/I)$ independent of $\bar{x}\bar{y}$ over I, and let $\bar{x}_0$ be a maximal subset of $\bar{x}'$ independent of I over A. Then $\text{cl}_P(\bar{x}'A) = \text{cl}_P(\bar{x}_0 I A)$ by regularity of P. Put $B = \text{acl}(A\bar{x}_0)$ and $\bar{x}_1 = \bar{x}' - \bar{x}_0$. Then $B \underset{A}{\cup} \bar{x}\bar{y}$; furthermore $\bar{x}_1 \underset{A}{\cup} B$, again by regularity of P.

We consider the pre-geometry on the set X of realizations of types in P over B. Note that X contains $\bar{x}$, $\bar{y}$ and $\bar{x}_1$. Now $B \underset{I}{\cup} \bar{x}\bar{y}$ and $\text{cl}_P(I) =$

$\text{cl}_P(\bar{x}I) \cap \text{cl}_P(\bar{y}I)$, so $\text{cl}_P(\bar{x}B) \cap \text{cl}_P(\bar{y}B) = \text{cl}_P(IB) = \text{cl}_P(\bar{x}_1 B)$ by Lemma 4.0.4. This means $\bar{x}_1 \subseteq \text{cl}(\bar{x}) \cap \text{cl}(\bar{y})$, and therefore $\text{cl}_P(\text{cl}(\bar{x}) \cap \text{cl}(\bar{y}), B) = \text{cl}_P(\bar{x}B) \cap \text{cl}_P(\bar{y}B)$. So if the dimension equality $\dim(\bar{x}) + \dim(\bar{y}) - \dim(\bar{x}_1) = \dim(\bar{x}\bar{y})$ holds, we get $\bar{x} \underset{\text{cl}_P(IB)}{\downarrow} \bar{y}$. By Lemma 4.1.1 we have $\bar{x} \underset{I}{\downarrow} \bar{y}$, and by Proposition 4.1.5 this proves local modularity. $\square$

Proposition 4.1.8 *Let Q be an $\emptyset$-invariant subfamily of P, and suppose P is locally modular. Then Q is locally modular.*

Proof: Suppose $\text{tp}(a/A)$ and $\text{tp}(b/A)$ are Q-internal. Then they are P-internal, and a and b are independent over $\text{cl}_P(aA) \cap \text{cl}_P(bA)$ by the local modularity of P. Let $I = \text{cl}_Q(aA) \cap \text{cl}_Q(bA)$. Suppose $ab \underset{I}{\not\downarrow} \text{cl}_P(aA) \cap \text{cl}_P(bA)$. By Proposition 3.0.6 there is $c \in (\text{cl}_P(aA) \cap \text{cl}_P(bA)) - I$ such that $\text{tp}(c/I)$ is $\text{tp}(ab/I)$-internal. So $\text{tp}(c/I)$, and hence $\text{tp}(c/A)$, is Q-analysable; since $U_Q(c/aA) \leq U_P(c/aA) = 0$ and $U_Q(c/bA) \leq U_P(c/bA) = 0$, we obtain $c \in I$, a contradiction. Therefore $a \underset{I}{\downarrow} b$. $\square$

Definition 4.1.2 *Two families P and Q of types are *perpendicular* if every type in P is foreign to every type in Q, and vice versa. We write $P \perp Q$.*

Proposition 4.1.9 *Let P_1 and P_2 be two perpendicular $\emptyset$-invariant families of stationary types, and put $P = P_1 \cup P_2$. Suppose P_1 and P_2 are locally modular. Then P is locally modular.*

Proof: Suppose $\text{tp}(a/A)$ and $\text{tp}(b/A)$ are P-internal. By Proposition 4.1.5 we may assume that $a = \bar{x}_1 \bar{x}_2$ and $b = \bar{y}_1 \bar{y}_2$ for some realizations $\bar{x}_i, \bar{y}_i$ of types in P_i (for $i = 1, 2$). Put $I := \text{cl}_P(\bar{x}_1 \bar{x}_2 A) \cap \text{cl}_P(\bar{y}_1 \bar{y}_2 A)$.

As P_i is locally modular, for $i = 1, 2$ we get that $\bar{x}_i$ and $\bar{y}_i$ are independent over $I_i := \text{cl}_{P_i}(\bar{x}_i A) \cap \text{cl}_{P_i}(\bar{y}_i A)$. However, $\text{tp}(\bar{x}_i/I_i)$ is foreign to all types of U_{P_i}-rank zero by Lemma 4.0.2; as P_1 and P_2 are perpendicular, this implies $\bar{x}_1 \underset{I_1}{\downarrow} \bar{y}_1 I \bar{y}_2$ and $\bar{x}_2 \underset{I_2}{\downarrow} \bar{y}_2 I \bar{x}_1 \bar{y}_1$. Hence $\bar{x}_1 \underset{I}{\downarrow} \bar{y}_1 \bar{y}_2$ and $\bar{x}_2 \underset{I\bar{x}_1}{\downarrow} \bar{y}_1 \bar{y}_2$. The independence $\bar{x}_1 \bar{x}_2 \underset{I}{\downarrow} \bar{y}_1 \bar{y}_2$ follows. $\square$

Lemma 4.1.10 *Let P' be an $\emptyset$-invariant family of P-semi-regular types extending P. Then internality, analysability, closure, and rank zero are the same for P and for P'.*

Proof: Since any type in $P' - P$ is P-internal, P'-internality equals P-internality, and P'-analysability equals P-analysability. Furthermore, $U_{P'}(p) \geq U_P(p)$ for any type p, but if $U_P(p) = 0$, then p is co-foreign to all types in P and hence to all types in $P' - P$ as well by P-semi-regularity of $P' - P$. Therefore $U_{P'}(p) = 0$. It follows that P-closure equals P'-closure. $\square$

In fact it is easy to see that in this case U_P-rank equals $U_{P'}$-rank.

Corollary 4.1.11 *Let Q be an $\emptyset$-invariant family of P-semi-regular types. If P is locally modular, then Q is locally modular.*

Proof: Put $P' = P \cup Q$. Then P' is locally modular by Lemma 4.1.10, and Q by Proposition 4.1.8. $\square$

To finish this section, we shall prove some useful facts about small locally modular types and theories. Recall that a theory is small if $S_n(\emptyset)$ is countable for all $n < \omega$. Smallness plays an important rôle when considering the number of countable models of a first-order theory: if $S_n(\emptyset)$ is not countable, then it must have size continuum, and there are continuum many non-isomorphic countable models (any type is realized in some countable model, but any model can realize only countably many types). In particular, a theory with less than the maximal number of non-isomorphic countable models must be small.

Lemma 4.1.12 *Suppose T is small. Let $\mathcal{L}$ be the lattice of P-closed subclasses of $\mathrm{cl}_P(a)$. Then no dense order can be embedded into $\mathcal{L}$.*

Proof: Suppose $\{L_i : i \in \mathbb{Q}\}$ is a collection of P-closed subclasses of $\mathrm{cl}_P(a)$ with $L_i \subset L_j$ for $i < j$. For $r \in \mathbb{R}$ put $L_r = \bigcup_{i<r} L_i$. Then $p_r := \mathrm{tp}(a/L_r)$ has a non-forking extension to $\mathrm{cl}_P(a)$, realized by some element a_r. Let $A = \mathrm{cl}_P(a) \cap \mathrm{acl}(aa_r)$. Then $a_r \underset{A}{\downarrow} \mathrm{cl}_P(a)$ by Lemma 4.0.2, so $\mathrm{tp}(a_r/\mathrm{cl}_P(a))$, and hence p_r, is based on aa_r. But for $s > r$ there are rational i and j such that $r < i < j < s$ and $a \underset{L_i}{\not\downarrow} L_j$, so the bounds satisfy

$$\beta(p_r) \geq \beta(p_i) > \beta(p_j) \geq \beta(p_s).$$

Hence all the p_r are different, and there are continuum many types over finite sets, contradicting smallness. $\square$

Proposition 4.1.13 *Suppose T is small, and $p \in S(\emptyset)$ is locally modular. Then there is a p-semi-regular regular type q non-orthogonal to p, based on a finite set.*

Proof: Choose some realization $a \models p$ and consider the lattice $\mathcal{L}$ of p-closed subclasses of $\mathrm{cl}_p(a)$; by Lemma 4.1.12 it does not embed a dense linear order. Hence there are L and L' in $\mathcal{L}$ such that $L \subset L'$ and p is not foreign to $\mathrm{tp}(L'/L)$, but for any L'' with $L \subset L'' \subset L'$ either p is foreign to $\mathrm{tp}(L'/L'')$, or it is foreign to $\mathrm{tp}(L''/L)$. Choose some element $b \in L'$ such that p is not foreign to $\mathrm{tp}(b/L)$; since $b \in \mathrm{cl}_p(a)$, there is a p-analysis $(b_i : i \leq \alpha) \subset \mathrm{acl}(bL)$ of $\mathrm{tp}(b/L)$. As p is not foreign to $\mathrm{tp}(b/L)$, there must be a minimal $i \leq \alpha$ such that p is not foreign to $\mathrm{tp}(b_i/Lb_j : j < i)$; we may clearly replace b by b_i and L by $\mathrm{cl}_p(Lb_j : j < i)$ and assume that $\mathrm{tp}(b/L)$ is p-internal.

Now consider some $B \supseteq L$ such that p is non-orthogonal to $\mathrm{tp}(b/B)$, and take $B' = \mathrm{acl}(bB) \cap \mathrm{cl}_p(B)$. Then p is still non-orthogonal to $q := \mathrm{tp}(b/B')$; furthermore q is foreign to all types of U_p-rank zero. In particular, q is p-semi-regular and stationary, and based on $\mathrm{cl}_p(bL) \cap \mathrm{cl}_p(B')$ by local modularity. Increasing L again, we may assume that q is based on L.

Claim. q is regular.

Proof of Claim: Suppose $b \underset{L}{\not\smile} c$. By local modularity $\mathrm{tp}(b/\mathrm{cl}_p(Lc))$ is based on $L'' := \mathrm{cl}_p(bL) \cap \mathrm{cl}_p(cL) \supseteq L$. As b and c fork over L, we have $b \underset{L}{\not\smile} L''$; since clearly $L'' \subseteq \mathrm{cl}_p(bL) \subseteq L'$, either $\mathrm{tp}(L''/L)$ or $\mathrm{tp}(L'/L'')$ is co-foreign to p. Now in the first case q would be foreign to $\mathrm{tp}(L''/L)$, which contradicts $b \underset{L}{\not\smile} L''$. So p, and hence q, is foreign to $\mathrm{tp}(L'/L'')$, and therefore to $\mathrm{tp}(b/\mathrm{cl}_p(cL))$, and finally to $\mathrm{tp}(b/cL)$ by the p-semi-regularity of q. Since c was arbitrary, q is regular. $\square$

Now let b' be a realization of q independent of a over L, and put $A = \mathrm{acl}(ab') \cap \mathrm{cl}_p(a)$. Then $b' \underset{A}{\smile} \mathrm{cl}_p(a)$ by Lemma 4.0.2, so $\mathrm{tp}(b'/\mathrm{cl}_p(A))$, and hence q, is based on the finite set ab'. $\square$

Corollary 4.1.14 *Let T be a small one-based theory, and p a type. Then there is a regular type q non-orthogonal to p.*

Proof: If a realizes p, then p is based on a, and is clearly p-semi-regular. By Proposition 4.1.13 there is a p-semi-regular regular type q non-orthogonal to p. $\square$

Definition 4.1.3 Let p be a type over A. The *weight* $w(p)$ *of p is less than* κ if there do not exist a superset $B \supseteq A$, a realization a of a non-forking extension of p to B, and a set $\{b_i : i < \kappa\}$ independent over B, such that $a \underset{B}{\not\smile} b_i$ for all $i < \kappa$.
  The weight of p is κ if it is less than κ^+, but not less than κ.

Clearly for all $i < \kappa$ we have $a \underset{B \cup \{b_j : j < i\}}{\not\smile} b_i$. So by Corollary 0.2.25 the weight is at most $|T|$. However, there are types of weight less than ω which do not have finite weight.

Definition 4.1.4 An element a *dominates* an element b over a set A if whenever $a \underset{A}{\smile} c$, we have $b \underset{A}{\smile} c$. A stationary type p dominates a type q if there are A and realizations a and b of non-forking extensions of p and q such that a dominates b over A. Two types are *domination-equivalent* if there are realizations of non-forking extensions of them which dominate each other.

Lemma 4.1.15 *Let p, q and r be stationary types.*

1. If p is regular, $w(p) = 1$.

2. If $w(p) = w(q) = 1$ and $p \not\perp q$, then p and q are domination-equivalent.

3. Domination is preserved under non-forking extensions. Furthermore, if p dominates q and q dominates r, then p dominates r.

4. If p dominates q, then $p \otimes r$ dominates $q \otimes r$.

Proof:

1. Suppose $p = \operatorname{tp}(a/A)$, $b \underset{A}{\downarrow} c$ and $a \underset{A}{\not\downarrow} b$. Replacing b by an initial segment $(b_i : i \leq n)$ of a Morley sequence in $\operatorname{stp}(a/Ab)$ independent of c over bA, we may choose n such that $a \underset{A}{\downarrow} (b_i : i < n)$ but $a \underset{A \cup (b_i : i < n)}{\not\downarrow} b_n$. Then $c \underset{A}{\downarrow} (b_i : i \leq n)$, and $b_n \underset{A}{\downarrow} (b_i : i < n)$ by regularity of p, so $b_n \underset{A}{\downarrow} c(b_i : i < n)$. Suppose $c \underset{A}{\not\downarrow} a$. Again by regularity, $a \underset{Ac(b_i : i < n)}{\downarrow} b_n$, but this yields $b_n \underset{A}{\downarrow} ac(b_i : i < n)$, a contradiction. Hence $w(p) = 1$.

2. Let a and b realize non-forking extensions of p and q over some set A with $a \underset{A}{\not\downarrow} b$. If $c \underset{A}{\downarrow} a$, then $w(b/A) = 1$ and $b \underset{A}{\not\downarrow} a$ together imply $b \underset{A}{\downarrow} c$, whence a dominates b. The reverse follows by symmetry.

3. Suppose a and b are realizations of non-forking extensions of p and q, respectively, over some set A, such that a dominates b over A. Suppose $B \underset{A}{\downarrow} ab$, and $c \underset{B}{\downarrow} a$. Then $cB \underset{A}{\downarrow} a$, whence $cB \underset{A}{\downarrow} b$ and $c \underset{B}{\downarrow} b$, so a dominates b over B, and domination is preserved under non-forking extensions. Hence if p dominates q and q dominates r, there are a set A and realizations a, b, c of non-forking extensions of p, q and r, respectively, such that a dominates b and b dominates c over A. So if $d \underset{A}{\downarrow} a$, then $d \underset{A}{\downarrow} b$ and hence $d \underset{A}{\downarrow} c$, and a dominates c over A.

4. Let a and b be realizations of non-forking extensions of p and q over some set A such that a dominates b over A, and c a realization of a non-forking extension of q over A with $c \underset{A}{\downarrow} ab$. Suppose $d \underset{A}{\not\downarrow} bc$. Then either $d \underset{A}{\not\downarrow} c$, so $d \underset{A}{\not\downarrow} ac$, or $d \underset{Ac}{\not\downarrow} b$, whence $dc \underset{A}{\not\downarrow} b$. As a dominates b we get $dc \underset{A}{\not\downarrow} a$, whence $d \underset{Ac}{\not\downarrow} a$, and $d \underset{A}{\not\downarrow} ac$. Therefore ac dominates bc. $\square$

Theorem 4.1.16 Let T be small, and $p \in S(\emptyset)$ be locally modular. Then p is domination-equivalent to a finite product of regular types. In particular, p has finite weight.

Proof: Let a realize p. First suppose p has infinite weight, as exemplified by a set A independent of a and some sequence $\{b_i : i \in \mathbb{Q}\}$ independent over A such that $a \not\perp_A b_i$ for all $i \in \mathbb{Q}$. Then $a \perp A$ implies $a \perp_{\mathrm{cl}_p(\emptyset)} \mathrm{cl}_p(A)$, and since trivially $a \perp \mathrm{cl}_p(\emptyset)$, even $a \perp \mathrm{cl}_p(A)$. By Lemma 4.0.3 the family $\{\mathrm{cl}_p(Ab_i) : i \in \mathbb{Q}\}$ is independent over $\mathrm{cl}_p(A)$, and clearly $a \not\perp_{\mathrm{cl}_p(A)} \mathrm{cl}_p(Ab_i)$ for all $i \in \mathbb{Q}$. But this yields $a \not\perp_{\mathrm{cl}_p(Ab_i : i \leq q)} \mathrm{cl}_p(Ab_i : i \leq r)$ for any rational $q < r$. However, by local modularity, $\mathrm{tp}(a/\mathrm{cl}_p(Ab_i : i \leq q))$ is based on $\mathrm{cl}_p(a) \cap \mathrm{cl}_p(Ab_i : i \leq q)$. As these types have different bounds, the intersections $\mathrm{cl}_p(a) \cap \mathrm{cl}_p(Ab_i : i \leq q)$ must all be different for varying $q \in \mathbb{Q}$, contradicting Lemma 4.1.12.

Claim. If $\mathrm{tp}(b/\mathrm{cl}_p(\emptyset))$ is p-internal with $w(b/\mathrm{cl}_p(\emptyset)) \geq n$, then there is a sequence $(b_i : i < n)$ in $\mathrm{cl}_p(b)$, independent and p-semi-regular over $\mathrm{cl}_p(\emptyset)$.

Proof of Claim: We should note that any p-internal b is p-semi-regular over $\mathrm{cl}_p(\emptyset)$, and locally modular by Corollary 4.1.11.

Consider $A \supseteq \mathrm{cl}_p(\emptyset)$ independent of b over $\mathrm{cl}_p(\emptyset)$ and elements $(b_i' : i < n)$ independent over A, such that b forks with each of them over A. Then $\mathrm{cl}_p(b) \perp_{\mathrm{cl}_p(\emptyset)} \mathrm{cl}_p(A)$, and for $i < n$ we have $b \not\perp_{\mathrm{cl}_p(\emptyset)} Ab_i'$ and $\mathrm{tp}(b/\mathrm{cl}_p(Ab_i'))$ is based on $\mathrm{cl}_p(b) \cap \mathrm{cl}_p(Ab_i')$ by local modularity (over $\mathrm{cl}_p(\emptyset)$). So in particular there is $b_i \in \mathrm{cl}_p(b) \cap \mathrm{cl}_p(Ab_i')$ with $b \not\perp_{\mathrm{cl}_p(\emptyset)} b_i$; as $\mathrm{tp}(b/\mathrm{cl}_p(\emptyset))$ is p-internal, we may assume that $\mathrm{tp}(b_i/\mathrm{cl}_p(\emptyset))$ is p-internal (and hence p-semi-regular by Proposition 3.0.6). Now $(b_i : i < n) \perp_{\mathrm{cl}_p(\emptyset)} \mathrm{cl}_p(A)$ and $(\mathrm{cl}_p(Ab_i') : i < n)$ is an independent sequence over $\mathrm{cl}_p(A)$ by Lemma 4.0.3; therefore $(b_i : i < n)$ is independent over $\mathrm{cl}_p(A)$ and hence over $\mathrm{cl}_p(\emptyset)$. $\square$

By the first paragraph of the main proof, there exists a maximal sequence $(a_i : i < n)$ of elements in $\mathrm{cl}_p(a)$, independent over $\mathrm{cl}_p(\emptyset)$, such that $\mathrm{tp}(a_i/\mathrm{cl}_p(\emptyset))$ is p-semi-regular for all $i < n$. By the claim and maximality, $w(a_i/\mathrm{cl}_p(\emptyset)) = 1$ for all $i < n$, since otherwise we could split up a_i into two elements a_i^1 and a_i^2 in $\mathrm{cl}_p(a_i)$ and preserve independence of the sequence by p-semi-regularity.

Claim. $(a_i : i < n)$ dominates a over $\mathrm{cl}_p(\emptyset)$.

Proof of Claim: Suppose not, so there is a' with $a' \not\perp_{\mathrm{cl}_p(\emptyset)} a$, but $a' \perp_{\mathrm{cl}_p(\emptyset)} (a_i : i < n)$. We may then replace a' by some element $a_n \in \mathrm{cl}_p(a) \cap \mathrm{cl}_p(a')$ by local modularity, which we may choose to be p-internal and of weight one over $\mathrm{cl}_p(\emptyset)$ by the above argument. Then a_n is independent of $(a_i : i < n)$; this contradicts the maximality of that sequence. $\square$

Conversely, if $a' \perp_{\mathrm{cl}_p(\emptyset)} a$, then $a' \perp_{\mathrm{cl}_p(\emptyset)} (a_i : i < n)$ by Lemma 4.0.3, since $a_i \in \mathrm{cl}_p(a)$ for all $i < n$. Hence a and $(a_i : i < n)$ are domination-equivalent over $\mathrm{cl}_p(\emptyset)$.

Claim. If $(b_i : i < m)$ is another maximal sequence, then $m = n$.

Proof of Claim: To simplify notation, we add $\mathrm{acl}(a_i, b_j : i < n, j < m) \cap \mathrm{cl}_P(\emptyset)$ to the language. Let $I \subset n$ be maximal such that $b_0 \perp \{a_i : i \in I\}$.

Suppose there are two elements $i_1 \neq i_2$ in $n - I$. Then a_{i_1} and a_{i_2} are independent over $\{a_i : i \in I\}$; since $w(b_0/a_i : i \in I) = w(b_0) = 1$, either a_{i_1} or a_{i_2} must be independent of b_0 over $\{a_i : i \in I\}$, contradicting maximality of I. Therefore there is a unique $j \in n - I$, and $\{b_0, a_i : i \in I\}$ is an independent set of size n.

Suppose $c \downarrow \{b_0, a_i : i \in I\}$. Since $w(a_j/a_i : i \in I) = w(a_j) = 1$, either c or b_0 must be independent of a_j over $\{a_i : i \in I\}$. The latter case is impossible by our choice of j, so $c \downarrow \{a_i : i < n\}$, and $\{b_0, a_i : i \in I\}$ dominates $\{a_i : i < n\}$, and hence a, over $cl_P(\emptyset)$. As a clearly dominates $\{b_0, a_i : i \in I\}$, we may now successively exchange the a_i for b_j and retain an independent set of size n which is domination-equivalent to a (note that we never have to throw an element b_j out again). Therefore $m = n$. $\square$

In particular, $w(a) = w(a/cl_p(\emptyset)) = n$ by the first claim.

Now for all $i < n$ the type $tp(a_i/cl_p(\emptyset))$ is based on $acl(a_i) \cap cl_p(\emptyset)$ and hence on a_i. So by Proposition 4.1.13 there is a p-semi-regular regular type q_i, based on some finite set, that is non-orthogonal to $tp(a_i/cl_p(\emptyset))$. By Lemma 4.1.15, the product $q_1 \otimes \cdots \otimes q_n$ is domination-equivalent to p. $\square$

Corollary 4.1.17 *Let T be small and one-based. Then every type has finite weight, and is domination-equivalent to a finite product of regular types.*

Proof: Any type in a one-based theory is finitely based; we may therefore apply Theorem 4.1.16. $\square$

4.2 Locally Modular Groups

In this section, we shall assume that P is locally modular.

Proposition 4.2.1 *Let G be a P-internal group type-definable over $\emptyset$. Suppose H is a locally connected relatively definable subgroup. Then H is relatively definable over $cl_P(\emptyset)$.*

Proof: Suppose u is the canonical parameter for H. Let h realize the principal generic type for H over u, and g realize the principal generic type for G over u, h. So $tp(hg/g, u)$ is generic for Hg over u, g, and $tp(hg/u, h)$ is principal generic for G.

Now let v be the canonical parameter for the coset Hg. If $(h_i : i < \omega)$ is a Morley sequence in $stp(hg/g, u)$, then h_i is generic in Hg for all $i < \omega$. But local connectivity of H means that any coset of a model-theoretic conjugate of H intersects Hg in a coset of infinite index and cannot contain infinitely many generic elements, so v is definable over $(h_i : i < \omega)$. As $v \in dcl(g, u)$, we get

$v \in \mathrm{Cb}(hg/g, u)$. On the other hand, local modularity implies $\mathrm{Cb}(hg/g, u) \subset \mathrm{cl}_P(hg)$.

As $H = (Hg)(Hg)^{-1}$, the canonical parameter u is definable over v, whence $u \in \mathrm{cl}_P(hg)$. But $hg \underset{}{\downarrow} u$, so $\mathrm{cl}_P(hg) \underset{\mathrm{cl}_P(\emptyset)}{\downarrow} u$ by Lemma 4.0.3, and therefore $u \in \mathrm{cl}_P(\emptyset)$. $\square$

Theorem 4.2.2 *Suppose G is P-internal. Then G^P is central in G^0. In particular, a P-semi-regular group is abelian-by-finite.*

Proof: We may assume that G is centralizer-connected. Consider, for $g \in G$, the subgroup $H_g = \{(h, h^g) : h \in G\} < G^2$. Then H_g is locally connected and definable over $\mathrm{cl}_P(\emptyset)$ by Proposition 4.2.1. But there is a definable bijection between cosets $gZ(G)$ of the centre and groups of the form H_g, for $g \in G$. Hence $U_P(G/Z(G)) = 0$, which means that $G^P \leq Z(G)$ must be abelian. Finally, G is P-semi-regular iff it is P-internal and $G^0 = G^P$. $\square$

Corollary 4.2.3 *A one-based group is abelian-by-finite.*

Proof: If P is the family of all types, then $G^P = G^0$ is abelian by Theorem 4.2.2, and contained in a definable abelian supergroup of finite index in G. $\square$

Theorem 4.2.4 *Suppose G is P-analysable. Then G^P is nilpotent.*

Proof: By Lemma 3.1.3 there is a sequence $(G_i : i \leq \alpha)$ of type-definable normal subgroups of G, continuous at limits, with $G = G_0$ and $G_\alpha = \{1\}$, such that every quotient G_i/G_{i+1} is P-internal.
Claim. There is a descending sequence $\alpha = k_0 \geq k_1 \geq k_2 \geq \cdots$ such that $C_G^i(G^P) \geq G_{k_i}^0$ for all $i < \omega$, and either $k_i > k_{i+1}$ or $k_i = 0$.
Proof of Claim: We use induction on i, the case $i = 0$ being trivial. So assume that $k \leq k_i$ is minimal such that $C_G^i(G^P)$ contains G_k^0. Note that k cannot be a limit ordinal: if $C_G^i(G^P) \geq G_k^0 = (\bigcap_{j<k} G_j)^0 = \bigcap_{j<k} G_j^0$, then by compactness and relative definability of $C_G^i(G^P)$ we get that $C_G^i(G^P) \geq G_j^0$ for some $j < k$, contradicting the choice of k.

Suppose $k > 0$. Put $j := k - 1$, and let H be the centralizer-connected component of G_j/G_k. For $g \in G^P$, consider the group $H_g := \{(h, h^g) : h \in H\}$. As H is P-internal and H_g is locally connected and definable relative to H^2, H_g is $\mathrm{cl}_P(\emptyset)$-definable by Proposition 4.2.1; since the different H_g are in definable bijection with the cosets of $C_{G^P}(H)$ in G^P, it follows that $U_P(G^P/C_{G^P}(H)) = 0$. By P-connectivity, H is centralized by G^P, and $[G_j^0, G^P] \leq G_k$. But this commutator group is connected by Lemma 1.1.7, whence $[G_j^0, G^P] \leq C_G^i(G^P)$. So we may take $k_{i+1} \leq j < k \leq k_i$. $\square$

Now any strictly descending sequence of ordinal numbers is finite. Therefore G^P must be nilpotent. $\square$

Proposition 4.2.5 *Suppose G is P-internal, and $p \in S_1(G)$ is a P-semi-regular stationary type of elements of G. Then p is a translate of the generic type of some $\mathrm{cl}_P(\emptyset)$-type-definable connected subgroup of G.*

Proof: We may assume that G is saturated. Let $\mathcal{G}$ be a saturated elementary superstructure of G containing a realization g of $\mathrm{gen}(G)$. If a realizes the non-forking extension p' of p to $\mathcal{G}$, put $q = \mathrm{tp}(ga/\mathcal{G})$. Note that $\mathrm{stab}(p') = \mathrm{stab}(p) =: S$ is type-definable over G. Now if σ is any automorphism of $\mathcal{G}$ fixing G, then $q = \sigma(q)$ iff $gp' = \sigma(gp')$ iff $gp' = \sigma(g)p'$ iff $\sigma(g)^{-1}gp' = p'$ iff $\sigma(g)^{-1}g \in \mathrm{stab}(p')$ iff $\sigma(g)^{-1}gS = S$ iff $\sigma(g)S = gS$ iff $\sigma(gS) = gS$. So if C is the canonical parameter for gS, then $\mathrm{Cb}(q) \subset \mathrm{acl}(G, C)$, and therefore $ga \downarrow_{G,C} \mathcal{G}$. Furthermore q is a translate of the P-semi-regular type p' and therefore P-semi-regular itself, whence $ga \downarrow_{\mathcal{G}} \mathrm{cl}_P(\mathcal{G})$. As $\mathrm{tp}(ga)$ is P-internal, q must be based on $\mathrm{cl}_P(ga)$ by local modularity, so C is definable over $G \cup \mathrm{cl}_P(ga)$ as the coset $\{s \in \mathcal{G} : sp' = q\}$. As $a \downarrow_G ga$ and $\mathrm{tp}(a/G) = p$ is foreign to all types of U_P-rank zero, we get $a \downarrow_G ga, C$, whence $ga \downarrow_{G,C} a$. Therefore $\mathrm{tp}(ga/\mathcal{G})$ and $\mathrm{tp}(ga/G, C, a)$ are both non-forking extensions of $\mathrm{stp}(ga/G, C)$; because $ga \in (gS)a$ and gS is both C- and $\mathcal{G}$-definable, we can find $a' \in \mathcal{G}$ with $ga \in (gS)a'$ by saturation of $\mathcal{G}$. Then $a \in Sa'$, and $a \downarrow_G a'$. By saturation of G we find $h \in G$ with $a \in Sh$, so $p(x) \vdash x \in Sh$.

Now $S = \mathrm{stab}(ph^{-1})$ as well, and ph^{-1} is a generic type for S by Proposition 2.1.24. So S is connected by Proposition 2.1.20, hence $\mathrm{cl}_P(\emptyset)$-definable by Proposition 4.2.1, and p is the unique generic type of Sh. $\square$

Under suitable circumstances, we may interpret these subgroups as endomorphisms and use them to describe the structure of forking on generic elements. We shall first have to generalize the notion of an endomorphism.

Definition 4.2.1 Let A be a connected abelian group, and $\mathfrak{S}$ a family of relatively definable subgroups of A closed under finite sums and finite extensions. An $\mathfrak{S}$-endomorphism of A, or *endogeny*, is an endomorphism of $A/\bigcup\mathfrak{S}$.

Remark 4.2.1 Note that $\bigcup\mathfrak{S}$ is a subgroup of A, which in a saturated model is proper iff $A \notin \mathfrak{S}$. We shall call an endogeny r *definable* if there are a supergroup A_0 of A with $A = A_0^0$, a definable subgroup $S \leq A_0$ with $S \cap A \in \mathfrak{S}$, and a definable endomorphism $r_0 : A_0 \to A_0/S$ which induces r; if A_0 and S are only type- or relatively definable, so is the endogeny r.

Theorem 4.2.6 *Suppose G is connected, with locally modular regular generic type p. Let $\mathfrak{D}$ be the collection of relatively definable proper subgroups of G, and R the ring of $\mathfrak{D}$-endogenies of G which are type-definable over $\mathrm{cl}_p(\emptyset)$. Then R is a division ring, and for every $\bar{g} = (g_0, g_1, \ldots, g_n)$ in G, the tuple is dependent iff there are $r_i \in R$, not all zero, with $\sum_{i=0}^n r_i g_i \subset \mathrm{cl}_p(\emptyset)$.*

Proof: Note first that G is abelian by Corollary 3.6.7. Furthermore, any subgroup $H \in \mathfrak{D}$ has U_p-rank zero, H^{lc} is $\mathrm{cl}_p(\emptyset)$-definable by Proposition 4.2.1, and $\mathfrak{D}$ is closed under sums and finite extensions. Hence $\mathfrak{D}$-endogenies are well-defined, and are either trivial or surjective, hence invertible. It follows that R is a division ring. Clearly, if $\sum r_i g_i \subset \mathrm{cl}_P(\emptyset)$ for some non-trivial endogenies r_i, then $\bar{g}$ is dependent.

For the other direction, suppose $\bar{g}$ is dependent; we may assume that every proper subtuple of $\bar{g}$ is independent. Consider $\mathrm{tp}(\bar{g}/\mathrm{cl}_P(\emptyset))$. This type is P-semi-regular; by Proposition 4.2.5 there is a $\mathrm{cl}_P(\emptyset)$-type-definable connected subgroup H of G^{n+1} such that $\mathrm{tp}(\bar{g})$ is the generic type of a coset $H + \bar{h}$, for some $\bar{h} \perp \bar{g}$. Put

$$r_i = \{(x_i, x_n) : (0, \dots, 0, x_i, 0, \dots, 0, x_n) \in H\}.$$

Claim. r_i induces an $\mathfrak{D}$-endomorphism.

Proof of Claim: As $(g_0, \dots, g_{n-1})$ is independent over $\bar{h}$, for any i with $0 \le i \le n - 1$ and any independent generic $g \in G$ there is $g' \in G$ with $(g_0, \dots, g_{i-1}, g, g_{i+1}, \dots, g_{n-1}, g') \in H + \bar{h}$, whence $(0, \dots, 0, g_i - g, 0, \dots, 0, g_n - g') \in H$. As $g_i - g$ is again generic for G, we see that every r_i projects surjectively onto the first co-ordinate. Similarly, it projects surjectively onto the second co-ordinate and must be non-zero. In fact, since $(g_0, \dots, g_{n-1})$ is independent, the projection of H to the first n co-ordinates is surjective, so if $r_i = G \times G$ for some $i < n$, then $H = G^{n+1}$ and $\bar{g}$ would be independent. Therefore $r_i(0)$ is a proper subgroup of G which is contained in a relatively $\mathrm{cl}_P(\emptyset)$-definable proper subgroup S, and r_i is a homomorphism from G to G/S. $\square$

Note that $S = \{x \in G : (\bar{0}, x) \in H\}$ does not, in fact, depend on i. Let $r_n(x) = -x$, and put $h = \sum_{i=0}^n r_i(h_i)$. Then $\bar{g} \in H + \bar{h}$ implies

$$\left[\sum_{i=0}^n r_i(g_i)\right] - h = \sum_{i=0}^n r_i(g_i - h_i) = S \subset \mathrm{cl}_p(\emptyset);$$

as $h \perp_{\mathrm{cl}_p(\emptyset)} \bar{g}$, we get $h \in \mathrm{cl}_p(\emptyset)$. $\square$

Together with the construction of a group for a non-trivial locally modular family P in the next section, this may be used to describe forking on an arbitrary regular locally modular non-trivial type.

For groups, this yields a characterization of one-basedness:

Proposition 4.2.7 *A group G is one-based iff every definable subset of G^n (for all $n < \omega$) is a Boolean combination of cosets of $\mathrm{acl}(\emptyset)$-definable subgroups.*

Hence a one-based group is an abelian structure.

Proof: Suppose every definable subset is a Boolean combination of cosets of $\mathrm{acl}(\emptyset)$-definable groups. Consider a model $\mathfrak{M}$ and a type $p \in S_n(\mathfrak{M})$ realized by $\bar{a}$, and put $A = \mathrm{acl}(\bar{a}) \cap \mathrm{acl}(\mathfrak{M})$. Let H be an $\mathrm{acl}(\emptyset)$-definable subgroup of G^n. If $\bar{a}H$ intersects $\mathfrak{M}^n$ in $\bar{m}$, then clearly $\bar{m}H \in A$. Therefore the set $d_p\bar{x}\,\bar{x}H = \bar{y}H$ is A-definable (either as $\emptyset$ or as $\bar{m}H$). As every formula is a Boolean combination of cosets, p is definable over A and the theory is one-based.

Conversely, suppose that G is one-based. We may assume that G is $|T|^+$-saturated, and consider a type $p \in S_n(G)$. Then p is a translate of the generic type of some $\mathrm{acl}(\emptyset)$-definable connected subgroup S by Proposition 4.2.5. Therefore, if $p, q \in S_1(G)$ agree on all cosets of $\mathrm{acl}(\emptyset)$-definable subgroups, then $p = q$. But this implies that every formula is a Boolean combination of cosets of $\mathrm{acl}(\emptyset)$-definable subgroups of G. $\square$

We shall now come back to Example 0.3.1 of the introduction and prove that every abelian structure is one-based.

Theorem 4.2.8 *Suppose A is an abelian structure. Then A is one-based.*

Proof: We define a *positive primitive formula* to be of the form

$$\varphi(\bar{x}, \bar{a}) := \exists \bar{y}\ (\bar{x}\bar{y}) \in (H + \bar{a}),$$

where H is an $\mathrm{acl}(\emptyset)$-definable subgroup of A^n for some appropriate $n < \omega$. Clearly $\varphi(\bar{x}, 0)$ defines a subgroup of A^m (where m is the length of the tuple $\bar{x}$) over $\mathrm{acl}(\emptyset)$, and $\varphi(\bar{x}, \bar{a})$ defines a coset of that subgroup. We claim that every formula is equivalent to a Boolean combination of positive primitive formulæ. Inductively, it is enough to show that we can eliminate a single existential quantifier. First we shall prove Neumann's Lemma:

Claim. Suppose a group H is covered by finitely many cosets C_i of subgroups H_i, say $H = \bigcup_{i<n} C_i$. Let $I \subseteq n$ be the set of indices such that H_i has finite index in H. Then $H = \bigcup_{i \in I} C_i$, and $\sum_{i \in I}(1/|H : H_i|) \geq 1$.

Proof of Claim: First we prove by induction on n that we may omit all cosets of subgroups of infinite index. Clearly, the base cases $n = 1, 2$ are trivial. So consider a covering $H = \bigcup_{i<n} C_i$. Then every coset C of H_n different from C_n is covered by $\bigcup_{i<n}(C_i \cap C)$, and by the inductive hypothesis we may omit (in the covering of C, and consequently in the covering of H) all those cosets C_i such that $H_i \cap H_n$ has infinite index in H_n. If we can omit something, we are done by the inductive hypothesis; otherwise the index $H_i \cap H_j$ must be finite in H_i and H_j for all $i, j \leq n$ by symmetry. But then the intersection $N := \bigcap_{i<n} H_i$ has finite index in H_i for all $i \leq n$, so finitely many cosets of N cover H. But this means that N, and consequently all H_i, must have finite index in H.

Second, if H is covered by cosets C_i of subgroups H_i of finite index, we consider the number n_i of cosets of $N := \bigcap_{i \in I} H_i$ in every C_i, namely $n_i =$

$|H : N|/|H : H_i|$. As we need at least $|H : N|$ cosets of N to cover H, the inequality follows. $\Box$

Claim. A conjunction of positive primitive formulae is equivalent to a positive primitive formula.
Proof of Claim: Consider the formula

$$\bigwedge_{i \in I} \exists \bar{y}_i \ (\bar{x}, \bar{y}_i) \in (H_i + \bar{z}_i).$$

This is logically equivalent to

$$(\exists \bar{y}_i)_{i \in I} \ (\bar{x}, \bar{y}_i)_{i \in I} \in \bigoplus_{i \in I}(H_i + \bar{z}_i). \ \Box$$

Claim. The formula $\vartheta(\bar{x}) := \exists v \, \varphi(v, \bar{x}) \wedge \bigwedge_{i<n} \neg \psi_i(v, \bar{x})$, where φ and the ψ_i (for $i < n$) are positive primitive, is equivalent to a Boolean combination of positive primitive formulæ.
Proof of Claim: We have to find conditions on $\bar{x}$ such that the cosets ψ_i do not cover φ. Clearly, we may assume that the ψ_i define subcosets of φ. By Neumann's Lemma, we may omit all those ψ_i such that the corresponding subgroup H_i has infinite index in the subgroup H corresponding to φ (this is independent of $\bar{x}$ and depends only on the theory of A and the formulæ involved). If $N = \bigcap_{i<n} H_i$, then for every non-empty $I \subset n$ there is some $n_I < \omega$, again independent of $\bar{x}$, such that $\bigcap_{i \in I} \psi_i$ either is empty, or contains exactly n_I cosets of N; the required positive primitive condition on $\bar{x}$ now is that the $\psi_i(v, \bar{x})$ are not sufficiently disjoint to contain $|H : N|$ distinct cosets of N. Put $\alpha_I(\bar{x}) = 0$ if $\bigcap_{i \in I} \psi_i(v, \bar{x})$ is empty, and $\alpha_I(\bar{x}) = 1$ otherwise. Then the $\psi_i(v, \bar{x})$ cover $\varphi(v, \bar{x})$ iff

$$\sum_{\emptyset \neq I \subset n} (-1)^{|I|+1} \alpha_I(\bar{x}) n_I = |H : N|.$$

Let S be the set of sequences $(\alpha_I : \emptyset \neq I \subset n)$ with $\sum_{\emptyset \neq I \subset n}(-1)^{|I|+1}\alpha_I n_I = |H : N|$. Then ϑ is equivalent to

$$\neg \bigvee_{(\alpha_I) \in S} \Big[\bigwedge_{\alpha_I=1} \exists v \bigwedge_{i \in I} \psi_i(v, \bar{x}) \wedge \bigwedge_{\alpha_I=0} \neg \exists v \bigwedge_{i \in I} \psi_i(v, \bar{x}) \Big],$$

which is a Boolean combination of positive primitive formulæ by the second claim. $\Box$

As $\exists v \, (\varphi(v) \vee \psi(v))$ is equivalent to $\exists v \, \varphi(v) \vee \exists v \, \psi(v)$, this proves the quantifier elimination. So every formula is equivalent to a Boolean combination of cosets of $\mathrm{acl}(\emptyset)$-definable subgroups, and A is one-based by Proposition 4.2.7. $\Box$

4.3 The Group Configuration

In this section, we shall describe the construction of a group from geometric data. P will denote an $\emptyset$-invariant family of stationary types.

Definition 4.3.1 P is *trivial* if over any P-closed set of parameters any three pairwise independent P-analysable elements are independent. A theory is *trivial* if the family of all types is trivial.

Note that the generic type of a stable group is never trivial: if x and y are two independent realizations of the principal generic type, then x, y and xy are pairwise independent, but not independent.

Lemma 4.3.1 *Suppose P is locally modular and non-trivial. Then over some parameter set there are sets a_1, a_2, a_3, b_1, b_2 and b_3 such that*

- *in the triples (a_1, a_2, a_3), (a_1, b_2, b_3), (b_1, a_2, b_3) and (b_1, b_2, a_3) any co-ordinate is in the P-closure of the other two,*

- *all other triples and all pairs of sets are independent, and*

- *$\operatorname{tp}(a_1 a_2 a_3 b_1 b_2 b_3)$ is P-analysable of non-zero U_P-rank, and foreign to all types of U_P-rank zero.*

Note that these conditions imply that every co-ordinate of the 6-tuple has non-zero U_P-rank: if $\operatorname{tp}(a/A)$ is P-analysable, then $U_P(a/A) = 0$ iff $a \in \operatorname{cl}_P(A)$; as soon as one of the co-ordinates of the 6-tuple has non-zero U_P-rank, the pairwise independence together with the P-closure conditions implies that none of the others has U_P-rank zero either. The conditions are best visualized in a diagram, where lines indicate dependence:

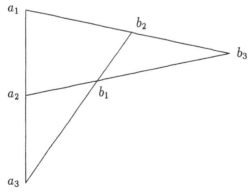

Proof of Lemma: Consider three elements a, b, c which are P-analysable over some P-closed set X, and pairwise independent but not independent. Let $(a_i : i \leq \alpha) \subset \operatorname{acl}(aX)$ be a P-analysis of a. Let i be minimal such that

$(a_j : j \leq i) \not\perp_X bc$. Then $(a_j : j < i) \perp_X bc$ and hence $X' := \mathrm{cl}_P(Xa_j : j < i) \perp_X bc$ by Lemma 4.0.3 and the P-closedness of X. Therefore $a_i \perp_{X'} b$, $b \perp_{X'} c$ and $c \perp_{X'} a_i$, but $a_i \not\perp_{X'} bc$. We may thus replace X by X' and a by a_i, and assume that $\mathrm{tp}(a/X)$ is P-internal, whence P-semi-regular. Similarly, we may suppose $\mathrm{tp}(b/X)$ and $\mathrm{tp}(c/X)$ are P-semi-regular. We incorporate X – or rather a base for $\mathrm{tp}(abc/X)$ – into the language.

Put $A := \mathrm{cl}_P(bc) \cap \mathrm{cl}_P(a)$, $B := \mathrm{cl}_P(ca) \cap \mathrm{cl}_P(b)$ and $C := \mathrm{cl}_P(ab) \cap \mathrm{cl}_P(c)$. Then by local modularity, $a \perp_A bc$, and by P-semi-regularity $a \perp_A \mathrm{cl}_P(b)c$. Let $B' := \mathrm{cl}_P(cA) \cap \mathrm{cl}_P(b)$; then $b \perp_{B'} cA$. Furthermore $a \perp_{AB'c} b$, whence $b \perp_{B'} ca$; by Lemma 4.0.3 we get $\mathrm{cl}_P(b) \perp_{\mathrm{cl}_P(B')} \mathrm{cl}_P(ca)$, and as B' is P-closed, we must have $B' = B$. Similarly $C = \mathrm{cl}_P(c) \cap \mathrm{cl}_P(Ab)$. So we may replace a by A and retain the intersections of the P-closures. It follows that in the triple (A, B, C) any co-ordinate is in the P-closure of the other two, and none is in $\mathrm{cl}_P(\emptyset)$, as this would contradict the original dependence of (a, b, c). Furthermore $\mathrm{tp}(ABC)$ is P-analysable, and (A, B, C) is pairwise independent over $\mathrm{cl}_P(\emptyset)$, since e.g. $a \perp b$ implies $\mathrm{cl}_P(a) \perp_{\mathrm{cl}_P(\emptyset)} \mathrm{cl}_P(b)$.

Let $a^0 \in A - \mathrm{cl}_P(\emptyset)$, $b^0 \in B - \mathrm{cl}_P(\emptyset)$ and $c^0 \in C - \mathrm{cl}_P(\emptyset)$. Suppose we have subsets $a^i \subset A$, $b^i \subset B$ and $c^i \subset C$ for $i \leq n$ such that $a^i \in \mathrm{cl}_P(b^{i+1}c^{i+1})$ (and similarly for the other co-ordinates) for all $i < n$; then clearly we can extend this sequence one further step, and hence find such a sequence of length ω. Put $a_1 = \bigcup_{i<\omega} a^i$, $a_2 = \bigcup_{i<\omega} b^i$, and $a_3 = \bigcup_{i<\omega} c^i$, and add a base for $\mathrm{tp}(a_1a_2a_3/\mathrm{cl}_P(\emptyset))$ to the language. Then the triple (a_1, a_2, a_3) is P-analysable, independent, not in $\mathrm{cl}_P(\emptyset)$, and any co-ordinate is in the P-closure of the other two.

Suppose (b_2, b_3) realizes $\mathrm{stp}(a_2, a_3/a_1)$ independently of a_2a_3. Then $a_2 \perp_{a_1} b_2b_3$ and $a_2 \perp a_1$ imply $a_2 \perp b_2b_3$; similarly $a_3 \perp b_2b_3$, and also a_2a_3 is independent of b_2 and of b_3. Put $B'' := \mathrm{cl}_P(a_3b_2) \cap \mathrm{cl}_P(a_2b_3)$; as $a_2a_3b_2b_3$ is not independent, $B'' \not\subseteq \mathrm{cl}_P(\emptyset)$. Then $b_3 \perp b_2a_3$, and Lemma 4.0.3 implies $\mathrm{cl}_P(b_3) \perp_{\mathrm{cl}(\emptyset)} B''$. Furthermore $a_2a_3 \perp b_3$ and $a_2 \perp a_3$ yield $a_3b_3 \perp_{b_3} a_2b_3$ and hence $a_3b_3 \perp_{\mathrm{cl}_P(b_3)} B''$, and $b_2b_3 \perp_{b_3} a_2b_3$ yields $b_2b_3 \perp_{\mathrm{cl}_P(b_3)} B''$ (again using Lemma 4.0.3), so B'' is independent of a_3b_3 and of b_2b_3 over $\mathrm{cl}_P(\emptyset)$, and hence over $\emptyset$ by P-semi-regularity of a_3b_3 and of b_2b_3. The independences $B'' \perp a_2b_2$ and $B'' \perp a_2a_3$ follow similarly; P-semi-regularity and $a_1 \subset \mathrm{cl}_P(a_2a_3) \cap \mathrm{cl}_P(b_2b_3)$ yield the remaining independence requirements.

Put $b_1 = \mathrm{Cb}(a_1a_2a_3b_2b_3/B'')$. Since $a_3b_2 \perp_{B''} a_2b_3$ by Lemma 4.1.3, we get $a_3b_2 \perp_{b_1} a_2b_3$ and hence $\mathrm{cl}_P(a_3b_2b_3) \perp_{\mathrm{cl}_P(b_1b_3)} a_2$ by Lemma 4.0.3. Since $a_1 \in \mathrm{cl}_P(b_2b_3)$ and $a_2 \in \mathrm{cl}_P(a_1a_3)$, we have $a_2 \in \mathrm{cl}_P(a_3b_2b_3)$ and hence $a_2 \in \mathrm{cl}_P(b_1b_3)$. The other dependences follow similarly, so we may add $\mathrm{acl}(a_1a_2a_3b_1b_2b_3) \cap \mathrm{cl}_P(\emptyset)$ to the language, and obtain a 6-tuple satisfying all the requirements. $\square$

Note that we can actually get another element c' in $\mathrm{cl}_P(a_1b_1) \cap \mathrm{cl}_P(a_3b_3)$ (with all other triples independent): since $a_1b_1 \not\perp a_3b_3$, the intersection $C'' := \mathrm{cl}_P(a_1b_1) \cap \mathrm{cl}_P(a_3b_3)$ is non-trivial. As above for B'', the triples (a_1, b_1, C'') and (a_3, b_3, C'') have pairwise independent co-ordinates, and any one co-ordinate is in the P-closure of the other two. All remaining triples are independent, and we may again replace C'' by a subset c.

In the group example from above, consider a third principal generic element z independent of x and y. Then $(a_1, a_2, a_3, b_1, b_2, b_3)$ corresponds to (x, y, xy, zxy, z, zx); in the abelian case (and a locally modular group is abelian-by-finite) the additional element c corresponds to yz.

Definition 4.3.2 A 6-tuple $(a_1, a_2, a_3, b_1, b_2, b_3)$ forms a *group configuration* with respect to P, or a *P-configuration*, if

1. any one of a_1, a_2, a_3 is in the P-closure of the other two,

2. $b_i \in \mathrm{cl}_P(a_j, b_k)$ for $\{i, j, k\} = \{1, 2, 3\}$,

3. all other triples and all pairs are independent, and

4. the type $\mathrm{tp}(a_1a_2a_3b_1b_2b_3)$ is P-analysable, foreign to all types of U_P-rank zero, and every co-ordinate has non-zero U_P-rank.

Lemma 4.3.2 *Suppose* $(a_1, a_2, a_3, b_1, b_2, b_3)$ *forms a P-configuration. Then there is a P-configuration* $(a'_1, a'_2, a'_3, b'_1, b'_2, b'_3)$ *over an independent parameter set* A *with* $\mathrm{cl}_P(a_iA) = \mathrm{cl}_P(a'_iA)$ *and* $\mathrm{cl}_P(b_iA) = \mathrm{cl}_P(b'_iA)$ *for* $i = 1, 2, 3$, *and such that* $b'_1 \in \mathrm{dcl}(a'_3b'_2)$, $b'_3 \in \mathrm{dcl}(a'_1b'_2)$, *and* $b'_2 \in \mathrm{dcl}(a'_3b'_1) \cap \mathrm{dcl}(a'_1b'_3)$.

Proof: First we consider some a^1 realizing $\mathrm{stp}(a_1)$ independently of $a_1a_3b_1b_2$. Since also a_1 is independent of $a_3b_1b_2$, there is a^2b^3 such that $\mathrm{stp}(a_1a_2a_3b_1b_2b_3) = \mathrm{stp}(a^1a^2a_3b_1b_2b^3)$. Note that $a_2 \perp a_3b_2$ implies $a_2 \perp a_3b_1b_2$, whence $a_2b_1 \perp_{b_1} a_3b_2$ and therefore $a_2b_3 \perp_{b_1} a_3b_2$. This yields $a^2b^3 \perp_{b_1} a_3b_2$, and $\mathrm{cl}_P(a^2b^3) \cap \mathrm{cl}_P(a_3b_2) = \mathrm{cl}_P(b_1)$ by Lemma 4.0.3.

Let $b' = \mathrm{Cb}(b_1/a^2a_3b_2b^3)$, and let b^1 be the set of conjugates of b' over $a^2a_3b_2b^3$. Then $b' \subset \mathrm{acl}(a^2a_3b_2b^3)$ (i.e. every element in b' is algebraic over $a^2a_3b_2b^3$), and $b^1 \subset \mathrm{dcl}(a^2a_3b_2b^3)$. Furthermore, b' is algebraic in a Morley sequence in $\mathrm{stp}(b_1/a^2a_3b_2b^3)$, and every conjugate of b' is algebraic in a conjugated Morley sequence. Hence b^1 is algebraic in an independent sequence I of realizations of $\mathrm{tp}(b_1/a^2a_3b_2b^3)$. As $b' \in \mathrm{acl}(b^1)$, we have $b_1 \perp_{b^1} a^2a_3b_2b^3$, so $b_1 \in \mathrm{cl}_P(a_3b_2)$ implies $b_1 \in \mathrm{cl}_P(b^1)$ by Lemma 4.0.3. On the other hand, every element of I lies in $\mathrm{cl}_P(a^2b^3) \cap \mathrm{cl}_P(a_3b_2) = \mathrm{cl}_P(b_1)$, so $b^1 \in \mathrm{cl}_P(b_1)$. As a^1 was independent of the original group configuration and $b^3 \in \mathrm{cl}_P(a^1b_2)$ and $a^2 \in \mathrm{cl}_P(a^1a_3)$, it follows that over $\mathrm{cl}_P(a^1)$ the 6-tuple $(a_1, a_2, a_3a^2, b^1, b_2b^3, b_3)$ forms a P-configuration whose sets have the same P-closure (over $\mathrm{cl}_P(a^1)$) as the original sets, and which in addition satisfies $b^1 \in \mathrm{dcl}(a_3 a^2, b_2 b^3)$. We may also replace the class $\mathrm{cl}_P(a^1)$ by the subset $\mathrm{cl}_P(a^1) \cap \mathrm{acl}(a^1 a_1 a_2 a_3 a^2 b^1 b_2 b^3 b_3)$.

Note that this construction left b_3 unchanged. Hence a similar construction (with indices 1 and 3 exchanged) yields a parameter set A and a group configuration $(a_1^0, a_2^0, a_3^0, b_1^0, b_2^0, b_3^0)$ over A with $\mathrm{cl}_P(a_i A) = \mathrm{cl}_P(a_i^0 A)$ and $\mathrm{cl}_P(b_i A) = \mathrm{cl}_P(b_i^0 A)$ for $i = 1, 2, 3$, and such that $b_1^0 \in \mathrm{dcl}(a_3^0 b_2^0)$ and $b_3^0 \in \mathrm{dcl}(a_1^0 b_2^0)$. We incorporate A into the language and omit the superscript ^{0}.

Let $b' = \mathrm{Cb}(b_2 / a_1 a_3 b_1 b_3)$ and b_2' be the set of conjugates of b' over $a_1 a_3 b_1 b_3$. Then $b_2' \subset \mathrm{dcl}(a_1 a_3 b_1 b_3)$, and b_2' must be algebraic in an independent sequence I of realizations of $\mathrm{tp}(b_2 / a_1 a_3 b_1 b_3)$. Now every element of I lies in $\mathrm{cl}_P(a_1 b_3) \cap \mathrm{cl}_P(a_3 b_1) = \mathrm{cl}_P(b_2)$, whence $b_2' \subset \mathrm{cl}_P(b_2)$. On the other hand, $b' \subset \mathrm{acl}(b_2')$, whence $b_2 \underset{b'}{\downarrow} a_1 b_3$, and $b_2 \subset \mathrm{cl}_P(a_1 b_3)$ yields $b_2 \subset \mathrm{cl}_P(b_2')$ by Lemma 4.0.3. Since $b_1 \subset \mathrm{dcl}(b_2 a_3)$, Lemma 0.2.23 yields $b_1 \subset \mathrm{dcl}(b' a_3)$. As b_2' is a set of conjugates of b' over $a_3 b_1$, clearly $b_1 \subset \mathrm{dcl}(b_2' a_3)$; similarly we get $b_3 \subset \mathrm{dcl}(b_2' a_1)$.

Now let a_2^1 be a realization of $\mathrm{stp}(a_2)$ independent of the configuration. Then there are $a_1^1 b_3^1$ and $a_3^1 b_1^1$ such that

$$\mathrm{stp}(a_1 a_2 a_3 b_1 b_2' b_3) = \mathrm{stp}(a_1^1 a_2^1 a_3 b_1 b_2' b_3^1) = \mathrm{stp}(a_1 a_2^1 a_3^1 b_1^1 b_2' b_3).$$

Put $a_1' = a_1 a_3^1$, $a_2' = a_2$, $a_3' = a_3 a_1^1$, $b_1' = b_1 b_3^1$, and $b_3' = b_3 b_1^1$. Then it is easy to check – and follows immediately from conditions 1. and 2. – that over $\mathrm{cl}_P(a_2^1)$ the new elements have the same P-closures as the old ones, so they form a P-configuration. Furthermore

- $b_1 \subset \mathrm{dcl}(a_3 b_2')$ and $b_3^1 \subset \mathrm{dcl}(a_1^1 b_2')$, whence $b_1' \subset \mathrm{dcl}(a_3' b_2')$,

- $b_3 \subset \mathrm{dcl}(a_1 b_2')$ and $b_1^1 \subset \mathrm{dcl}(a_3^1 b_2')$, whence $b_3' \subset \mathrm{dcl}(a_1' b_2')$,

- $b_2' \subset \mathrm{dcl}(a_1 a_3^1 b_1^1 b_3) = \mathrm{dcl}(a_1' b_3')$ and $b_2' \subset \mathrm{dcl}(a_1^1 a_3 b_1 b_3^1) = \mathrm{dcl}(a_3' b_1')$.

We add $\mathrm{acl}(a_1' a_2' a_3' b_1' b_2' b_3' a_2^1) \cap \mathrm{cl}_P(a_2^1)$ to the language and obtain a P-configuration with all the required properties. $\square$

Theorem 4.3.3 *Let $(a_1, a_2, a_3, b_1, b_2, b_3)$ be a P-configuration. Then there is a type-definable P-analysable P-connected group G acting transitively on a set X, such that for a generic element $x \in X$ (with respect to that action) there is b realizing $\mathrm{stp}(b_1)$ such that (possibly over some additional parameters) $\mathrm{cl}_P(x) \subseteq \mathrm{cl}_P(b)$.*

Proof: We may assume that the group configuration satisfies the conclusion of Lemma 4.3.2. Let $p_i = \mathrm{stp}(a_i)$ and $q_i = \mathrm{stp}(b_i)$ for $i = 1, 2, 3$. For any $a \models p_3$ there is an invertible function $f_a : q_1 | a \to q_2 | a$, and for every $b \models p_1$

there is an invertible function $g_b : q_2|b \to q_3|b$, such that if (b, c, a, x) has the same strong type as (a_1, a_2, a_3, b_1), then $(b, c, a, x, f_a(x), g_b(f_a(x)))$ has the same strong type as $(a_1, a_2, a_3, b_1, b_2, b_3)$.

Recall that two functions on a strong type have the same *germ* if they agree on some (any) element independent of their canonical parameters, and that having the same germ is a congruence with respect to composition, as long as the functions have the property that if x is independent of f, so is $f(x)$. Let F be the set of germs of functions of the form f_a for $a \models p_3$, and F' the set of germs of functions of the form g_b for $b \models p_1$. Then these functions are type-definable (on an infinite sequence of variables) and have the required property; furthermore the equivalence relations which yield the germs, namely $d_{q_1} x \, f_a(x) = f_{a'}(x)$ and $d_{q_2} x \, g_b(x) = g_{b'}(x)$, are also type-definable.

Claim.

1. If $f \in F$ and $g \in F'$ are independent, then $g \circ f$ is independent of f and of g.

2. If f and f' are independent in F, then $f^{-1} \circ f'$ is independent of f and of f'.

3. If $f_0, f_1, f_2, f_3 \in F$ are independent, then there are independent $f_4, f_5 \in F$ such that $f_0^{-1} \circ f_1 \circ f_2^{-1} \circ f_3 = f_4^{-1} \circ f_5$, and $f_4^{-1} \circ f_5$ is independent of both $f_0^{-1} \circ f_1$ and $f_2^{-1} \circ f_3$.

Proof of Claim:

1. Suppose f corresponds to a_3 and g to a_1 in a group configuration $(a_1, a_2, a_3, b_1, b_2, b_3)$. Then for any generic $b \models q_1|a_1 a_3$ we have $(g \circ f)(b) \in \mathrm{cl}_P(ba_2)$. Now if $(b^i : i < \omega)$ is a Morley sequence in q_1 over $a_1 a_3$, then $g \circ f \in \mathrm{acl}(b^i, (g \circ f)(b^i) : i < \omega) \subseteq \mathrm{cl}_P(a_2, b^i : i < \omega)$. But $a_1 a_3 \underset{a_2}{\bigcup} (b^i : i < \omega)$, whence $g \circ f \underset{\mathrm{cl}_P(a_2)}{\bigcup} \mathrm{cl}_P(a_2, b^i : i < \omega)$ and $g \circ f \in \mathrm{cl}_P(a_2)$. The first assertion now follows from the independence of $\mathrm{cl}_P(a_2)$ of a_1 and of a_3.

2. Let $g \in F'$ be independent of f, f'. Then $f^{-1} \circ g^{-1}$ is independent of g, f', whence $f^{-1} \circ g^{-1} \underset{g \circ f'}{\bigcup} f'$. Furthermore $g \circ f' \bigcup f'$, and hence $f^{-1} \circ g^{-1}, g \circ f' \bigcup f'$, so $f^{-1} \circ f' \bigcup f'$. The other independence follows similarly.

3. Let f be independent of f_0, f_1, f_2, f_3. Then by 2. there are f_4 and f_5 independent of f such that $f_0^{-1} \circ f_1 = f_4^{-1} \circ f$ and $f_2^{-1} \circ f_3 = f^{-1} \circ f_5$; as $f_0^{-1} \circ f_1 \underset{f}{\bigcup} f_2^{-1} \circ f_3$, we may also choose $f_0, f_1, f_4 \underset{f}{\bigcup} f_2, f_3, f_5$. Thus

$$f_0^{-1} \circ f_1 \circ f_2^{-1} \circ f_3 = f_4^{-1} \circ f \circ f^{-1} \circ f_5 = f_4^{-1} \circ f_5,$$

and clearly $f_4 \perp f_5$. Furthermore f_4 is independent of $f_4^{-1} \circ f = f_0^{-1} \circ f_1$. As $f_5 \perp f, f_0, f_1, f_4$ we get $f_0^{-1} \circ f_1 \perp f_4, f_5$, and the independences follow. $\square$

Put $F'' = \{f^{-1} \circ f : f, f' \in F \text{ independent}\}$. Then F'' is (the set of realizations of) a complete stationary type, and closed under inverse and generic composition. Let $G = \{h \circ h' : h, h' \in F''\}$. If $h, h', h'' \in F''$, then we choose any $g \in F''$ independent of h, h', h''. By 3. there are independent $g, g' \in F''$ with $h' = g \circ g'$ and h' independent of g and of g'. We may in addition choose it such that $g, g' \perp_{h'} h, h''$, whence $g \perp h$ and $g' \perp h''$. By 3. again $h \circ g$ and $g' \circ h''$ are both in F'', and $h \circ h' \circ h'' = h \circ g \circ g' \circ h'' \in G$. It follows that G is closed under composition and inverses, and thus a type-definable group. We shall now omit $\circ$.

Now consider independent $h \in F''$ and $g \in G$, say $g = h_1 h_2$ with $h_1, h_2 \in F''$. We may assume that $h \perp h_1, h_2$, so by 3. first $hh_1 \in F''$ and $hh_1 \perp h_1$; since $hh_1 \perp_{h_1} h_2$ we get $hh_1 \perp h_1, h_2$, and by 3. again $hh_1 h_2 = hg \in F''$ with $hg \perp h_2$. As $hh_1 h_2 \perp_{h_2} h_1$, we get $hg \perp h_1, h_2$, whence $hg \perp g$. So F'' could be called "generic" for G.

Fix some realization $b_1 \models q_1$, and put $H = \{g \in G : d_{F''} x \, (xg)b_1 = xb_1\}$. This is well-defined, as elements of F'' act on independent realizations of q_1. If $h, h' \in H$ and $f \in F''$ is independent of h, h', then fh is in F'' and independent of h', whence $(f(hh'))b_1 = ((fh)h')b_1 = (fh)b_1 = fb_1$ and $hh' \in H$; similarly fh^{-1} is in F'' and independent of h, whence $fb_1 = ((fh^{-1})h)b_1 = (fh^{-1})b_1$ and $h^{-1} \in H$. So H is a subgroup, and G acts on $X := G/H$ by left translation.

Suppose $h \in F''$ is independent of b_1. There are independent $f, f' \in F$, independent of b_1 with $f^{-1} \circ f' = h$; since $f'(b_1) =: b_2$ realizes q_2 independently of b_1, f and $f^{-1}(b_2) = hb_1$ realizes q_1 independently of b_1, we get that the action of F'' on q_1 is generically transitive. In particular, we have a generic embedding from q_1 to X, given by $hb_1 \mapsto hH$ (for independent $h \in F''$).

We shall now deal with the fact that the domain of G consists of infinite tuples. Let π_k be the projection to the first k co-ordinates, and put

$$N_k := \{g \in G : d_{F''} x \, d_{F''} x' \, \pi_k(xgx') = \pi_k(xx')\}.$$

If $n, n' \in N_k$ are independent of the independent pair $f, f' \in F''$, then $f, nf' \in F''$ are independent and independent of n, n', whence $\pi_k(fn'n^{-1}(nf')) = \pi_k(fn'f')) = \pi_k(ff') = \pi_k(f(nf'))$, and $n'n^{-1} \in N_k$. Furthermore, if $g \in G$, $n \in N_k$, and if $f, f' \in F''$ are independent, then $fg, g^{-1}f' \in F''$ are independent and independent of g, n, whence

$$\pi_k((fg)n^g(g^{-1}f')) = \pi_k(fnf') = \pi_k(ff') = \pi_k((fg)(g^{-1}f')).$$

It follows that N_k is a normal subgroup of G. Clearly $\bigcap_{k<\omega} N_k = \{1\}$, and G is equal to the inverse limit of the system given by $G_k := G/N_k$. Furthermore, as $\pi_k(fgf')$ depends only on $\pi_i(g)$ for some $i < \omega$ for fixed f, f', already $\pi_i(g^{-1}g') = \pi_i(1)$ implies $g^{-1}g' \in N_k$; since $\pi_i(g^{-1}g')$ depends only on a finite part of g and of g', the quotient G_k is a type-definable group whose domain are finite tuples. Since it is translation-invariant and consists of a single strong type, F''/N_k is the unique generic type for G_k by Corollary 2.1.23. Hence G_k is connected; since F'' is P-analysable and foreign to all types of zero U_P-rank, G_k is P-connected and P-analysable, and acts transitively on $G/HN_k =: X_k$. Finally, if $x \in X_k$ is generic for that action, then there are $b \models q_1$ and $f \in F''$ with $x = fHN_k$ and $b = fb_1$, and $\mathrm{cl}_P(x) \subseteq \mathrm{cl}_P(b)$. $\square$

We should note that G acts faithfully on X, since it is a group of germs of translations.

Remark 4.3.1 In fact, any group in a stable theory given by a partial type in infinitely many variables is the inverse limit of a directed system of type-definable groups. In order to prove this, we would have to define and prove the existence of generic types in that set-up; then the last paragraph of the above proof will easily generalize.

Corollary 4.3.4 *Suppose P is locally modular and non-trivial. Then there is a definable abelian group.*

Proof: By Lemma 4.3.1 there is a P-configuration; now Theorem 4.3.3 yields a type-definable P-connected P-analysable group. By Theorem 3.1.1 it has a P-internal quotient, which is abelian-by-finite by Theorem 4.2.2; by Theorem 2.1.17 it is contained in a definable abelian supergroup. $\square$

If $(a_1, a_2, a_3, b_1, b_2, b_3)$ is a P-configuration and $U_P(b_1)$ is finite, it is easy to see that $U_P(b_1) = U_P(b_2) = U_P(b_3)$, and that by choosing k sufficiently large, we get $U_P(X_k) = U_P(b_1)$.

Hrushovski has classified the possible cases for $U_P(b_1) = 1$; little is known about bigger values.

Theorem 4.3.5 *Suppose the P-connected group G acts faithfully and transitively on a set X of U_P-rank one. Then one of the following holds:*

1. $U_P(G) = 1$, G is abelian and the action is simply transitive.

2. $U_P(G) = 2$, X is the affine line over a type-definable algebraically closed field K and G acts as $\mathrm{AGL}_1(K)$.

3. $U_P(G) = 3$, X is the projective line over a type-definable algebraically closed field K and G acts as $\mathrm{PSL}_2(K)$.

Proof: If $U_P(G) = 1$, then G is abelian by Corollary 3.6.7. Let $x \in X$ and $g \in C_G(x)$. For any $y \in X$ there is $h \in G$ with $y = hx$, but then $gy = ghx = hgx = hx = y$, and $g \in C_G(y)$ as well. So g stabilizes the whole of X, and must be trivial by faithfulness.

Next assume $U_P(G) = 2$, so G is soluble by Corollary 3.8.2. Consider any $x \in X$ and its stabilizer $C := C_G(x)$. As G acts transitively on X, the quotient G/C is isomorphic to X and has U_P-rank one. Therefore $U_P(C) = 1$, and C^P is abelian. Suppose C^P is normal, and consider any $y \in X$ and $g \in G$ with $y = gx$. For any $c \in C^P$ also $c^g \in C^P$, and $cy = cgx = gc^g x = gx = y$; as G acts faithfully, C^P must be trivial, a contradiction. It follows that G is not abelian, and $N := \mathrm{dc}(G')$ is a normal abelian subgroup of G, which is P-connected by Lemma 3.6.3 and hence has U_P-rank one. $U_P(N \cap C)$ cannot be one, as otherwise $N = C^P$ would be normal. Therefore $U_P(N \cap C) = 0$, so $U_P(NC^P) = 2$ and $NC^P = G$ by P-connectivity.

Consider the orbit Nx. Then

$$U_P(Nx) = U_P(N/C_N(x)) = U_P(N) - U_P(C_N(x)) = U_P(N) - 0 = 1 = U_P(X);$$

as X is in bijection with G/C and G is P-connected, there is a unique type in X of U_P-rank one, and any forking extension of it must have U_P-rank zero. It follows that Nx is a generic subset for X for any $x \in X$ and any two such subsets intersect: X is a unique orbit under the action of N, and N acts simply transitively by 1. In particular, $C_N(x) = \{1\}$, so $G = N \rtimes C$, and C is P-connected.

Now we consider the action of C on N by conjugation. Suppose $c \in C_C(N)$. For any $y \in X$ there is $n \in N$ with $nx = y$, and $cy = cnx = ncx = nx = y$ as before, whence $c = 1$ and C acts faithfully on N. On the other hand, for every $n \in N - Z(G)$ we have $C_C(n) < C$ and so $U_P(C_C(n)) = 0$ by P-connectivity. So a non-trivial orbit has rank

$$U_P(n^C) = U_P(C/C_C(n)) = U_P(C) - U_P(C_C(n)) = U_P(C) - 0 = 1 = U_P(N),$$

and as N is P-connected, n^C is generic and any two non-trivial orbits must intersect: there is just one non-trivial orbit. Furthermore, every element $m \in N$ is the sum of two generic elements n^{c_1} and n^{c_2}, with $c_1, c_2 \in C$.

Let R be the ring of endomorphisms of N generated by C. Then R is commutative; we claim that R is a definable field. Firstly, given any $r \in R$, the image $\mathrm{im}(r)$ is again P-connected, and therefore either N or zero since N is P-connected of U_P-rank one: R is an integral domain. Secondly, if $r(n) = m = n^{c_1} + n^{c_2}$, then $r - c_1 - c_1$ is an endomorphism of N whose kernel contains n and is non-trivial; therefore $r = c_1 + c_2$. It follows that R is type-definable as C^2/E, where $(c, c')E(d, d')$ iff $n^c + n^{c'} = n^d + n^{d'}$. But R is a field by Lemma 1.0.1, and algebraically closed by Theorem 3.7.2; clearly $K^+ \cong N$, and $K^\times \cong C$ by P-connectivity (and C must act on N without fixed points). Now fixing any element $x \in X$ will yield an isomorphism between N and $Nx = X$ which turns X into the affine line over K on which G acts as the group of linear transformations. Note that G is sharply 2-transitive on X.

Now assume $U_P(G) = 3$, fix $x \in X$, and consider $C := C_G(x)$. It has rank $U_P(C) = U_P(G) - U_P(G/C) = U_P(G) - U_P(X) = 3 - 1 = 2$. If for some $x' \in X$ the orbit $C^P x'$ has U_P-rank zero, then $U_P(C_{C^P}(x')) = U_P(C^P) - U_P(C^P x') = U_P(C^P)$, so x' is centralized by the whole of C^P. This cannot happen for the whole of X, since G acts faithfully, so there is one generic orbit O of U_P-rank one, and only one, since X is isomorphic to $G/C_G(x)$ and G is P-connected. Clearly, C^P acts transitively on O, and faithfully as well (since the points in $X - O$ are already stabilized by C^P and G acts faithfully on X). So O has the structure of an affine line over a field K on which C^P acts as $\mathrm{AGL}_1(K)$.

Let $D(y)$ denote the smallest intersection of centralizers which contains $C_G(y)^P$; if $g \in G$ is such that $gy = x$, then $D(y) = D(x)^g$ and $D(.)$ is uniformly definable. Consider $y \in X - O$. As $C_G(y) \geq C^P = C_G(x)^P$ and hence $D(y) \geq D(x)$, Lemma 1.0.2 implies $D(x) = D(y)$. But $X - O$ constitutes the class of x modulo the definable equivalence relation xEy iff $D(x) = D(y)$; as E is clearly G-invariant and G acts transitively, every element of x has a class of U_P-rank zero. In particular, O cannot consist of a single class. However, C^P acts 2-transitively on O, which implies that every class modulo E is reduced to a single point. Therefore $O = X - \{x\}$, and G acts sharply 3-transitively on X.

Now choose three points $0, 1, \infty$ on X such that $0, 1$ yield a parametrization of the affine line $X - \{\infty\}$ by the field K. As G is 3-transitive, there is an element $g \in G$ with $g(\infty, 0, 1) = (0, \infty, 1)$. Clearly g^2 fixes three points and must be trivial. Furthermore, g normalizes $C_G(\infty, 0) \cong K^\times$; as $K^\times$ is P-connected of U_P-rank one, g is either the identity or the map $x \mapsto 1/x$ by Corollary 3.6.7. But if $g = \mathrm{id}_{K^\times}$, then for any $k \in K^\times$ we get

$$g(k \cdot 1) = (gk) \cdot 1 = (k^{g^{-1}} g) \cdot 1 = (kg) \cdot 1 = k(g \cdot 1) = k \cdot 1,$$

so g stabilizes all but two points and must be trivial, a contradiction. Therefore g is the map $x \mapsto 1/x$. But now g and C^P generate $\mathrm{PSL}_2(K)$, which is exactly 3-transitive on the projective line $O \cup \{\infty\}$, and must thus equal G.

Finally, suppose $U_P(G) = n \geq 4$. If $x \in X$, then $U_P(C_G(x)) = n - 1$ as before, and not all orbits of $C_G(x)^P =: C$ can have U_P-rank zero: if Cy is such an orbit of rank zero, then C must centralize y, but C cannot centralize the whole of X. So there is a unique generic orbit O on which C acts faithfully and transitively. Taking iterated centralizers (and reducing X accordingly), we can reduce to the case $U_P(G) = 4$; then $U_P(C) = 3$. Considering the equivalence relation $D(x) = D(y)$ as before and using the fact that C acts sharply 3-transitively on O, we see that $O = X - \{x\}$ and G acts sharply

4-transitively.

Now consider four points $0, 1, x, y$. Then there are involutions $g, h \in G$ such that $g(0, 1, x, y) = (x, 1, 0, y)$ and $h(0, 1, x, y) = (y, 1, x, 0)$, and both g and h act on the other points (which we identify with $K^\times \cdot 1$) as inversions. Hence gh stabilizes all other points and must be trivial. But this means $g = h$, yielding the final contradiction. $\square$

Corollary 4.3.6 *A simple P-connected group of finite U_P-rank n with a subgroup of U_P-rank $n - 1$ has rank 3 and is isomorphic to $PSL_2(K)$ for some algebraically closed field K.*

Proof: G acts faithfully and transitively by translation on the right quotient G/H, and $U_P(G/H) = 1$. $\square$

4.4   Quasi-endomorphisms

In this section, we shall generalize Theorem 4.2.6 to the non-regular context. In order to do this, we shall first look at quasi-endomorphisms from a different angle.

Definition 4.4.1 Let A be a stable (type-)definable abelian group and $\mathfrak{S}$ a set of relatively definable subgroups of A of infinite index closed under finite sums and (relatively definable) subgroups. An $\mathfrak{S}$-*endomorphism* of A, or *endogeny*, is a relatively definable homomorphism r from a subgroup A_r of finite index in A to A/S_r, where $S_r \in \mathfrak{S}$.

We may also view an endogeny as a relatively definable subgroup E of $A \times A$ with $\mathrm{pr}_1(E) = A_r$ (the *domain* of E) and $S_r = \{a \in A : (0, a) \in E\}$ (the *co-kernel* of E). We shall write aE for $\{b \in A : (a, b) \in E\}$ and Ea for $\{b \in A : (b, a) \in E\}$. If E and E' are two endogenies, we define sum, difference, and product as follows:

$$
\begin{aligned}
a(E + E') &= aE + aE', \\
a(E - E') &= aE - aE', \\
a(EE') &= (aE)E'.
\end{aligned}
$$

The domain of $E \pm E'$ is $A_E \cap A_{E'}$, the domain of EE' is $A_E \cap E^{-1}(A_{E'})$.

We shall now define an equivalence relation on the set of endogenies.

Definition 4.4.2 An endogeny E is *trivial* if there is some $S_E \in \mathfrak{S}$ with $E \leq A_E \times S_E$. Two endogenies E and E' are $\mathfrak{S}$-*equivalent*, denoted by $E \equiv^{\mathfrak{S}} E'$, if there is some trivial endogeny S with $E + S = E' + S$. We call $\mathfrak{S}$ *closed* if it is closed under subgroups, finite sums, finite extensions and $\mathfrak{S}$-endomorphisms (i.e. for any $\mathfrak{S}$-endomorphism r and any element S in $\mathfrak{S}$, the image Sr also lies in S).

Lemma 4.4.1 *For an abelian group A, let $\mathfrak{F}$ be the set of its finite subgroups, $\mathfrak{D}$ the set of its relatively definable subgroups to which $\mathrm{gen}(A)$ is foreign, and $\mathfrak{A}$ the set of its relatively definable subgroups B such that for any relatively definable subgroup C of infinite index in A the index of $B+C$ in A is infinite. Then $\mathfrak{F}$, $\mathfrak{D}$ and $\mathfrak{A}$ are closed.*

Remark 4.4.1 In the case of a connected group with regular generic type, $\mathfrak{D}$ is just the family of relatively definable proper subgroups, so this definition agrees with that given in Theorem 4.2.6.

Proof: Trivial for $\mathfrak{S} = \mathfrak{F}$. For $\mathfrak{S} = \mathfrak{D}$, if $\mathrm{gen}(A)$ is foreign to S_1 and S_2, then it is also foreign to $S_1 + S_2$. If r is an $\mathfrak{D}$-endomorphism with co-kernel S_2, s is an element of $S_1 r$ and x realizes $\mathrm{gen}(A)$, then we first choose an element t in S_1 such that (t, s) lies in r. So t is independent of x. Now we choose s' in tr independent of x over t. Then $s - s'$ lies in S_2, whence x is independent of $s - s'$ over $\{t, s'\}$, so x is independent of s. Therefore $S_1 r$ lies in $\mathfrak{D}$.

Finally, if A_1 and A_2 are elements of $\mathfrak{A}$ and B is a relatively definable subgroup of infinite index in A, then $B + A_1$ has infinite index A, and so does $B + A_1 + A_2$, whence $A_1 + A_2$ again lies in $\mathfrak{A}$. Suppose now that r is an $\mathfrak{A}$-endomorphism with co-kernel A_2, A_1 lies in $\mathfrak{A}$ and B is a relatively definable subgroup of infinite index in A, with $A_1 r + B = A$. Clearly, we may assume that A_2 is contained in B. Then B intersects $A_1 r$ in a subgroup of infinite index, so $B' = r^{-1}B = \{a \in A : \exists b \in B \, (a, b) \in r\}$ must also have infinite index in A. But now for any a in A there are a_1 in A_1 and b in B with $a_1 r + (b + A_2) = ar$, so there is b' in B' with $a_1 + b' \equiv a$ modulo A_2, that is $A_1 + A_2 + B' = A$, a contradiction.

As for closure under finite extensions, this is obvious in all cases. $\square$

For weakly minimal A, all three notions agree, and for A with regular generic type, $\mathfrak{D} = \mathfrak{A}$. But in general different $\mathfrak{S}$ may well lead to different $\mathfrak{S}$-endomorphisms; and while on the one hand increasing $\mathfrak{S}$ increases the number of $\mathfrak{S}$-endomorphisms, on the other it also coarsens the $\mathfrak{S}$-equivalence.

Lemma 4.4.2 *Let $\mathfrak{S}$ be closed and E, E' and E'' be endogenies.*

1. *$E + E'$ and $E - E'$ are again endogenies.*

2. *EE' is an endogeny.*

3. *$A \times \{0\}$ is additive, id_A multiplicative identity.*

4. *Addition is associative and commutative.*

5. *Multiplication is associative.*

6. $E(E' + E'') \leq EE' + EE''$, and $E'E + E''E = (E' + E'')E$ (at least on the intersection of the domains).

7. $\equiv$ is an equivalence relation.

8. Addition and subtraction are well-defined modulo $\equiv$.

9. If E' is trivial, then $E \equiv E + E'$.

10. $E - E \equiv 0$.

11. $E(E' + E'') \equiv EE' + EE''$.

12. Multiplication is well-defined modulo equivalence.

Therefore the endogenies modulo $\equiv$ form a ring.

Proof: 1. and 2.: co-ker$(E \pm E')$ is in $\mathfrak{S}$ because $\mathfrak{S}$ is closed under addition; co-ker(EE') because $\mathfrak{S}$ is closed. It is obvious that the respective domains have finite index in A.
3., 4., 5. and 6. are obvious.
7., 8. and 9. follow from the fact that the sum of two trivial endogenies is again trivial.
10. uses that $E - E$ is trivial.
11.: Consider $S := A \times 0EE''$. Then S is trivial, and $EE' + EE'' \leq E(E' + E'') + S$. The equivalence follows from 6.
12. If $T = E + S$ and $T' = E' + S'$ for trivial S and S', then $TT' = (E + S)(E' + S') \equiv EE' + ES' + SE' + SS'$, but the last three summands are all trivial. $\square$

Theorem 4.4.3 Let G be a type-definable group over A, and $r_i : G \to H_i/K_i$ ($i < \omega$) be a sequence of surjective homomorphisms type-definable over A, such that $H_0 \geq H_1 \geq \cdots \geq H_i \geq \cdots \geq K_i \geq \cdots \geq K_1 \geq K_0$. If the sequence of the H_i is properly descending, then there are continuum many types over A.

Proof: Let $H := \bigcap_{i < \omega} H_i$. By thinning out the sequence, we may assume that $\sum(1/|H_i : H_i \cap H_{i+1}^*|) < 1$ for some definable supergroups H_{i+1}^* of H_{i+1}. We shall consider the cosets of H in H_0. If N is a proper subgroup of H_i/K_i of index k, then $r_i^{-1}N$ is a proper subgroup of G of index k. By Neumann's Lemma and compactness the type $\{r_i x \in (H_i - H_{i+1}^*)/K_i : i < \omega\}$ is consistent. If $g \in G$ satisfies this type, then there are continuum many (partial) types over g: choose any subset I of ω and consider the partial type

$$p_I = \{ x \cdot \prod_{i \in I \cap j} r_i g \leq H_j : j < \omega \}$$

(we identify j with the set $\{0, 1, \ldots, j-1\}$).

As $r_i g \in H_i/K_i$, this is a consistent set. If J is another subset and j is minimal in $(I-J) \cup (J-I)$ (say j lies in $I-J$), then for realizations x_I of p_I and x_J of p_J we have $x_I \cdot \prod_{I \cap j} r_i g \leq H_{j+1}(r_j g)^{-1} \neq H_{j+1}$, but $x_J \cdot \prod_{I \cap j} r_i g \leq H_{j+1}$, so p_I and p_J are disjoint. $\Box$

In particular we may apply this lemma to a sequence of homomorphisms $H_i/K_i \to H_{i+1}/K_{i+1}$.

Corollary 4.4.4 *An abelian group A type-definable over a finite set in a small theory is the sum of a divisible group and one of finite exponent.*

Thus a small group is ABD.

Proof: We take $r_i : x \mapsto i!x$. The sequence of images $i!A$ has to become stationary at some i_0. Then $i_0!A$ is divisible, and $A = i_0!A + A[i_0!]$ as in Lemma 1.5.2. $\Box$

From now on let $\mathfrak{G}$ be closed and contained in $\mathfrak{A}$. (If there were some subgroup B in $\mathfrak{G} - \mathfrak{A}$, then we might just consider quasi-endomorphisms of any relatively definable subgroup C of infinite index in A with $A = B + C$.)

Lemma 4.4.5 *Let A be an ABD abelian group and r an $\mathfrak{G}$-endomorphism of A. Then there is an equivalent endogeny r' with domain A.*

Proof: If $A = D + B$, where B has finite exponent and D is divisible, then D is connected and hence contained in the domain of r. Let $a_1, \ldots, a_n \in B$ be a system of representatives for A/A_r, and consider the definable supergroup $r' := \langle r, (a_1, 0), \ldots, (a_n, 0) \rangle$. As the a_i have finite orders, this is a finite extension and its co-kernel is a finite extension of the co-kernel of r. Therefore r' is an $\mathfrak{G}$-endomorphism, and must be $\mathfrak{G}$-equivalent to r. Furthermore, the domain of r' is the whole of A. $\Box$

Note that we may have to enlarge the co-kernel: if r is not defined on a but is defined on pa for some prime p, and $(pa, b) \in r$ but b is not p-divisible, then any extension of r to a puts $b - pr'(a)$ into the co-kernel.

Lemma 4.4.6 *Let Φ be a set of $\emptyset$-definable subsets of a small group G. Assume that Φ is closed under disjunction and division (i.e. for any two sets φ and ψ in Φ both $\varphi \vee \psi$ and $\exists y, z\, [\varphi(y) \wedge \psi(z) \wedge x = yz^{-1}]$ are in Φ). Then there is some φ in Φ such that $\varphi(G) \cdot G^0$ is maximal.*

Proof: Choose a descending sequence of $\emptyset$-definable normal subgroups $(G_i : i < \omega)$ whose intersection is G^0. We may assume that G is saturated, and that for any formula φ in Φ, if some element x lies in $\varphi(G)$, then so do x^{-1} and 1. We shall write φ^2 for the formula $\exists y, z\, [\varphi(y) \wedge \varphi(z) \wedge x = yz]$. Suppose that the conclusion of the lemma does not hold.

Claim. There are a sequence $i_0 < i_1 < \cdots$ and a sequence $(\varphi_i : i < \omega)$ of formulæ in Φ such that

1. φ_j^2 is contained in φ_{j+1} for all $j < \omega$, and

2. $(\varphi_j^2(G)G_{i_{j+1}}) \cap G_{i_j}$ is properly contained in $(\varphi_{j+1}(G)G_{i_{j+1}}) \cap G_{i_j}$, for all $j < \omega$.

Proof of Claim: We use induction on j. So suppose we have already found $\varphi_0, \ldots, \varphi_n$ and $i_0, \ldots, i_n$ as required. By assumption, there is an increasing sequence $(\psi_j)_{j<\omega}$ of formulæ in Φ such that $\varphi_n^2 = \psi_0$ and $\psi_j^2(G)G^0$ is properly contained in $\psi_{j+1}(G)G^0$. Therefore we can choose a G_{i_n}-coset X and $k < l$ such that $(\psi_k(G)G^0) \cap X$ is non-empty, and $(\psi_k^2(G)G^0) \cap X$ is properly contained in $(\psi_l(G)G^0) \cap X$.

By saturation and compactness there is an m with $i_n < m < \omega$ such that $(\psi_k^2(G)G_m) \cap X$ is properly contained in $(\psi_l(G)G_m) \cap X$. As $XG_m = X$, we can then find elements $x \in \psi_k(G) \cap X$ and $y \in \psi_l(G) \cap X$ with $y \notin \psi_k^2(G)G_m$; then xy^{-1} lies in $\psi_l^2(G)$, but not in $\psi_k(G)G_m$. Moreover, xy^{-1} is an element of G_{i_n}. We conclude that $(\psi_k(G)G_m) \cap G_{i_n}$ is a proper subset of $(\psi_l^2(G)G_m) \cap G_{i_n}$. We put $\varphi_{n+1} := \psi_l^2$ and $i_{n+1} := m$, and are done. $\square$

For any set I of even natural numbers, put

$$p_I := \{x \in \varphi_j(G)G_{i_{j+1}} : j \in I\} \cup \{x \notin \varphi_j(G)G_{i_{j+1}} : j \notin I\}.$$

Claim. p_I is consistent.
Proof of Claim: By the first claim, we can choose elements g_j in $\varphi_{j+1}(G) \cap G_{i_j}$ but not in $\varphi_j^2(G)G_{i_{j+1}}$, for $j < \omega$. By property 1., for any even $k < \omega$, the product $\prod_{j \in I \cap k} g_j$ lies in $\varphi_k(G)$ and therefore also in $\varphi_k(G)G_{i_{k+1}}$. Moreover, g_j lies in $G_{i_{k+1}}$ for any $j > k$. On the other hand, g_k does not lie in $\varphi_k^2(G)G_{i_{k+1}}$. Hence for any initial segment J of I and even $k \leq \max\{J\}$, the product $\prod_J g_j$ lies in $\varphi_k(G)G_{i_{k+1}}$ iff $k \notin I$. $\square$

This yields continuum many types over $\emptyset$ and contradicts smallness. $\square$

Theorem 4.4.7 *Let A be a small stable definable abelian group. Then either the ring of definable endogenies of A is locally finite, or there is a definable subgroup A_0 of infinite index such that the quotient A/A_0 is connected-by-finite.*

Proof: By Corollary 4.4.4 there are a definable divisible subgroup D of A and a definable subgroup B of bounded exponent such that $A = D + B$. If B has infinite index, we may choose $A_0 = B$ and A/B is divisible, hence connected. Therefore we may assume that A has finite exponent, and also that A^0 is not definable. Suppose $r_1, \ldots, r_n$ are endogenies such that $R = \langle r_1, \ldots, r_n \rangle$ is infinite. By Lemma 4.4.5 we may assume that all r_i have domain A and are $\emptyset$-definable (by adding finitely many parameters).

The set of co-kernels of the endogenies in R satisfies the assumptions of Lemma 4.4.6. Hence there is a co-kernel K such that $A^0 + K$ is maximal. If the index of $A^0 + K$ in A is finite, then A/K is connected-by-finite (K must have infinite index) and we are done. Otherwise modulo K all co-kernels are contained in A^0, and we may consider the endogenies as endomorphisms of A/K modulo the union of all the co-kernels. So from now on we assume $A = A/K$, and thus all co-kernels are contained in the connected component.

Claim. A subring S of finite index in a finitely generated ring R (with unit) is finitely generated.

Proof of Claim: Let $\bar{e}$ be a system of representatives of R/S with $0, 1 \in \bar{e}$, and $\bar{s}$ a finite tuple in S such that $\bar{e}\bar{s}$ generates R and for any $e_0, e_1 \in \bar{e}$ there are $e_2, e_3 \in \bar{e}$ with $e_0 + e_1 - e_2 \in \bar{s}$ and $e_0 e_1 - e_3 \in \bar{s}$. Let $\bar{s}' \in S$ be the set of elements of the form $ese' - e''$, with $e, e', e'' \in \bar{e}$ and $s \in \bar{s}$ (for any e, e', s there is a unique $e'' \in \bar{e}$ such that $ese' - e'' \in S$). We claim that $\bar{s}\bar{s}'$ generates S. Let S_0 be the subring of S generated by $\bar{s}\bar{s}'$; we show that $\bar{e}$ is a set of representatives of R/S_0.

Since $R_0 := \bigcup_{e \in \bar{e}} e + S_0$ is an additive subgroup of R and $\bar{e}\bar{s}$ generates R, it is sufficient to show that every finite product of elements in $\bar{e}\bar{s}$ lies in R_0. We use induction on the number of factors; the case of a single factor is obvious.

So suppose the assertion is true for products of length at most n, and consider a product $p_0 \cdots p_n$ of length $n + 1$. Assume $p_n \in \bar{s}$. By the inductive hypothesis, $p_0 \cdots p_{n-1} = e + s_0$ for some $e \in \bar{e}$ and $s_0 \in S_0$. Therefore

$$p_0 \cdots p_n = (e + s_0)p_n = ep_n 1 + s_0 p_n = e' + s' + s_0 p_n \in e' + S_0,$$

for some $e' \in \bar{e}$, $s' \in \bar{s}'$. On the other hand, if $p_n \in \bar{e}$, then $p_{n-1} \in \bar{s}$ (since $n > 0$). By the inductive hypothesis $p_0 \cdots p_{n-2} = e + s$ for some $e \in \bar{e}$ and $s \in S_0$. By construction, $1 p_{n-1} p_n = s' + e'$ for some $s' \in \bar{s}'$ and $e' \in \bar{e}$, and by the inductive hypothesis again $p_0 \cdots p_{n-2} e' = (s + e)e' \in R_0$. Hence

$$\begin{aligned} p_0 \cdots p_n &= (e + s)p_{n-1}p_n = ep_{n-1}p_n + s(1 p_{n-1}p_n) = ep_{n-1}p_n + s(s' + e') \\ &= ep_{n-1}p_n + ss' + (s + e)e' - ee' \in R_0. \ \square \end{aligned}$$

Therefore there are only countably many subrings of finite index in R and there is a descending sequence $(I_i : i < \omega)$ of right ideals of finite index such that every right ideal of finite index contains I_i for some $i < \omega$. Let $(A_i : i < \omega)$ be a descending sequence of $\emptyset$-definable subgroups of A with $\bigcap_{i<\omega} A_i = A^0$. If I is a right ideal of finite index in R, then $I^{-1}A^0$ is type-definable and contains A^0; it is hence definable iff it has finite index. Suppose there is no right ideal I of finite index in R such that $I^{-1}A^0$ is definable. Replacing $(A_i : i < \omega)$ by a sub-sequence, we may assume that $\sum_{i<\omega}(1/|A : I_i^{-1}A_i|) < 1$; by Neumann's Lemma and compactness (since I_i is finitely generated, $I_i^{-1}A_i$ is definable) there is some element a in $A - \bigcup_{i<\omega} I_i^{-1}A_i$, so the right ideal $I_a := \{r \in R : ar \in A^0\}$ has infinite index in R. Hence aR/A^0 is infinite.

Claim. If $\bar{a}$ is a tuple of elements such that $\bar{a}R/A^0$ is infinite, then there are continuum many cosets of A^0 definable over $\bar{a}$.

Proof of Claim: For every i we find $r_i \in R$ and $a_i \in \bar{a}$ with $a_i r_i \in A_i$, but $a_i r_i \notin A_j$ for sufficiently big j (which we rename to be A_{i+1}). Then for any subset S of ω the (partial) types

$$p_S = \{x + \sum_{i \in S \cap j} a_i r_i \leq A_j : j < \omega\}$$

are consistent and pairwise contradictory. $\square$

It follows that aR/A^0 cannot be infinite and there is some right ideal I of finite index in R such that $A_I := I^{-1}A^0$ is definable. We consider a set $\bar{a}$ of representatives of A/A_I; the claim now implies that $AI/A^0 = \bar{a}I/A^0$ is finite, and I has a subideal of finite index which maps A to A^0.

Therefore the set of endogenies r in R with $Ar \leq A^0$ forms an ideal I of finite index (it is necessarily two-sided). Now R has finite characteristic $e(A)$ and I is finitely generated; there must therefore be an infinite word $w = \cdots r_2 r_1$ in the generators of I whose end segments are non-trivial. By Theorem 4.4.3 there is an end segment $r = r_n \cdots r_1$ such that $A^0 r$ is minimal. Then $A^0 r = A^0 r_{n+1} r$ is infinite and connected. But $A^0 r = A^0 r_{n+1} r \leq A r_{n+1} r \leq A^0 r$, so equality holds. So for $A_0 := \ker(r_{n+1}r)$, the quotient A/A_0 is definably isomorphic to $Ar_{n+1}r = A^0 r$ and therefore infinite and connected. $\square$

Note that all we needed stability for was to get the connected component to be the intersection of $\emptyset$-definable groups. Thus the theorem holds also for small abelian groups without the independence property.

If $\mathfrak{S}$ is the set of all subgroups of infinite index in A (e.g. for $\mathfrak{S} = \mathfrak{D}$ and a regular generic type), then every non-surjective $\mathfrak{S}$-endomorphism is equivalent to a trivial one and surjective $\mathfrak{S}$-endomorphisms are invertible. Thus R is a division ring. And of course, if R is locally finite and without zero-divisors, then it is a locally finite field by Wedderburn's Theorem. Note that if $A^0 r$ is a minimal image of A^0 under an $\emptyset$-definable endogeny (which exists by Theorem 4.4.3), then the ring of $\emptyset$-definable endogenies of $A^0 r$ (which in particular comprises the $\emptyset$-definable elements of Rr) has no zero-divisors.

Lemma 4.4.8 *Let A be a small stable abelian group without infinite definable connected-by-finite quotient. Then for any $\mathfrak{S}$-endomorphism r with domain A there is $n < \omega$ such that $Ar^n = Ar^{n+1}$. We then have $A = \ker(r^n)+\mathrm{im}(r^n)$, and some power r^{n_0} of r is the projection to Ar^n.*

Proof: By Theorem 4.4.3 there is a smallest image Ar^n, and as Ar^{n+1} is contained in Ar^n, we must have equality. If a is an element of A, then ar^n

lies in $Ar^n = Ar^{2n}$, so there is an element a' in Ar^n with $a'r^n = ar^n$, whence $(a - a')r^n = 0$ (i.e. $(a - a', 0) \in r^n$), and $A = \ker(r^n) + \operatorname{im}(r^n)$. Secondly, $\langle r \rangle$ is finite by Theorem 4.4.7, so there are $m > k \geq n$ with $r^m \equiv^{\mathfrak{S}} r^k$. Hence r^{m-k} is equivalent to the identity on $Ar^k = Ar^n$. If n_0 is a multiple of $m - k$ greater than or equal to n, then r^{n_0} is the projection to Ar^n. □

Corollary 4.4.9 *Let A be a small stable abelian group without infinite definable connected-by-finite quotient, and X an arbitrary set of parameters. Then the ring R of X-definable $\mathfrak{A}$-endomorphisms of A has a locally nilpotent ideal $I = \{r \in R : \operatorname{im}(r) \in \mathfrak{A}(X)\}$.*

Note that we consider endogenies up to $\mathfrak{A}$-equivalence, not $\mathfrak{A}(X)$-equivalence, which would be coarser. In particular, $Ar \in \mathfrak{A}(X)$ does not imply $r \equiv^{\mathfrak{A}} 0$ even for X-definable r, as Ar might have a supplement needing more parameters for its definition.

Proof: Clearly I is an additive subgroup of R; it is a right ideal since $\mathfrak{A}(X)$ is closed under X-definable $\mathfrak{A}$-endomorphisms, and a left ideal as $Aqr \leq Ar$ for any $q \in R$. By Lemma 4.4.8, any $r \in I$ must be nilpotent and I is nil.

Claim. A finite nil ring is nilpotent.

Proof of Claim: Let R be a minimal counter-example, and $r \in R$ arbitrary. As $r^n = 0$ for some $n < \omega$, we have $rR < R$ and rR must be nilpotent by the inductive hypothesis, say of class k_r. But this means that any product in R involving at least k_r factors r is zero. Taking $k =: 1 + \sum_{r \in R}(k_r - 1)$, any product in R must have at least k_r factors r for some $r \in R$, and hence be zero. So R is nilpotent of class k. □

Since I is locally finite by Theorem 4.4.7, it is locally nilpotent. □

It is unknown whether I must be nilpotent.

Theorem 4.4.10 *Let A be a small stable abelian group without infinite definable connected-by-finite quotient. Then the following are equivalent for an $\mathfrak{S}$-endomorphism r:*

1. $\operatorname{im}(r)$ has finite index in A,

2. $\ker(r) \in \mathfrak{S}$,

3. r is invertible,

4. r is left invertible, and

5. r is right invertible.

Proof: 1. ⇒ 2.: Consider n_0 as given by Lemma 4.4.8. Then $\operatorname{im}(r^{n_0})$ must have finite index in A, so $r^{n_0} \equiv^{\mathfrak{S}} 1$ and r is invertible.

2. $\Rightarrow$ 1.: By Lemma 4.4.8 for some $n < \omega$ we have $A^0 \leq \ker(r^n) + \operatorname{im}(r^n)$. But $\ker(r) \in \mathfrak{S}$ implies $\ker(r^n) \in \mathfrak{S}$, hence $\operatorname{im}(r^n)$ has finite index in A, and so has $\operatorname{im}(r)$.

3. $\Rightarrow$ 4. & 5. is obvious, as are 4. $\Rightarrow$ 1. and 5. $\Rightarrow$ 2. And 1. & 2. $\Leftrightarrow$ 3. $\square$

Definition 4.4.3 Let A be an abelian group, and $\mathfrak{S}$ a closed family of subgroups. A sum $\sum_{i=1}^{n} A_i$ of subgroups of A is $\mathfrak{S}$-*direct*, or *almost direct*, if all intersections $A_j \cap \sum_{i \neq j} A_i$ are in $\mathfrak{S}$. If $\sum_{i=1}^{n} A_i \geq A^0$, then it is called an $\mathfrak{S}$-*decomposition* of A. If B is a subgroup of A, a subgroup H with $B \cap H \in \mathfrak{S}$ and $B + H \geq A^0$ is called an $\mathfrak{S}$-*complement* for B in A; we say that B is $\mathfrak{S}$-*irreducible* if every proper (relatively definable) subgroup of infinite index is in $\mathfrak{S}$.

Our aim now is to decompose a group A into a finite $\mathfrak{S}$-direct sum of $\mathfrak{S}$-irreducible subgroups A_i. In that case, the ring of endogenies of A can be retrieved as a finite direct product of finite-dimensional matrix rings over some fields. However, there are two main obstacles:

1. A might not have enough $\mathfrak{S}$-irreducible subgroups. This will happen if there is an infinite descending chain $(A_i : i < \omega)$ of subgroups not in $\mathfrak{S}$, but without $\mathfrak{S}$-complements.

2. A might only decompose as an infinite sum of $\mathfrak{S}$-irreducible subgroups. In this case, there is a family $(A_i : i < \omega)$ of irreducible subgroups of A with $\mathfrak{S}$-complements $(H_i : i < \omega)$ such that $A_i \leq H_j$ for all $i \neq j$.

In order to overcome these difficulties, we shall have to look at the parameters needed for the $\mathfrak{S}$-endomorphisms. Note first that we may, in Definition 4.4.1, consider only families $\mathfrak{S}$ of subgroups defined over some set X of parameters. In this case, we shall talk about $\mathfrak{S}(X)$-endomorphisms. Note that if $X \subseteq X'$, then $\mathfrak{F}(X) \subseteq \mathfrak{F}(X')$ and $\mathfrak{O}(X) \subseteq \mathfrak{O}(X')$, but not necessarily $\mathfrak{A}(X) \subseteq \mathfrak{A}(X')$, as it might happen that there is a relatively X-definable subgroup B of infinite index which does not have a relatively X-definable supplement in A^0 – whence $B \in \mathfrak{A}(X)$ – but which does have a relatively X'-definable supplement, thus removing B from $\mathfrak{A}(X')$.

Lemma 4.4.11 *Let A be a small stable abelian group, and put $\mathfrak{S} = \mathfrak{A}(\operatorname{acl}(\emptyset))$. Then A has a finite $\mathfrak{S}$-decomposition into $\mathfrak{S}$-irreducible subgroups. Moreover, there is a finite bound $n < \omega$ on the length of any sequence of $\operatorname{acl}(\emptyset)$-definable subgroups A_i with $A_i^0 \not\leq \sum_{j \neq i} A_j$ and $A^0 \leq \sum_i A_i$.*

Proof: For any $\operatorname{acl}(\emptyset)$-definable subgroup we add a predicate P_S for the orbit S of the canonical parameter of that subgroup under automorphisms of the structure. (So in particular P_S will be a finite set.) We consider the many-sorted structure which consists of A and all the P_S (S an orbit), together with the structure induced by addition and the subgroups G_s for $s \in P_S$. This structure is $\emptyset$-definable and inherits smallness. But over $\mathrm{acl}(\emptyset)$ it is just an abelian structure and thus one-based by Theorem 4.2.8, which it remains even without naming $\mathrm{acl}(\emptyset)$. By Corollary 4.1.17 the generic type of A has finite weight n. Now suppose $(A_i : i < m)$ is a family of relatively $\mathrm{acl}(\emptyset)$-definable subgroups of A with $A^0 \leq \sum_{i<m} A_i$ and $A_i^0 \nleq \sum_{j\neq i} A_j$ for all $i < m$, then consider independent principal generic elements a_i of A_i for $i < m$. Their sum $a = \sum_{i<m} a_i$ is a principal generic element of A and forks with every a_i, for $i < m$. So $m \leq n$.

Now suppose $\sum_{i<m} A_i \geq A^0$ has maximal length m. We claim that this is an $\mathfrak{S}$-decomposition into $\mathfrak{S}$-irreducible components. First, suppose $I := A_0 \cap \sum_{j\neq 0} A_j$ were not in $\mathfrak{A}(\mathrm{acl}(\emptyset))$. Then we could find an $\mathrm{acl}(\emptyset)$-definable group B of infinite index in A such that $I + B \geq A^0$. We may assume that $\sum_{i<m} A_i = A$ and $I + B = A$ (replacing A by a subgroup of finite index).

Put $B_0 := A_0 \cap B$ and $B_i := (A_i + I) \cap B$ for $0 < i < m$. Then $B_i + I = A_i + I$, hence $A = I + \sum_{i<m} B_i$. We claim that $I^0 \nleq \sum_{i<m} B_i$ and $B_i^0 \nleq I + \sum_{j\neq i} B_j$, contradicting the maximality of m.

As B_i is contained in B for all $i < m$ it follows immediately from $A = I + B$ and $B^0 < A^0$ that $I^0 \nleq \sum_{i=1}^{n} B_i$. Furthermore $B_i \leq A_i + I$; since $A_0 = B_0 + I$ and $I + \sum_{j\neq 0} B_j \leq \sum_{j\neq 0} A_j$ does not contain A_0^0, it cannot contain B_0^0 either. Finally, $I \leq A_0$ and $B_i + I = A_i + I$, so for $i \neq 0$ we have that $A_i^0 \nleq \sum_{j\neq i} A_j$ implies $A_i^0 \nleq I + \sum_{j\neq i} B_j$, whence $B_i^0 \nleq I + \sum_{j\neq i} B_j$.

Let now B be an $\mathrm{acl}(\emptyset)$-definable subgroup of infinite index in A_i and suppose $B \notin \mathfrak{S}$. Then there is an $\mathrm{acl}(\emptyset)$-definable subgroup B' of infinite index in A such that $B + B' \geq A^0$. We may assume that both B and B' contain $A_i \cap \sum_{j\neq i} A_j$. Put $B'' := B' \cap A_i$, then $A_i^0 \leq B + B''$, and B'' has infinite index in A_i. Furthermore A^0 is contained in $B + B'' + \sum_{j\neq i} A_j$, $B^0 \nleq B'' + \sum_{j\neq i} A_j$, $(B'')^0 \nleq B + \sum_{j\neq i} A_j$, and easily $A_k^0 \nleq B + B'' + \sum_{j\neq i,k} A_j$, contradicting the maximality of m. $\square$

Theorem 4.4.12 *Suppose A is a small stable abelian group such that no quotient of A is connected-by-finite, and let R be the ring of $\mathrm{acl}(\emptyset)$-definable $\mathfrak{A}(\mathrm{acl}(\emptyset))$-endomorphisms of A. Then R decomposes as a finite direct sum of finite-dimensional matrix rings over locally finite fields.*

Proof: By Lemma 4.4.11 there is an $\mathfrak{A}(\mathrm{acl}(\emptyset))$-decomposition $\sum_{i<n} A_i$ of A. Then $A_i \cap \sum_{j\neq i} A_j$ is in $\mathfrak{A}(\mathrm{acl}(\emptyset))$ for $i \neq j$, so $S := \sum_{i<n}(A_i \cap \sum_{j\neq i} A_j)$ is also in that family. Modulo S the sum of the A_i is direct. We replace A by A/S and A_i by $(A_i + S)/S$. Then any $\mathrm{acl}(\emptyset)$-definable subgroup of infinite index in some A_i is in $\mathfrak{A}(\mathrm{acl}(\emptyset))$, hence (if π_i denotes the projection to A_i) for any $\mathfrak{A}(\mathrm{acl}(\emptyset))$-endomorphism r in R, the endogeny $\pi_i r \pi_j$ is either invertible as $\mathfrak{S}$-homomorphism from A_i^0 to A_j^0, or zero. Furthermore, R is locally finite

by Theorem 4.4.7. It follows that $\pi_i R \pi_i$ is a locally finite integral domain, hence a locally finite field. The result follows. $\square$

Note that the projections π_i are elements of R.

Lemma 4.4.13 *If all* $\mathrm{acl}(\emptyset)$-*conjugates of some endogeny* r *are* $\mathfrak{D}$-*equivalent, then* r *is* $\mathfrak{D}$-*equivalent to an* $\mathrm{acl}(\emptyset)$-*definable* $\mathfrak{G}$-*endomorphism.*

Proof: Suppose that $r' \equiv^{\mathfrak{D}} r$ for any $\mathrm{acl}(\emptyset)$-conjugate r' of r. Then for a principal generic element a of A there are c and x such that $(a, c) \in r$, $(a, c + x) \in r'$, and x does not depend on the choice of a. Hence for an independent principal generic element b there is d such that $(b, d) \in r$, $(b, d + x) \in r'$. But now $a - b$ is again principal generic and $(a - b, c - d)$ lies in $r \cap r'$, so $r \cap r'$ is also an $\mathfrak{G}$-endomorphism, which must necessarily be equivalent to r. It is now easy to show that $\bigcap\{r' : r'$ conjugate to r over $\mathrm{acl}(\emptyset)\}$ is an $\mathfrak{G}$-endomorphism (using that this is a finite subintersection), definable over $\mathrm{acl}(\emptyset)$ and equivalent to r. $\square$

Lemma 4.4.14 *Let* A *be a small abelian group without infinite definable connected-by-finite quotient. Then any two* $\mathrm{acl}(\emptyset)$-*conjugates* r *and* r' *of some* $\mathfrak{A}$-*endomorphism of* A *are* $\mathfrak{A}$-*equivalent.*

Proof: As $\mathrm{stp}(r) = \mathrm{stp}(r')$, for any two independent generic elements a_1 and a_2 with $a_2 \in a_1 r + A^0$, there is an automorphism fixing $\mathrm{acl}(\emptyset) \cup \{a_1, a_2\}$ and mapping r to r'. Hence $a_1 r'$ and a_2 lie in the same coset of A^0, and modulo the co-kernel $r - r'$ maps A into A^0. But then $A' = A(r - r')$ must be in $\mathfrak{A}$: otherwise there is a subgroup B of infinite index in A – which we may assume to contain co-ker$(r - r')$ – such that $A' + B$ has finite index in A; hence $(A' + B)/B \cong A^0/(B \cap A^0)$ is infinite and connected, contradicting the hypothesis on A. $\square$

Theorem 4.4.12 now yields a characterization of the ring of endogenies in the case $\mathfrak{A} = \mathfrak{D}$:

Proposition 4.4.15 *Let* A *be a small stable abelian group without infinite definable connected-by-finite quotient. If* $\mathfrak{D} = \mathfrak{A}$, *then the ring* R *of* $\mathfrak{D}$-*endomorphisms has a locally nilpotent ideal* I, *and* R/I *decomposes as a finite subdirect sum of finite-dimensional matrix rings over locally finite fields.*

Proof: If $\mathfrak{D} = \mathfrak{A}$, then by Lemmas 4.4.13 and 4.4.14 any $\mathfrak{D}$-endomorphism is equivalent to one defined over $\mathrm{acl}(\emptyset)$. Let I be the set of $(\mathrm{acl}(\emptyset)$-definable) $\mathfrak{D}$-endomorphisms r with $\mathrm{im}(r) \in \mathfrak{A}$. It is locally nilpotent by Corollary 4.4.9. Since every $\mathrm{acl}(\emptyset)$-definable $\mathfrak{D}$-endomorphism can be viewed as an $\mathfrak{A}(\mathrm{acl}(\emptyset))$-endomorphism (but not necessarily *vice versa*), we may apply Theorem 4.4.12 and get a homomorphism φ from R to a direct sum of finite-dimensional matrix rings over locally finite fields. Clearly $\ker(\varphi) = \{r \in R : \operatorname{im}(r) \in \mathfrak{A}(\operatorname{acl}(\emptyset))\} = I$. However, φ need not be surjective; in particular the projections to the summands in the irreducible $\mathfrak{A}(\operatorname{acl}(\emptyset))$-decomposition need not be in R, and so R/I is only a subdirect sum. $\square$

Remark 4.4.2 Alternatively, we might view R as the inverse limit of finite sub-direct sums of matrix rings R/I_i, for a chain (or, more generally, a directed system) of finite sets $(P_i : i < \omega)$ of parameters such that $\bigcup_{i<\omega} P_i$ contains parameters for all supplements of subgroups in $\mathfrak{A}(\operatorname{acl}(\emptyset)) - \mathfrak{A}$, where I_i is the ideal $\{r \in R : \operatorname{im}(r) \in \mathfrak{A}(P_i)\}$. Note that $(I_i : i < \omega)$ forms a descending sequence of ideals with $\bigcap_{i<\omega} I_i = (0)$. As a nil ideal in R/I_i is nilpotent, I is also residually nilpotent.

For an arbitrary small stable group without connected-by-finite quotient we may consider the ring R of $\mathfrak{A}$-endomorphisms and a directed system $(P_i : i \in I)$ of finite sets of parameters such that every $r \in R$ has a representative over some P_i by Lemma 4.4.14. Let R_i be the subring of P_i-definable $\mathfrak{A}$-endomorphisms, and $I_i := \{r \in R_i : \operatorname{im}(r) \in \mathfrak{A}(P_i)\}$. Then $R = \bigcup_{i \in I} R_i$, I_i is a locally nilpotent ideal in R_i, and R_i/I_i is a subdirect finite sum of finite-dimensional matrix rings over locally finite fields.

The following example shows that in general the nilexponent need not be bounded by the weight of a generic type:

Example 4.4.1 Let V be a vector space over $\mathbb{Q}$, together with a nilpotent endomorphism σ of class n, with $\ker(\sigma) = \operatorname{im}(\sigma^{n-1})$, and a complement U for $\ker(\sigma)$ in V. This structure (as a module) has Morley rank n, σ is nilpotent of nilexponent n (even modulo $\mathfrak{A}$-equivalence), and $w(\operatorname{gen}(V)) = 2$.

4.5 CM-Triviality

Recently, Hrushovski [57] has introduced a new geometric property, CM-triviality (*CM* standing for *curve memory*, or *Cohen-Macaulay*), which holds for the stable structures obtained via the amalgamation process mentioned after Example 0.3.4. Again, P will denote an $\emptyset$-invariant family of stationary types.

Definition 4.5.1 The family P is *CM-trivial* if for any expansion of T by naming parameters we have:
whenever $A \subset B$ and c are such that

- $c \underset{A}{\downarrow} \operatorname{cl}_P(A)$ and $c \underset{B}{\downarrow} \operatorname{cl}_P(B)$,

- $\operatorname{tp}(c)$ is P-internal, and

- $\mathrm{cl}_P(cA) \cap \mathrm{cl}_P(B) = \mathrm{cl}_P(A)$,

then $\mathrm{Cb}(c/A) \subset \mathrm{cl}_P(\mathrm{Cb}(c/B))$.
A theory is CM-trivial if the family of all types is CM-trivial.

We should note that CM-triviality of P is preserved under naming parameters.

Proposition 4.5.1 *We may omit the condition $c \underset{B}{\downarrow} \mathrm{cl}_P(B)$ in Definition 4.5.1. If $U_P(c)$ is finite, we may also omit $c \underset{A}{\downarrow} \mathrm{cl}_P(A)$.*

Proof: Assume that P is CM-trivial, and suppose that A, B and c are such that $\mathrm{tp}(c)$ is P-internal, $c \underset{A}{\downarrow} \mathrm{cl}_P(A)$, and $\mathrm{cl}_P(cA) \cap \mathrm{cl}_P(B) = \mathrm{cl}_P(A)$. Let $B' = \mathrm{acl}(Bc) \cap \mathrm{cl}_P(B)$. Then $A \subseteq B \subseteq B'$, and $c \underset{B'}{\downarrow} \mathrm{cl}_P(B')$. Then $\mathrm{Cb}(c/B') \subset \mathrm{cl}_P(\mathrm{Cb}(c/B))$ by Lemma 4.0.5, and clearly $\mathrm{cl}_P(B') = \mathrm{cl}_P(B)$. Hence
$$\mathrm{cl}_P(cA) \cap \mathrm{cl}_P(B') = \mathrm{cl}_P(cA) \cap \mathrm{cl}_P(B) = \mathrm{cl}_P(A).$$
By CM-triviality of P we get $\mathrm{Cb}(c/A) \subset \mathrm{cl}_P(\mathrm{Cb}(c/B')) \subseteq \mathrm{cl}_P(\mathrm{Cb}(c/B))$.

For the second assertion, suppose that A, B, c are such that $\mathrm{cl}_P(cA) \cap \mathrm{cl}_P(B) = \mathrm{cl}_P(A)$, and $\mathrm{tp}(c)$ is P-internal of finite U_P-rank. Let $B' = \mathrm{acl}(Bc) \cap \mathrm{cl}_P(B)$ as above, so $\mathrm{cl}_P(B') = \mathrm{cl}_P(B)$ and $\mathrm{Cb}(c/B') \subset \mathrm{cl}_P(\mathrm{Cb}(c/B))$. If $A' = \mathrm{acl}(Ac) \cap \mathrm{cl}_P(A)$, then $c \underset{A'}{\downarrow} \mathrm{cl}_P(A')$ and $\mathrm{cl}_P(A') = \mathrm{cl}_P(A)$; by Lemma 4.0.5 we have $Cb(c/A) \subset \mathrm{cl}_P(Cb(c/A'))$. Note that $A' \subseteq B'$, and
$$\mathrm{cl}_P(cA') \cap \mathrm{cl}_P(B') = \mathrm{cl}_P(cA) \cap \mathrm{cl}_P(B) = \mathrm{cl}_P(A) = \mathrm{cl}_P(A').$$
CM-triviality of P now implies $\mathrm{Cb}(c/A') \subset \mathrm{cl}_P(\mathrm{Cb}(c/B'))$, whence
$$Cb(c/A) \subset \mathrm{cl}_P(Cb(c/A')) \subseteq \mathrm{cl}_P(\mathrm{Cb}(c/B')) \subseteq \mathrm{cl}_P(\mathrm{Cb}(c/B)). \quad \square$$

Proposition 4.5.2 *CM-triviality of P is preserved under forgetting parameters. That is, if A, B and c are a counter-example to the CM-triviality of P in T and $D \underset{}{\downarrow} ABc$, then A, B and c are a counter-example to the CM-triviality of P in $T(D)$.*

Proof: The assumptions imply that $c \underset{A}{\downarrow} D$ and $c \underset{B}{\downarrow} D$. As $Bc \underset{A}{\downarrow} D$, Lemma 4.0.4 yields (with $C = Ac$) that $\mathrm{cl}_P(cAD) \cap \mathrm{cl}_P(BD) = \mathrm{cl}_P(AD)$. However, since $\mathrm{Cb}(c/AD) \subseteq \mathrm{acl}(A)$ and $\mathrm{Cb}(c/B) = \mathrm{Cb}(c/BD) \subseteq \mathrm{acl}(B)$, the independence $AB \underset{}{\downarrow} D$ implies that $\mathrm{Cb}(c/AD)$ and $\mathrm{Cb}(c/BD)D$ are independent over $\mathrm{Cb}(c/B)$. Hence
$$\mathrm{Cb}(c/AD) \underset{\mathrm{cl}_P(\mathrm{Cb}(c/B))}{\downarrow} \mathrm{cl}_P(\mathrm{Cb}(c/BD)D)$$
by Lemma 4.0.3, and as $\mathrm{Cb}(c/AD) = \mathrm{Cb}(c/A)$ is not in $\mathrm{cl}_P(\mathrm{Cb}(c/B))$, it cannot be in $\mathrm{cl}_P(\mathrm{Cb}(c/BD)D)$ either. Thus A, B and c are a counter-example to the CM-triviality of P in $T(D)$. $\quad \square$

Proposition 4.5.3 *In Definition 4.5.1 we may in addition require c to be a tuple realizing types in P, and tp(AB) to be P-internal, or A and B to be models.*

We might not be able to require A and B to be P-internal and models of T at the same time, as a P-internal model need not exist.

Proof: Suppose A, B, c are a counter-example to the CM-triviality of P. By P-internality of $\mathrm{tp}(c)$, we find some $D \downharpoonleft c$ and a tuple $\bar{d}$ of realizations of types in P over D, such that c is definable over $D\bar{d}$. We may choose $D \downharpoonleft ABc$. Then A, B and c are a counter-example to CM-triviality in $T(D)$ by Proposition 4.5.2. From now on assume D named.

We may suppose $\bar{d} \downharpoonleft_c B$. Put $A' = \mathrm{acl}(A\bar{d}) \cap \mathrm{cl}_P(A)$ and $B' = \mathrm{acl}(B\bar{d}) \cap \mathrm{cl}_P(B)$. Then $c \downharpoonleft_A A'$ and $c \downharpoonleft_B B'$; furthermore $A' \subseteq B'$, and $\bar{d} \downharpoonleft_{A'} \mathrm{cl}_P(A')$ and $\bar{d} \downharpoonleft_{B'} \mathrm{cl}_P(B')$. By Lemma 4.1.4 we get

$$\mathrm{cl}_P(\mathrm{Cb}(c/A)) = \mathrm{cl}_P(\mathrm{Cb}(c/\mathrm{cl}_P(A))) = \mathrm{cl}_P(\mathrm{Cb}(\bar{d}/\mathrm{cl}_P(A))) = \mathrm{cl}_P(\mathrm{Cb}(\bar{d}/A'))$$

and

$$\mathrm{cl}_P(\mathrm{Cb}(c/B)) = \mathrm{cl}_P(\mathrm{Cb}(c/\mathrm{cl}_P(B))) = \mathrm{cl}_P(\mathrm{Cb}(\bar{d}/\mathrm{cl}_P(B))) = \mathrm{cl}_P(\mathrm{Cb}(\bar{d}/B')).$$

Furthermore, since $A\bar{d} \downharpoonleft_{Ac} B$, Lemma 4.0.3 implies that $\mathrm{cl}_P(A\bar{d})$ is independent of $\mathrm{cl}_P(B)$ over $\mathrm{cl}_P(Ac)$, so

$$\mathrm{cl}_P(A'\bar{d}) \cap \mathrm{cl}_P(B') = \mathrm{cl}_P(Ac) \cap \mathrm{cl}_P(B) = \mathrm{cl}_P(A) = \mathrm{cl}_P(A').$$

It follows that A', B', $\bar{d}$ are a counter-example to the CM-triviality of P (over the additional parameters D).

We now replace A' by $A'' := \mathrm{Cb}(\bar{d}/A')$ and B' by $B'' := \mathrm{Cb}(\bar{d}/B') \cup A''$. As any element in a canonical base is algebraic in a finite part of a Morley sequence in the strong type, and since $\mathrm{tp}(\bar{d})$ is P-internal, both A'' and B'' have P-internal types over $\emptyset$. Furthermore, $\mathrm{tp}(\bar{d}/A'')$ is P-analysable and a non-forking restriction of $\mathrm{tp}(\bar{d}/A')$, hence orthogonal to all types of U_P-rank zero, and $\bar{d} \downharpoonleft_{A''} \mathrm{cl}_P(A'')$; similarly $\bar{d} \downharpoonleft_{B''} \mathrm{cl}_P(B'')$. Since $\bar{d} \downharpoonleft_{A''} A'$, Lemma 4.0.3 yields $\mathrm{cl}_P(\bar{d}A'') \downharpoonleft_{\mathrm{cl}_P(A'')} \mathrm{cl}_P(A')$. Now trivially $e \in \mathrm{cl}_P(\bar{d}A'') \cap \mathrm{cl}_P(B'')$ implies $e \in \mathrm{cl}_P(\bar{d}A') \cap \mathrm{cl}_P(B') = \mathrm{cl}_P(A')$, whence $e \in \mathrm{cl}_P(\bar{d}A'') \cap \mathrm{cl}_P(A') = \mathrm{cl}_P(A'')$. Thus $\mathrm{cl}_P(\bar{d}A'') \cap \mathrm{cl}_P(B'') = \mathrm{cl}_P(A'')$, and A'', B'' and $\bar{d}$ are a counter-example to the CM-triviality of P.

Now let $\mathfrak{M} \supseteq A'$ be a model with $\mathfrak{M} \downharpoonleft_{A'} B'\bar{d}$. By Lemma 4.0.4 we get $\mathrm{cl}_P(\mathfrak{M}\bar{d}) \cap \mathrm{cl}_P(\mathfrak{M}B') = \mathrm{cl}_P(\mathfrak{M})$ (putting $C = A'\bar{d}$ and $D = \mathfrak{M}$). Next consider a model $\mathfrak{N} \supseteq \mathfrak{M}B'$ with $\bar{d} \downharpoonleft_{\mathfrak{M}B'} \mathfrak{N}$. We claim that

$$\mathrm{cl}_P(\mathfrak{M}\bar{d}) \cap \mathrm{cl}_P(\mathfrak{N}) = \mathrm{cl}_P(\mathfrak{M}).$$

Take any $e \in \mathrm{cl}_P(\mathfrak{M}\bar{d}) \cap \mathrm{cl}_P(\mathfrak{N})$. Then $\mathrm{cl}_P(\mathfrak{M}\bar{d})$, and hence e, is independent of $\mathrm{cl}_P(\mathfrak{N})$ over $\mathrm{cl}_P(\mathfrak{M}B')$. Hence $e \in \mathrm{cl}_P(\mathfrak{M}B') \cap \mathrm{cl}_P(\mathfrak{M}\bar{d}) = \mathrm{cl}_P(\mathfrak{M})$.

As $\bar{d} \underset{A'}{\downarrow} \mathfrak{M}$ and $\bar{d} \underset{B'}{\downarrow} \mathfrak{N}$, the triple $\mathfrak{M}$, $\mathfrak{N}$ and $\bar{d}$ yields a final counter-example to the CM-triviality of P. $\square$

Proposition 4.5.4 *Let P be CM-trivial and $Q \subset P$ be an $\emptyset$-invariant subfamily. Then Q is CM-trivial.*

Proof: Suppose A, B and c are such that $c \underset{A}{\downarrow} \mathrm{cl}_Q(A)$, $c \underset{B}{\downarrow} \mathrm{cl}_Q(B)$, $\mathrm{tp}(c)$ is Q-internal, and $\mathrm{cl}_Q(cA) \cap \mathrm{cl}_Q(B) = \mathrm{cl}_Q(A)$. First, note that U_Q-rank is smaller than or equal to U_P-rank and $\mathrm{tp}(\mathrm{acl}(cA)/A)$ is Q-analysable. Hence $\mathrm{cl}_P(A) \cap \mathrm{acl}(cA) \subset \mathrm{cl}_Q(A) \cap \mathrm{acl}(cA)$; Lemma 4.0.2 yields $c \underset{A}{\downarrow} \mathrm{cl}_P(A)$. Similarly, $c \underset{B}{\downarrow} \mathrm{cl}_P(B)$.

Let $e \in \mathrm{cl}_P(cA) \cap \mathrm{cl}_P(B)$, and $e_0 \in \mathrm{Cb}(c/eA)$. As e_0 is definable in a Morley sequence in $\mathrm{stp}(c/eA)$, $\mathrm{tp}(e_0/A)$ is $\mathrm{tp}(c/A)$-internal and hence Q-analysable; since $U_Q(e_0/cA) \leq U_P(e_0/cA) \leq U_P(e/cA) = 0$ and similarly $U_Q(e_0/B) = 0$, we get $e \in \mathrm{cl}_Q(cA) \cap \mathrm{cl}_Q(B) = \mathrm{cl}_Q(A)$. Therefore $c \underset{A}{\downarrow} e_0$ and hence $c \underset{A}{\downarrow} e$, whence $e \in \mathrm{cl}_P(A)$. CM-triviality of P now implies $\mathrm{Cb}(c/A) \subset \mathrm{cl}_P(\mathrm{Cb}(c/B))$. But as $\mathrm{tp}(c)$ is Q-internal, so is $\mathrm{Cb}(c/A)$. Hence $\mathrm{Cb}(c/A) \subset \mathrm{cl}_Q(\mathrm{Cb}(c/B))$. $\square$

Proposition 4.5.5 *Let P_1 and P_2 be two perpendicular $\emptyset$-invariant families of types, and put $P = P_1 \cup P_2$. Suppose both P_1 and P_2 are CM-trivial. Then P is CM-trivial.*

Proof: Let A and B be sets and c an element such that $c \underset{A}{\downarrow} \mathrm{cl}_P(A)$ and $c \underset{B}{\downarrow} \mathrm{cl}_P(B)$, $\mathrm{tp}(c)$ is P-internal, and $\mathrm{cl}_P(cA) \cap \mathrm{cl}_P(B) = \mathrm{cl}_P(A)$ (having possibly added some parameters to the language). By Proposition 4.5.3 we may assume that $c = c_1 c_2$, where c_i is a tuple of realizations of P_i over $\emptyset$ (after adding some more parameters).

Since P_2 is foreign to P_1, the P_1-closure of any set is contained in its P-closure. Therefore $c_1 \underset{A}{\downarrow} \mathrm{cl}_{P_1}(A)$ and $c_1 \underset{B}{\downarrow} \mathrm{cl}_{P_1}(B)$; furthermore $\mathrm{tp}(c_1)$ is clearly P_1-internal. Let $e \in \mathrm{cl}_{P_1}(c_1 A) \cap \mathrm{cl}_{P_1}(B)$. Then $e \in \mathrm{cl}_P(cA) \cap \mathrm{cl}_P(B) = \mathrm{cl}_P(A)$, and $U_{P_1}(e/A) \leq U_P(e/A) = 0$. As both $\mathrm{tp}(e/c_1 A)$ and $\mathrm{tp}(c_1/A)$ are P_1-analysable, so is $\mathrm{tp}(e/A)$, whence $e \in \mathrm{cl}_{P_1}(A)$. CM-triviality of P_1 now implies

$$\mathrm{Cb}(c_1/A) \subset \mathrm{cl}_{P_1}(\mathrm{Cb}(c_1/B)) \subseteq \mathrm{cl}_P(\mathrm{Cb}(c_1/B));$$

we similarly obtain $\mathrm{Cb}(c_2/A) \subset \mathrm{cl}_P(\mathrm{Cb}(c_2/B))$.

Put $X = \mathrm{Cb}(c_1/A) \cup \mathrm{Cb}(c_2/A)$, a subset of $\mathrm{acl}(A)$. Then $c_1 \underset{X}{\downarrow} A$; as $c_1 \underset{A}{\downarrow} \mathrm{cl}_{P_1}(A)$, by Lemma 4.0.2 the type $\mathrm{tp}(c_1/A)$ is orthogonal to all types of U_{P_1}-rank zero. As P_1 and P_2 are perpendicular and $\mathrm{tp}(c_2)$ is P_2-internal, we get $c_1 \underset{A}{\downarrow} c_2$, whence $c_1 \underset{X}{\downarrow} Ac_2$, and $c_1 \underset{Xc_2}{\downarrow} A$. Since also $A \underset{X}{\downarrow} c_2$, we get $A \underset{X}{\downarrow} c_1 c_2$, and thus $\mathrm{Cb}(c/A) \subseteq X$; the inclusion $X \subseteq \mathrm{Cb}(c/A)$

is obvious. Similarly, $\mathrm{Cb}(c/B) = \mathrm{Cb}(c_1/B) \cup \mathrm{Cb}(c_2/B)$, and therefore $\mathrm{Cb}(c/A) \subset \mathrm{cl}_P(\mathrm{Cb}(c/B))$. $\square$

Proposition 4.5.6 *Let Q be an $\emptyset$-invariant family of P-semi-regular types. Then if P is CM-trivial, so is Q.*

Proof: Put $P' = P \cup Q$. Then internality, analysability and closure agree for P and P' by Lemma 4.1.10, so P' is CM-trivial. Therefore Q is CM-trivial by Proposition 4.5.4. $\square$

Problem 4.5.1   1. Find a theory T with CM-trivial types p and q such that $\{p, q\}$ is not CM-trivial.

2. If every type p is CM-trivial, is the whole theory CM-trivial (i.e. is the family of all types CM-trivial)?

3. If T is superstable and all regular types are CM-trivial, is T CM-trivial? What if T has finite rank?

4. If T is superstable and the family P of all regular types is CM-trivial, is T CM-trivial?

Note that for a theory of finite rank, questions 3. and 4. are the same by Proposition 4.5.5. Also, as every type in a superstable theory is analysable in the regular types, question 4. really asks for conditions which allow us to replace P-*internality* of tp(c) by P-*analysability*. My feeling is that the answer to questions 2.–4. is probably negative (as it is for the analogous questions about local modularity); one should then look for additional constraints on the theory under which local CM-triviality implies global CM-triviality.

4.6   CM-Trivial Groups

In this section we shall analyse CM-trivial groups, and show that certain structures, notably fields and bad groups, are not CM-trivial.

Lemma 4.6.1 *Let F be an infinite type-definable stable field F, let a, b and c be independent realizations of the generic type p of F, and put $d = ca + b$. Then $\mathrm{cl}_p(a, b) \cap \mathrm{cl}_p(c, d) = \mathrm{cl}_p(\emptyset)$.*

Proof: We may assume that F is type-defined over $\emptyset$. Choose a_1 generic and independent of a, b, c and put $b_1 = d - ca_1 = c(a - a_1) + b$. As for two independent generic elements x and y both the sum $x + y$ and the product xy are generic over x and over y, it can be immediately seen that c, d, a and c, d, a_1 are independent generic triples, and $a, b \underset{\smile}{\bigcup} a_1, b_1$. Note that $b = d - ca$.

By Lemma 4.0.3 we get $\mathrm{cl}_p(a, b) \mathop{\downarrow}_{\mathrm{cl}_p(\emptyset)} \mathrm{cl}_p(a_1, b_1)$. Both a and a_1 realize p over (c, d) and, by stationarity of p and independence, also over $\mathrm{cl}_p(c, d)$; therefore

$$\mathrm{tp}(a/\mathrm{cl}_p(c, d)) = \mathrm{tp}(a_1/\mathrm{cl}_p(c, d)) = p|\mathrm{cl}_p(c, d),$$

which yields $\mathrm{tp}(a, b/\mathrm{cl}_p(c, d)) = \mathrm{tp}(a_1, b_1/\mathrm{cl}_p(c, d))$. Now consider $e \in \mathrm{cl}_p(a, b) \cap \mathrm{cl}_p(c, d)$. Then $e \in \mathrm{cl}_p(a_1, b_1)$, and therefore $e \in \mathrm{cl}_p(a, b) \cap \mathrm{cl}_p(a_1, b_1)$, but these two closures are independent over $\mathrm{cl}_p(\emptyset)$. So $e \in \mathrm{cl}_p(\emptyset)$. $\square$

Proposition 4.6.2 *The generic type p of an infinite field F interpretable in a stable theory is not CM-trivial.*

Proof: We may assume that F is type-definable over $\emptyset$. Let a, b, c, d, e, f be independent generic elements. We consider the plane P in F^3 given by $z = ax + by + c$, the line $l \subset P$ given by $y = dx + e$, and the point $q = (f, g, h) \in l$. We identify P with its canonical parameter (which is the ordered triple (a, b, c)), and similarly l with its canonical parameter. Clearly l is inter-definable with (d, e) over P; as the line l is generic in P and the point q is generic both in l and in P with $q \mathop{\downarrow}_l P$, we get $l = \mathrm{Cb}(q/l, P)$ and $P = \mathrm{Cb}(q/P)$. Actually, l is given by the two equations $z = ax+b(dx+e)+c = (a+bd)x + (be+c)$ and $y = dx + e$, so l is equivalent to the ordered quadruple $(d, e, a + bd, be + c)$. Clearly a (say) is generic over l, whence $P \not\in \mathrm{cl}_p(l)$.

By Lemma 4.6.1 (with P named), $\mathrm{cl}_p(q, P) \cap \mathrm{cl}_p(l, P) = \mathrm{cl}_p(P)$. Clearly $\mathrm{tp}(q)$ is p-internal. Furthermore, q is inter-definable with f over l and $f \models p|l$, so $q \mathop{\downarrow}_l \mathrm{cl}_p(l)$. Finally, q is inter-definable with (f, g) over P and $(f, g) \models p^{(2)}|P$, so $q \mathop{\downarrow}_P \mathrm{cl}_p(P)$. Hence P, $\{l, P\}$ and q yield a counter-example to the CM-triviality of p. $\square$

Lemma 4.6.3 *Let G be a type-definable stable connected centreless group, which is internal in some $\emptyset$-invariant family Q of types. Suppose there is a family $\mathfrak{B}$ of type-definable nilpotent connected subgroups of G such that for any $B \in \mathfrak{B}$ there is $q \in Q$ not foreign to B, the generic type of every $B \in \mathfrak{B}$ is foreign to every type of U_Q-rank zero, and every generic element of G is contained in some $B \in \mathfrak{B}$. Then Q is not CM-trivial.*

Proof: We may assume that G is saturated and type-definable over $\emptyset$. Take a group $B \in \mathfrak{B}$ containing a generic element of G and consider the family $\mathfrak{B}'$ consisting of all conjugates of B under model-theoretic automorphisms of G. By $\emptyset$-invariance of Q, the foreignness conditions in the assumption must still hold for $\mathfrak{B}'$, and by connectivity of G every generic element must be contained in some conjugate of B, whence in some element of $\mathfrak{B}'$. We may therefore replace $\mathfrak{B}$ by $\mathfrak{B}'$ and assume that all groups in $\mathfrak{B}$ are (model-theoretically) conjugate. Now let b be a canonical parameter for B, and c a generic element for G independent of b. Let l be the canonical parameter for the coset cB. Note that connectivity of B implies that cB is connected, i.e. there is no relatively definable subcoset of finite index, and cB has a unique generic type (the left translate by c of the generic type of B).

Claim. $\mathrm{cl}_Q(c) \cap \mathrm{cl}_Q(l) = \mathrm{cl}_Q(\emptyset)$.

Proof of Claim: B contains an element d generic for G and independent of c, so cB contains c and cd, which are two independent generic elements of G. Furthermore, both c and cd are generic elements of the coset cB over l. As cB is connected, $\mathrm{stp}(c/l) = \mathrm{stp}(cd/l)$; since this is a translate of the generic type of B which is foreign to all types of U_Q-rank zero, both c and cd are independent of $\mathrm{cl}_Q(l)$ over l, and $\mathrm{tp}(c/\mathrm{cl}_Q(l)) = \mathrm{tp}(cd/\mathrm{cl}_Q(l))$. So if $e \in \mathrm{cl}_Q(c) \cap \mathrm{cl}_Q(l)$, we have $e \in \mathrm{cl}_Q(cd)$ as well; but $c \underset{}{\bigcup} cd$ implies $\mathrm{cl}_Q(c) \underset{\mathrm{cl}_Q(\emptyset)}{\bigcup} \mathrm{cl}_Q(cd)$ by Lemma 4.0.3, whence $e \in \mathrm{cl}_Q(\emptyset)$. $\square$

Now take three generic independent elements a, c, d of G over b. Let $G_a :=$ $\{(g, g^a) : g \in G\} < G \times G$ and $B_a := \{(g, g^a) : g \in B\} < G_a$, and consider the coset $(c, d)G_a$ with canonical parameter P, and the coset $(c, d)B_a$ with canonical parameter l. Then (c, d) is a generic point of $(c, d)G_a$ over P and a generic point of $(c, d)B_a$ over l. Furthermore $c, d \underset{l}{\bigcup} P$, and $P = \mathrm{Cb}(c, d/P)$ and $l = \mathrm{Cb}(c, d/l) = \mathrm{Cb}(c, d/l, P)$, as a coset is the canonical base of its generic elements.

Claim. $\mathrm{cl}_Q(c, d, P) \cap \mathrm{cl}_Q(l, P) = \mathrm{cl}_Q(P)$.

Proof of Claim: We name P. Let (c', d') be generic points of $(c, d)G_a$ independent of b, c, d over P. We can now apply the first claim to the group G_a with the coset $(c'^{-1}c, d'^{-1}d)B_a$ (over P, c', d') and obtain

$$(\dagger) \qquad \mathrm{cl}_Q(c, d, P, c', d') \cap \mathrm{cl}_Q(l, P, c', d') = \mathrm{cl}_Q(P, c', d').$$

But $c', d' \underset{P}{\bigcup} b, c, d$, and $l \in \mathrm{dcl}(b, c, d, P)$. Therefore

$$\mathrm{cl}_Q(c', d', P) \underset{\mathrm{cl}_Q(P)}{\bigcup} \mathrm{cl}_Q(c, d, l, P),$$

and $(\dagger)$ implies

$$\mathrm{cl}_Q(c, d, P) \cap \mathrm{cl}_Q(l, P) \subseteq \mathrm{cl}_Q(P, c', d') \cap \mathrm{cl}_Q(l, P) = \mathrm{cl}_Q(P). \square$$

By Lemma 3.1.8, if a type $q_0 \in Q$ is not foreign to B, it cannot be foreign to $Z(B)$ either. We shall now show that $P \notin \mathrm{cl}_Q(l)$. Let a' be generic for $Z(B)$ over a, b, c, d, and put $a'' = a'a$. Then a'' is generic for G over b, c, d and there is an automorphism σ which fixes b, c, d and moves a to a''. Furthermore

$$\sigma(B) = B_{a''} = \{(g, g^{a''}) : g \in B\} = \{(g, g^a) : g \in B\} = B_a,$$

since $g^{a'} = g$ for all $g \in B$. Therefore σ must fix l. Let $P' := \sigma(P)$. So if $P \in \mathrm{cl}_Q(l)$, also $P' \in \mathrm{cl}_Q(l)$. Now G_a is definable over P (as PP^{-1}) and

$G_{a''} = \sigma(G_a)$ is definable over P'. But G is centreless, so $a \in \mathrm{dcl}(G_a)$ and $a'' \in \mathrm{dcl}(G_{a''})$, whence $a' \in \mathrm{dcl}(a, a'') \subseteq \mathrm{dcl}(P, P') \subset \mathrm{cl}_Q(l)$. However, a' is generic for $Z(B)$ over a, b, c, d, so

$$U_Q(a'/l) \geq U_Q(a'/a, b, c, d) = U_Q(Z(B)) > 0,$$

a contradiction.

Therefore $P \notin \mathrm{cl}_Q(l)$, and P, $\{l, P\}$ and (c, d) are a counter-example to CM-triviality of Q. $\square$

Theorem 4.6.4 *Let G be a type-definable stable soluble group such that every type is CM-trivial. Then G is nilpotent-by-finite.*

Proof: Suppose not, and let G be a counter-example of minimal derived length, which we may assume to be saturated. A nilpotent group is contained in a definable one, so if G^0 had a normal nilpotent subgroup N of finite index, then N would be contained in a definable G-invariant nilpotent group $\bar{N}$, and by saturation $\bar{N}$ must intersect G in a subgroup of finite index. We may therefore assume that G is connected. By Lemma 1.1.7 the definable hull $\mathrm{dc}(G')$ is connected, and by minimality of the derived length G' is contained in a relatively definable normal nilpotent subgroup N of G.

Since all the groups $H_i := (Z_{i+1}(N)/Z_i(N)) \rtimes (G/N)$ are interpretable in G and G is nilpotent iff all the H_i are nilpotent, our counter-example must have derived length 2 and we can choose N to be abelian. Furthermore, as the intersection of a chain of non-nilpotent connected groups (in a sufficiently saturated model) is again non-nilpotent and connected, we may assume that every proper relatively definable subgroup of G is nilpotent-by-finite.

Suppose there is $z \in Z_2(G) - Z_1(G)$ of finite order n modulo $Z_1(G)$. Then for any $g \in G$ we have

$$[g^n, z] = [g, z]^n = [g, z^n] = 1,$$

so $C_G(z)$ is a normal subgroup of G and $G/C_G(z)$ has exponent n. Since $C_G(z)$ is proper in G, it must be nilpotent-by-finite by minimality of G; but then G itself is nilpotent-by-finite by Corollary 1.2.18, a contradiction.

Suppose that $Z_3(G) > Z_2(G)$, and let $z \in Z_3(G)$ and $g \in G$ be such that $[z, g] \notin Z(G)$. Put $H_0 = \{h \in G : [h, g] \in Z_2(G)\}$ and $H := \{h \in H_0 : [h, H_0] \subseteq Z_2(G)\}$. Then H is a subgroup of G containing $Z_3(G)$ and g, and $H' \leq Z_2(G)$, so $H/Z(G)$ is nilpotent of class 2, with torsion-free commutator subgroup. By Corollary 1.3.10 there is an interpretable infinite field, contradicting Proposition 4.6.2. We may therefore divide out by the second centre and assume that G is centreless.

If a generic element $g \in G$ had finite order modulo N, then G/N would have finite exponent and Corollary 1.2.18 would imply that G is nilpotent-by-finite, contradicting our assumptions. Hence $C_G(g)N/N$ is infinite. Suppose

$C_G(g)N$ is a proper subgroup of G. Minimality of G implies that $C_G(g)N$ is nilpotent-by-finite and contains a maximal normal nilpotent subgroup M_g of finite index, which must be contained in the Fitting subgroup $F(G)$. Now some n-th power of g lies in M_g and thus in $F(G)$; this holds for every generic element, and $G/F(G)$ has finite exponent at most n. But $F(G)$ is nilpotent by Theorem 1.2.11, so by Corollary 1.2.18 again, G is nilpotent-by-finite, a contradiction.

Therefore $C_G(g)N = G$ for any generic g. If $C_G(g) \cap N$ contains a non-trivial element h, then $C_G(h)$ is a proper subgroup of G containing g and N, and has a maximal normal nilpotent subgroup M_g of finite index. Then M_g contains N and is normal in G and we get a contradiction as above. So $C_G(g) \cap N = \{1\}$ and G is the semi-direct product $N \rtimes C_G(g)$; connectivity of G now implies that $C_G(g)$ is connected and hence nilpotent. Furthermore, the generic type p of G is equivalent to the product of the generic types of N and of $C_G(g)$; in particular p cannot be foreign to $C_G(g)$, and the generic type of $C_G(g)$ is foreign to all types to which p is foreign.

Let $\mathfrak{B}$ be the family of centralizers of generic elements of G. Then $(G, \mathfrak{B})$ satisfies the assumptions of Lemma 4.6.3 with $Q = \{p\}$, contradicting CM-triviality of p. This final contradiction proves the theorem. $\square$

Theorem 4.6.5 *Let G be a type-definable stable group such that every type is CM-trivial. Suppose every type in G is non-orthogonal to some regular type. Then G is nilpotent-by-finite.*

Proof: Suppose G is a counter-example to the theorem. As in the proof of Theorem 4.6.4 we may assume that G is connected and every proper relatively definable subgroup of G is nilpotent-by-finite. By Theorem 4.6.4, the connected component of every normal soluble connected subgroup S of G is nilpotent and hence contained in the Fitting subgroup $F(G)$, which is nilpotent by Theorem 1.2.11. Since G is connected, S must be centralized by G modulo $F(G)$. Now $\bar{G} := G/Z(G/F(G))$ cannot be nilpotent, since otherwise G would be soluble, and hence nilpotent by Theorem 4.6.4. Furthermore any proper relatively definable subgroup of $\bar{G}$ must still be nilpotent-by-finite, as its pre-image is proper in G. It follows that $\bar{G}$ is definably simple and $\text{gen}(\bar{G})$ is $\mathfrak{R}$-primary by Corollary 3.1.2.

If q is any type to which the generic type of $\bar{G}$ is non-orthogonal, then $\bar{G}$ is q-internal. In particular, since $\text{gen}(\bar{G})$ is non-orthogonal to some regular type p, it is p-internal, and foreign to all types of U_p-rank zero; in other words, $\text{gen}(\bar{G})$ is p-semi-regular. In particular, $\bar{G}$ is p-connected of finite U_p-rank.

By Theorem 3.8.1, there is a type-definable p-connected centreless section G_1 of $\bar{G}$ with a relatively definable p-connected nilpotent (Borel) subgroup B, such that the family $\mathfrak{B}$ of the conjugates of B in G_1 satisfies the assumptions in Lemma 4.6.3 with $Q = \{p\}$. So p is not CM-trivial, a contradiction. $\square$

Theorem 4.6.6 *Let* G *be a CM-trivial type-definable torsion-free small stable group. Then* G *is abelian and connected.*

Proof: If H is any type-definable subgroup of G containing some element h, and if $n > 1$ is any positive natural number, then the endomorphism $x \mapsto x^n$ of $Z(C_H(h))$ has trivial kernel, and must be surjective by Corollary 5.0.7. (The proof of Corollary 5.0.7 will not depend on the present section.) So any $h \in H$ is divisible in H, and must be contained in every subgroup of finite index. Hence every type-definable subgroup of G is connected, and clearly this still holds for all type-definable sections of G.

Second, by Proposition 4.6.2, Corollary 1.3.10 and Theorem 4.6.4 any soluble subgroup of G must be abelian-by-finite. As in the proof of Theorem 4.6.5 we may assume that a potential counter-example G is definably simple, and that every proper relatively definable subgroup is nilpotent-by-finite, hence abelian-by-finite, and connected by the argument above. In particular, for a generic element g the centralizer $C_G(g)$ is abelian and connected. If we take Q to be the family of all types, then the family $\mathfrak{B} := \{C_H(g) : g \text{ generic}\}$ satisfies the assumptions of Lemma 4.6.3. Therefore G is not CM-trivial, a contradiction. $\square$

Problem 4.6.1   1. Is any CM-trivial stable group nilpotent-by-finite?

    2. Suppose every type in the stable group G is CM-trivial. Is G nilpotent-by-finite?

    3. If G is analysable in a CM-trivial family P, is G^P nilpotent? Or soluble?

4.7   Dimensionality

Definition 4.7.1 A theory is *dimensional* if there is a set Q of stationary types such that any non-algebraic type is non-orthogonal to some type in Q. The theory is *finite-dimensional* if the set Q is finite.

Remark 4.7.1 Dimensional theories are also called *non-multidimensional*.

In a dimensional theory there is only a bounded number of types, up to non-orthogonality; as a consequence, forking is easier to control definably. If the dimensions are carried by regular types, one may also try to classify the models of a theory by the dimension of the set of realizations of these regular types (although the *real* classification – see Shelah [105] – is far more complicated...).

Proposition 4.7.1 *Suppose T is dimensional and every type is non-orthogonal to a regular type. Then there is a family of regular types carrying the dimensions.*

Proof: Let P be the collection of all regular types, and Q the family of types carrying the dimensions. Then every type in Q has weight at most $|T|$, and can be non-orthogonal to at most $|T|$ pairwise orthogonal types in P. As regular types are either orthogonal or domination-equivalent and any one is non-orthogonal to a type in Q, there is a set P' with $|P'| \leq |Q| + |T|$ of regular types such that every regular type is non-orthogonal to some type in P'. Now since every type is non-orthogonal to a regular type, it must be non-orthogonal to a type in P'. $\square$

Theorem 4.7.2 *Let G be a group of finite U_P-rank. Then there are finitely many regular types of U_P-rank one such that every type of non-zero U_P-rank is non-orthogonal to at least one of them.*

Proof: By Proposition 3.5.10 there is a regular type p_1, of U_P-rank one and foreign to all types of U_P-rank zero, which is non-orthogonal to $\mathrm{gen}(G^P)$. By Theorem 3.1.1 there is a p_1-internal quotient G^P/N for some G-invariant relatively definable subgroup N, with $U_P(G/N) > 0$. Then $U_P(N) < U_P(G)$, so we can repeat and obtain a sequence $G = G_0 \geq G_1 \geq \cdots \geq G_n = \{1\}$ of definable normal subgroups of G, and regular types $p_1, \ldots, p_n$ of U_P-rank one and foreign to all types of U_P-rank zero, such that the quotient G_{i-1}^P/G_i is p_i-internal for $1 \leq i \leq n$. It follows that every type is analysable in $Q := \{p_1, \ldots, p_n\}$ together with the collection of types of U_P-rank zero. We add the parameters needed for the types in Q to the language.

Now suppose that $U_P(a/A) > 0$. If we put $A' = \{a' \in \mathrm{acl}(Aa) : U_P(a'/A) = 0\}$, then $\mathrm{tp}(a/A')$ is foreign to all types of U_P-rank zero by Proposition 3.0.6. Therefore a must be non-orthogonal to a realization of some $p_i \in Q$, say b, with $b \underset{}{\downarrow} A'$ and $a \underset{A'}{\not\downarrow} b$. But this implies $b \underset{A}{\not\downarrow} a$, and we are done. $\square$

Corollary 4.7.3 *A group of finite Lascar rank is finite-dimensional.*

Proof: Take P to be the family of all types. Then Lascar rank equals U_P-rank, and a type is algebraic iff it has Lascar rank zero. $\square$

For the next couple of results, we need another model-theoretic notion.

Definition 4.7.2 A theory T does not have the *finite cover property (fcp)* if for every formula $\varphi(x, y)$ without parameters there is a natural number $n_\varphi < \omega$ such that for all a the set defined by $\varphi(x, a)$ is either infinite or has at most n_φ elements.

Proposition 4.7.4 *A dimensional theory does not have the finite cover property.*

Proof: Let Q be the set of types carrying the dimensions. We may assume that all the types in Q are over $\emptyset$. Now consider a formula $\varphi(x, y)$ such that for all $n < \omega$ there are some model $\mathfrak{M}_n$ and a parameter a_n such that $\varphi(x, a_n)$ is finite with at least n elements. Let $\mathfrak{N} := (\prod_\omega \mathfrak{M}_n)/\mathcal{U}$ be a non-principal ultraproduct, and put $\bar{a} := (a_n : n < \omega)/\mathcal{U}$. The set $\varphi(\mathfrak{N}, \bar{a})$ is clearly infinite and there must be a non-algebraic type q over $\mathfrak{N}$ containing this formula. But then q is non-orthogonal to some type p in P, so there are a superset B of $\mathfrak{N}$, a realization b of p independent of B, and a realization c of q, such that $b \not\!\perp_B c$. In particular, there is a formula $\psi(x, y, B)$ not in the bound $\beta(p)$, such that $\models \psi(b, c, B)$. (Of course, the formula only uses a finite part of B.) Therefore

$$\mathfrak{N} \models \exists Z \, d_p x \, \exists y \, \varphi(y, \bar{a}) \wedge \psi(x, y, Z).$$

It follows that there is some $I \in \mathcal{U}$ such that for all $n \in I$ we have

$$\mathfrak{M}_n \models \exists Z \, d_p x \, \exists y \, \varphi(y, a_n) \wedge \psi(x, y, Z).$$

This means that there is some set B' of parameters (containing a_n) such that there are $b' \models p|B'$ and $c' \models \varphi(x, a_n)$ with $b' \not\!\perp_{B'} c'$. But as $\varphi(x, a_n)$ is finite, this is impossible. $\square$

Definition 4.7.3 A formula $\varphi(x)$ is *minimal* in a model $\mathfrak{M}$ if it is infinite and there is no infinite, co-infinite $\mathfrak{M}$-definable subset of φ. The formula is *strongly minimal* if it is minimal in all elementary extensions of $\mathfrak{M}$ as well. A type is strongly minimal if it is the unique transcendental (i.e. non-algebraic) type in a strongly minimal formula.

So a type or formula is strongly minimal iff it has Morley rank and degree one.

Lemma 4.7.5 *A minimal formula in a theory without the finite cover property is strongly minimal.*

Proof: By the lack of the finite cover property and the fact that φ is minimal in $\mathfrak{M}$, given any formula $\psi(x, y)$ there is n such that for any $a \in \mathfrak{M}$ either $\psi(x, a) \wedge \varphi(x)$ or $\varphi(x) \wedge \neg\psi(x, a)$ has size n (and the other one is correspondingly infinite). But this is preserved in all elementary extensions of $\mathfrak{M}$. $\square$

Corollary 4.7.6 *A group G of finite Morley rank is finite-dimensional, and every dimension is associated with a strongly minimal type over an isolated parameter. There is a series $G = G_n \geq G_{n-1} \geq \cdots \geq G_1 \geq G_0 = \{1\}$ of normal subgroups definable over these parameters, such that every quotient G_i/G_{i-1} is in the algebraic closure of one of the dimensions.*

Proof: As $U(G) \leq RM(G) < \omega$, the group is finite-dimensional by Corollary 4.7.3, and does not have the finite cover property. Now G is ω-stable by Remark 0.2.9 and hence trivially small; by Proposition 0.2.7 there is a countable atomic model. Let H be a definable subgroup of $\mathfrak{M}$ of minimal Morley rank and Morley degree. Then H has no subgroup of infinite index (which would have smaller rank), and no subgroup of finite index either (which would have smaller degree), so H is a minimal definable subgroup of G.
Claim. There is a minimal formula $\varphi(x, m)$ over $\mathfrak{M}$ contained in H.
Proof of Claim: Suppose not. Then any infinite formula over $\mathfrak{M}$ contained in H can be split into two infinite halves. This yields an infinite 2-splitting tree, whose branches can be completed to continuum many types over $\mathfrak{M}$, contradicting ω-stability. $\square$

By Lemma 4.7.5 the formula $\varphi(x, m)$ is strongly minimal, and we may translate it inside H so that it contains 1. As it is contained in a minimal subgroup, it is indecomposable; by Theorem 3.6.11 the family of conjugates of φ must generate a normal subgroup G_1. Clearly G_1 is definable over m, and $G_1 \subseteq \text{acl}(\varphi(\mathfrak{M}, m))$. Now G/G_1 has smaller Lascar rank, and we finish by induction. $\square$

Definition 4.7.4 A theory T is *almost strongly minimal* if there is a strongly minimal set $\varphi(x, m)$ over an isolated parameter m such that every model $\mathfrak{M}$ containing m is in $\text{acl}(\varphi(\mathfrak{M}, m))$.

So the quotients G_i/G_{i-1} in Corollary 4.7.6 are almost strongly minimal.

Proposition 4.7.7 *An almost strongly minimal theory T is categorical in all powers greater than $|T|$.*

Proof: Suppose $\vartheta(y)$ isolates the type of a parameter m such that $\varphi(x, m)$ is strongly minimal, and $\text{acl}(\varphi^{\mathfrak{N}}) = \mathfrak{N}$ for every $\mathfrak{N}$ which contains m. By compactness, this means that there are finitely many formulæ $\varphi_1(x, \bar{y}), \ldots, \varphi_n(x, \bar{y})$ without parameters, such that for all tuples $\bar{n}$ of realizations of $\varphi(x, m)$ the set defined by $\varphi_i(x, \bar{n})$ is finite of bounded size, and

$$\models \forall x \bigvee_{i=1}^{n} \left[\exists \bar{y}\, \varphi_i(x, \bar{y}) \wedge \bigwedge_{y_j \in \bar{y}} \varphi(y_j, m) \right].$$

Let $\mathfrak{M}_1$ and $\mathfrak{M}_2$ be two uncountable models of the same cardinality κ. Then there is a realization $m_1 \in \mathfrak{M}_1$ and $m_2 \in \mathfrak{M}_2$ of ϑ, and $\mathfrak{M}_i \subseteq \text{acl}(\varphi(\mathfrak{M}_i, m_i))$ for $i = 1, 2$. Now for realizations of a strongly minimal set forking is the same as becoming algebraic; as forking does satisfy the exchange law by forking symmetry and algebraic closure is easily seen to satisfy the other axioms for a pre-geometry, $\varphi(\mathfrak{M}_i, m_i)$ has a basis B_i for $i = 1, 2$. Note that $\operatorname{acl}(B_i m_i) \supseteq \operatorname{acl}(\varphi(\mathfrak{M}_i, m_i)) \supseteq \mathfrak{M}_i$; as

$$|T| < \kappa = |\mathfrak{M}_i| \leq |\operatorname{acl}(B_i m_i)| \leq |B_i \cup T|,$$

we see that B_1 and B_2 must have the same cardinality κ. Furthermore m_1 and m_2 have the same type, so the map which sends $B_1 m_1$ to $B_2 m_2$ is a partial isomorphism from $\mathfrak{M}_1$ to $\mathfrak{M}_2$. But then this partial isomorphism can be extended to an isomorphism of the algebraic closures, and hence to an isomorphism from $\mathfrak{M}_1$ to $\mathfrak{M}_2$. $\square$

Proposition 4.7.8 *If G has finite Morley rank and $\operatorname{gen}(G)$ is strongly $\mathfrak{R}$-primary, then G is almost strongly minimal.*

Proof: Since G has finite Morley rank, it contains a strongly minimal subset $\varphi(x, m)$ defined over an isolated parameter, with transcendental type p. Furthermore $\operatorname{gen}(G)$ is non-orthogonal to a regular type q of Lascar rank one. So G is q-analysable, and p cannot be foreign to q. As forking extensions of q are algebraic, p is non-orthogonal to q. Hence q is dominated by p by Lemma 4.1.15, and $\operatorname{gen}(G)$ must be non-orthogonal to p as well. It follows that G is φ-internal, and hence almost strongly minimal. $\square$

So a simple group or a field of finite Morley rank is almost strongly minimal and hence uncountably categorical. An easy adaptation of the proof shows that a simple group or a field of finite Lascar rank is one-dimensional.

Proposition 4.7.9 *If G has finite Morley rank and no normal abelian subgroups, then G decomposes as a finite direct product of almost strongly minimal subgroups.*

Proof: Clearly we may assume that G is connected. By Theorem 3.4.6 the connected component G^0 (which equals the Φ-component) decomposes as a finite direct product of $\mathfrak{R}$-connected subgroups H_i; by Proposition 3.4.7 every H_i has a strongly $\mathfrak{R}$-primary generic type, and is almost strongly minimal by Proposition 4.7.8. $\square$

Finally, we shall connect Lascar and Morley ranks:

Proposition 4.7.10 *Suppose T is dimensional and all the dimensions are carried by strongly minimal types. Then Morley rank equals Lascar rank and is finite, and for every formula $\varphi(x, y)$ and every natural number $n < \omega$ the set $\{a : RM(\varphi(x, a)) = n\}$ is definable.*

Proof: We may assume that the dimensions are carried by the strongly minimal sets in the family $Q := \{\varphi_j(x) : j \in J\}$, which we assume to be

$\emptyset$-definable. Note firstly that the true formula $(x = x)$ is analysable in Q, so by Lemma 3.0.8 there is an analysis in finitely many steps. As $U(\varphi_j) = 1$ for $j \in J$, every type has finite Lascar rank, bounded by the number of steps used. In particular, every formula has a Lascar rank, namely the maximum of the Lascar ranks of the types it contains.

Claim. $\{a' : U(\varphi(x, a')) \geq n\}$ is open.

Proof of Claim: If $\models \varphi(b, a)$ and $U(b/a) \geq n$, then there are realizations b_i of p_i (where p_i is the transcendental type for some dimension in Q) and finite tuples $a \subseteq a_1 \subseteq \cdots \subseteq a_n$ with $b_i \in a_{i+1}$ for all $i < n$, such that $b_i \mathop{\smile}\limits_{a_i} a_i$ and $b_i \mathop{\smile}\limits_{a_i} b$. Then $b_i \in \mathrm{acl}(a_i b)$ via some formula $\varphi_i(b_i, a_i, b)$, and a satisfies

$$\vartheta(x) := \exists z_1\, d_{p_1} y_1 \cdots \exists z_n\, d_{p_n} y_n\, \exists y\, \varphi(y, x) \wedge \bigwedge_{i=1}^{n} \varphi_i(y_i, z_i, y).$$

Clearly for any $a' \models \vartheta$ we have $U(\varphi(x, a')) \geq n$. $\square$

Claim. If $U(p) = n$, then there is $\varphi \in p$ with $U(\varphi) = n$. Furthermore, for every formula $\varphi(x, y)$ and every $n < \omega$ there is an $\emptyset$-definable set $\vartheta(y)$ such that $\models \vartheta(a)$ iff $U(\varphi(x, a)) \leq n$.

Proof of Claim: We use induction on n. If $U(p) = 0$, then p is algebraic and contained in a finite formula of Lascar rank zero. By Proposition 4.7.4 a dimensional theory does not have the finite cover property, so there is some k such that $U(\varphi(x, a)) = 0$ iff $\varphi(x, a)$ has size at most k, and this is definable over $\emptyset$. So suppose now that the assertion holds for ranks at most n, and consider a type $\mathrm{tp}(a/A)$ of rank $n + 1$. Assume first that A is algebraically closed. Then there are $B \supseteq A$ with $B \mathop{\smile}\limits_A a$, and b realizing some $\varphi_j \in Q$ with $a \mathop{\smile}\limits_B b$; then $U(a/Bb) = n$ and $b \in \mathrm{acl}(Ba)$. By the inductive hypothesis there are a formula $\varphi(x, B, b) \in \mathrm{tp}(a/Bb)$ of Lascar rank n, and a formula $\vartheta(Z, z)$ which holds of $B'b'$ iff $U(\varphi(x, B', b')) \leq n$. Consider the formula

$$\psi(x, B) := \exists z\, \varphi_j(z) \wedge \vartheta(B, z) \wedge \varphi(x, B, z),$$

and the formula $\varphi'(x, A) := d_{\mathrm{tp}(B/A)} Z\, \psi(x, Z)$. Clearly, $\varphi'(x, A)$ is contained in $\mathrm{tp}(a/A)$ and has Lascar rank $n + 1$. (Of course, all of this only mentions a finite subtuple of B.) If A is not algebraically closed, then we just take the disjunction of the finitely many A-conjugates of $\varphi'(x, \mathrm{acl}(A))$.

If $(B_i : i < \omega)$ is a Morley sequence in $\mathrm{tp}(B/A)$ and $a \models \varphi'(x, A)$, then by Theorem 0.2.31 there is some $n < \omega$ such that a can fork with at most n of the B_i via the formula $\neg\psi(a, B_i)$. So

$$\vartheta(A) := \exists Z_0, \ldots, Z_n\, \forall x\, (\varphi'(x, A) \to \bigvee_{i=0}^{n} \psi(x, Z_i))$$

holds; if A is not algebraically closed and $\varphi'(x, A)$ has k conjugates over A, we need $k(n+1)$ different Z_i to take account of that. Clearly $\models \vartheta(A')$ implies

$U(\varphi'(x, A')) \leq n + 1$, whence $\varphi''(x, A) := \varphi'(x, A) \vee \neg \vartheta(A)$ is a formula in p with the property that $\models \vartheta(A')$ iff $U(\varphi''(x, A')) \leq n + 1$ (unless $n + 1$ is the maximal rank, in which case the assertion is trivial).

Now suppose $\varphi(x, a)$ is an arbitrary formula of Lascar rank $n + 1$. Then for every type p in $[\varphi]$ over a there are a formula $\varphi_p(x, a) \in p$ and a true sentence $\vartheta(a)$ such that $\models \vartheta_p(a')$ iff $U(\varphi_p(x, a')) \leq n + 1$. Then a finite family $(\varphi_p(x, a) : p \in P)$ covers $\varphi(x, a)$, and

$$\bigwedge_{p \in P} \vartheta_p(a') \wedge [\forall x\, \varphi(x, a') \rightarrow \bigvee_{p \in P} \varphi_p(x, a')]$$

is true for $a = a'$ and implies $U(\varphi(x, a')) \leq n + 1$. Therefore the set of a' such that $\varphi(x, a')$ has Lascar rank at most $n + 1$ is open. As it is also closed by the first claim, it is definable by compactness. $\square$

Claim. If $RM(p) \geq n$ then $U(p) \geq n$, for any type p.
Proof of Claim: This is clear for $n = 0$, as types of rank zero are algebraic types. So suppose it is true for n and assume $RM(a/A) = n + 1$. By going to a non-forking extension, we may assume that there is b realizing some formula $\varphi_j \in Q$ with $a \underset{A}{\not\smile} b$. Then $b \in \mathrm{acl}(Aa)$. As there is another type of Morley rank at least n in any neighbourhood of $\mathrm{tp}(a/A)$, there is a sequence $(a_i b_i : i < \omega)$ such that $\mathrm{tp}(a_i b_i/A)$ has limit $\mathrm{tp}(ab/A)$, the types $\mathrm{tp}(a_i/A)$ are all distinct of Morley rank at least n, and $b_i \models \varphi_j$ for all $i < \omega$.

If infinitely many of the b_i are non-algebraic over A, then they satisfy $\mathrm{tp}(b/A)$ and we may actually assume that $b_i = b$ for all $i < \omega$. Then $\mathrm{tp}(a_i/Ab)$ has limit $\mathrm{tp}(a/Ab)$. By the inductive hypothesis $U(a_i/A) \geq n$ whence $RM(a_i/Ab) \geq U(a_i/Ab) \geq n - 1$. Therefore $RM(a/Ab) \geq n$; by the inductive hypothesis again $U(a/Ab) \geq n$ and $U(a/A) \geq n + 1$.

Otherwise we may assume that all b_i are algebraic over A. Then $RM(a_i/Ab_i) = RM(a_i/A) \geq n$, whence $U(a_i/Ab_i) \geq n$ by the inductive hypothesis. Suppose $U(a/Ab) \leq n - 1$, and consider a formula $\varphi(a, b)$ which says so. For big enough i we get that $\varphi(x, b_i)$ has Lascar rank $n - 1$ and $\models \varphi(a_i, b_i)$, a contradiction. Hence $U(a/Ab) \geq n$ and $U(a/A) \geq n + 1$. $\square$

As $RM(p) \geq U(p)$ in any case, we are done. $\square$

Remark 4.7.2 By an argument similar to that used in the proof of Proposition 4.7.10, it is possible to show that a dimensional theory where all dimensions are associated to a type of Shelah rank one is superstable of finite and bounded U-rank, every type of Lascar rank n is contained in a formula of Lascar rank n (and so Lascar rank equals Shelah rank), and for all formulæ $\varphi(x, y)$ and all $n < \omega$ the set $\{a : U(\varphi(x, a)) = n\}$ is definable.

4.8 Binding Groups

We have already come across two group existence theorems: one in chapter 2 stating that a generically given group, or a definable group operation on an undefinable set, gives rise to a type-definable group, and the other in section 4.3 which constructs a type-definable group from geometric data, namely the group configuration. In this section, we shall encounter another group construction related to the forking geometry.

Definition 4.8.1 Let p and q be two stationary types over A. The *binding group* of p over q is the group of automorphisms of the realizations of p in the monster model $\mathfrak{C}$ which fixes all realizations of q.

Lemma 4.8.1 *Suppose p and q are two stationary types over a set A, and suppose that p is q-internal. Then there are an A-definable function $f(\bar{x}, \bar{y})$ and a tuple $\bar{a}$ of realizations of p, such that for any realization a of p there is some $\bar{b}$ realizing q with $a = f(\bar{a}, \bar{b})$.*

Such a tuple $\bar{a}$ is called a *fundamental system of solutions* of p relative to q.
Proof: Let B be a superset of A such that for all a realizing the non-forking extension of p to B there is a tuple $\bar{b}$ realizing q with $a \in \text{dcl}(B\bar{b})$. Let $\mathfrak{M}$ be a $|T \cup B|^+$-saturated model independent of $a, \bar{b}$ over B, and consider a Morley sequence $I := (a_i, \bar{b}_i : i < \omega)$ of $\text{stp}(a, \bar{b}/B)$ in $\mathfrak{M}$. Then $\text{Cb}(a, \bar{b}/B) \subset \text{dcl}(I)$, so Lemma 0.2.23 implies that $a \in \text{dcl}(I, \bar{b})$, and there are an A-definable function $g(\bar{u}, \bar{v})$ and tuples $\bar{a}$ realizing p and $\bar{b}'$ realizing q, with $\bar{a}\bar{b}' \in I$ and $a = g(\bar{a}, \bar{b}, \bar{b}')$. Note that $a \downarrow_B I$ and $a \downarrow_A B$; therefore a is independent of $\bar{a}$ over A.
We claim that $\bar{a}$ is a fundamental system of solutions of p relative to q. So consider any other realization a' of p, and $\bar{a}'$ a realization of $\text{tp}(\bar{a}/A)$ independent of $a', \bar{a}$ over A. Then since p is stationary, $\text{tp}(a\bar{a}/A) = \text{tp}(a'\bar{a}'/A)$ and there are realizations $\bar{b}_0$ and $\bar{b}'_0$ of q with $a' = g(\bar{a}', \bar{b}_0, \bar{b}'_0)$. On the other hand, every element $a_i \in \bar{a}'$ (say for $1 \leq i \leq n$) is independent of $\bar{a}$ over A and there are $\bar{b}_i$ and $\bar{b}'_i$ realizing q with $a_i = g(\bar{a}, \bar{b}_i, \bar{b}'_i)$; composing these function yields the required f with $a' = f(\bar{a}, \bar{b}_0, \ldots, \bar{b}_n, \bar{b}'_0, \ldots, \bar{b}'_n)$, and clearly f is A-definable and independent of the particular choice of a'. $\square$

Theorem 4.8.2 *Suppose p and q are stationary types over A and p is q-internal. Then the binding group of p over q is type-definable.*

Proof: Let $\bar{a}$ be a fundamental system of solutions of p relative to q, using the function $f(\bar{x}, \bar{y})$. Suppose σ is an automorphism of the monster model $\mathfrak{C}$ fixing the class B of all realizations of q. Then σ will map $\bar{a}$ to some tuple $\bar{a}'$ of realizations of p; this $\bar{a}'$ is again a fundamental system of solutions of p

relative to q, and $\text{tp}(\bar{a}/AB) = \text{tp}(\bar{a}'/AB) =: \bar{p}$. Since for every $a \models p$ there is $\bar{B} \in B$ with $a = f(\bar{a}, \bar{b})$, and σ must map a to $f(\bar{a}', \bar{b})$, the action of σ on $p^{\mathfrak{C}}$ is determined uniquely by $\bar{a}'$ (over $\bar{a}$).

Claim. If $\bar{a}'' \models \bar{p}$, then there is an automorphism of $\mathfrak{C}$ fixing AB and mapping $\bar{a}$ to $\bar{a}''$.

This is not immediately obvious from the saturation of $\mathfrak{C}$, since $|B| = |\mathfrak{C}|$.

Proof of Claim: It is sufficient to show that given any sets $X, X' \subset \mathfrak{C}$ with $\text{tp}(X/AB) = \text{tp}(X'/AB)$ and any $x \in \mathfrak{C}$ there is $x' \in \mathfrak{C}$ with $\text{tp}(Xx/AB) = \text{tp}(X'x'/AB)$. So consider a subset B' of B such that $\text{tp}(Xx/AB)$ is based on AB' and the set $\mathcal{X}$ of strong types of non-forking extensions of $\text{tp}(Xx/AB')$ to AB is minimal. Note that there are no forking extensions: if there is a tuple $\bar{b}$ (in an elementary extension of $\mathfrak{C}$) whose co-ordinates satisfy q with $Xx \underset{AB'}{\not\smile} \bar{b}$, then by saturation of $\mathfrak{C}$ there is such a tuple in $\mathfrak{C}$, hence in B. Therefore every AB-definable formula must either contain $\text{tp}(Xx/AB')$ or intersect it trivially. By saturation of $\mathfrak{C}$ there is some $x' \in \mathfrak{C}$ with $\text{tp}(X'x'/AB') = \text{tp}(Xx/AB')$. It follows that $\text{tp}(Xx/AB) = \text{tp}(X'x'/AB)$. $\square$

Now suppose $\bar{a}$ and $\bar{a}'$ both realize $\bar{p}$ and represent an automorphism σ of $p^{\mathfrak{C}}$ fixing AB, and $\bar{b}$ and $\bar{b}'$ are two tuples of realizations of q such that $f(\bar{a}, \bar{b}) = f(\bar{a}, \bar{b}')$. Then

$$f(\bar{a}', \bar{b}) = \sigma(f(\bar{a}, \bar{b})) = \sigma(f(\bar{a}, \bar{b}')) = f(\bar{a}', \bar{b}').$$

By compactness there are formulæ $\varphi(\bar{x}) \in \bar{p}$ and $\psi(\bar{y})$ in the appropriate power of q over AB, such that whenever $\bar{b}, \bar{b}' \models \psi$ and $\bar{a}, \bar{a}' \models \varphi$ with $f(\bar{a}, \bar{b}) = f(\bar{a}, \bar{b}')$, then $f(\bar{a}', \bar{b}) = f(\bar{a}', \bar{b}')$. It follows that the action of the automorphism σ represented by $\bar{a}'$ on a realization x of p is definable by the formula

$$\exists \bar{y} \, (f(\bar{a}, \bar{y}) = x \wedge f(\bar{a}', \bar{y}) = x' \wedge \psi(\bar{y})),$$

where $x' = \sigma(x)$. We shall denote this action by $\sigma_{\bar{a}'}$.

Define an equivalence relation E on $\bar{p}$ by $\bar{a}' E \bar{a}''$ iff $\sigma_{\bar{a}'}(\bar{a}) = \sigma_{\bar{a}''}(\bar{a})$. This clearly is an equivalence relation, and whenever $\bar{a}'$ and $\bar{a}''$ represent the same automorphism, then they are equivalent. Conversely, suppose $\bar{a}'$ and $\bar{a}''$ are equivalent. Then $\sigma_{\bar{a}'}^{-1}\sigma_{\bar{a}''}$ fixes the fundamental system $\bar{a}$ of solutions and hence all realizations of p, and $\sigma_{\bar{a}'}^{-1}\sigma_{\bar{a}''} = 1$. It follows that $\bar{p}/E$ is the domain of a type-definable group isomorphic to the binding group of p relative to q, with the group product given by $[\sigma_{\bar{a}_1}][\sigma_{\bar{a}_2}] = [\sigma_{\bar{a}_3}]$ (where $[.]$ denotes the equivalence class modulo E) iff

$$\sigma_{\bar{a}_1}\sigma_{\bar{a}_2}(\bar{a}) = \sigma_{\bar{a}_3}(\bar{a}),$$

and the action given by $[\sigma_{\bar{a}'}](x) := \sigma_{\bar{a}'}(x)$. $\square$

Remark 4.8.1 It can be shown that the binding group and its action on p are actually type-definable with parameters realizing p; an isomorphic copy of it is also definable with parameters from q alone, but in order to define the action on p we may need parameters from p.

Definition 4.8.2 Two types $p, q \in S(A)$ are *almost orthogonal* if any realization of p is independent of any realization of q over A.

Note that this does not mean that p and q are orthogonal, since they may have non-forking extensions which are no longer almost orthogonal, i.e. we need additional parameters to witness non-orthogonality.

Now suppose p and q are stationary and non-orthogonal; let B be a super-set of A, and a, b realize the non-forking extensions of p, q to B, with $a \not\!\!\downarrow_B b$. Then $\mathrm{tp}(a, b/B)$ is based on a Morley sequence $(a_i, b_i : i < \omega)$ in $\mathrm{stp}(a, b/B)$; in particular there are minimal m, n such that $(a_i < i < m)$ and $(b_j : j < n)$ are dependent over A. It follows that $p^{(m)}$ and $q^{(n)}$ are not almost orthogonal over A. For regular p we may now determine the maximal value of m (if we allow n to become arbitrarily large).

Proposition 4.8.3 *Let p and q be two non-orthogonal stationary types over A, and suppose p is regular. Then there are three independent realizations b_1, b_2, b_3 of p such that p and q are not almost orthogonal over $Ab_1b_2b_3$.*

Proof: Since p is non-orthogonal to q, for any realization a of p there is a q-internal $a_0 \in \mathrm{acl}(Aa) - \mathrm{acl}(A)$. Clearly $\mathrm{tp}(a_0/A)$ is also regular, and a_0 is domination-equivalent to a over A. We may hence replace p by $\mathrm{tp}(a_0/A)$ and assume that p is q-internal.

Let G be the binding group of p over q. If a realizes p, then G acts on a; we consider the orbit Ga. If it has U_p-rank zero, then it is stabilized by G^p. Therefore

$$U_p(Ga) = U_p(G/C_G(a)) \leq U_p(G/G^p) = 0,$$

so $U_p(a/Aq^{\mathfrak{C}}) = U_p(Ga) = 0$, and a forks with $q^{\mathfrak{C}}$ over A.

Otherwise $U_p(Ga) = 1$ and Ga must contain $p|a$ by regularity of p. Hence any two independent realizations of p are conjugate under G; if a, a' are any two realizations, we choose $a'' \models p|a, a'$ to see that a, a'', and a' are all conjugate under G. Therefore G acts transitively on $p^{\mathfrak{C}}$; by Theorem 4.3.5, there are three possibilities for the action of $(G/C_G(p^{\mathfrak{C}}))^p$ on $p^{\mathfrak{C}}$:

1. $U_p(G) = 1$, G^p is abelian and the action is simply transitive. If we fix one point $b_1 \models p$, then $C_{G^p}(b_1)$ is trivial, so any other realization of p lies in $\mathrm{cl}_p(b_1, A, q^{\mathfrak{C}})$. In particular, p and q^ω are not almost orthogonal over Ab_1.

2. $U_p(G) = 2$, and G^p acts sharply 2-transitively. So if we fix two independent realizations b_1 and b_2 of p, every other realization of p lies in $cl_p(b_1 b_2, A, q^c)$, and p and q^ω are not almost orthogonal over $Ab_1 b_2$.

3. $U_p(G) = 3$ and G^p acts sharply 3-transitively. Then if we fix three independent realizations b_1, b_2, b_3 of p, every other one lies in $cl_p(b_1 b_2 b_3, A, q^c)$, and p and q^ω are not almost orthogonal over $Ab_1 b_2 b_3$.

Furthermore, in cases 2. and 3. there is a definable field, and the action of G^p on p is well-determined. □

Definition 4.8.3 A theory is *unidimensional* if any two types are non-orthogonal.

Unidimensionality is a generalization of $|T|^+$-categoricity:

Proposition 4.8.4 *An uncountably categorical countable theory T is ω-stable and unidimensional.*

In fact, any $|T|^+$-categorical theory is unidimensional.
Proof: Any theory T has a model $\mathfrak{N}$ of cardinality ω^+ which realizes only countably many distinct types over any countable subset (see [94], Lemme 18.20). As this model is unique up to isomorphism, there can only be countably many distinct types over any countable subset and T is ω-stable.

Let now p be a type of Morley rank and degree one over a countable saturated model $\mathfrak{M}$ of T. We claim that no type q is orthogonal to p. So suppose otherwise. We may clearly assume that q is also over $\mathfrak{M}$.
Claim. If A is a superset of $\mathfrak{M}$, then there is an elementary superstructure $\mathfrak{N} \supseteq A$ of $\mathfrak{M}$ which is dominated by A over $\mathfrak{M}$.
Proof of Claim: Let $\exists x\, \varphi(x)$ be a true sentence with parameters from A. We have to find an element a realizing φ such that A dominates aA over $\mathfrak{M}$; adjoining successively witnesses for all the existential quantifiers, we shall obtain the required model $\mathfrak{N}$. Note that if $(A_i : i \in I)$ is a chain such that A dominates A_i over $\mathfrak{M}$ for all $i \in I$, then A dominates $\bigcup_{i \in I} A_i$ over $\mathfrak{M}$, since forking is finitary.

Take any realization a of φ. As the fundamental order is closed under union of chains, there is some $B \mathop{\smile}\limits_{\mathfrak{M}} A$ such that the bound $\beta(a/AB)$ is minimal; since, in a superstable theory, every type is based on some finite subset of its domain, we may assume that B is finite. There is a finite subset A' of A (containing the parameters of φ) such that $\operatorname{tp}(a/AB)$ is based on $A'B$, and $\operatorname{tp}(A'B/\mathfrak{M})$ is based on a finite subset $X \subset M$. By saturation there is $B' \subset M$ with $\operatorname{stp}(B'/X) = \operatorname{stp}(B/X)$; as $A' \mathop{\smile}\limits_X B$ and $A' \mathop{\smile}\limits_X B'$ we get $\operatorname{tp}(A'B) = \operatorname{tp}(A'B')$, and there is a' with $\operatorname{tp}(aA'B) = \operatorname{tp}(a'A'B')$. Then $a' \models \varphi(x)$; furthermore

$$\beta(a'/A) \leq \beta(a'/A'B') = \beta(a/A'B) = \beta(a/AB)$$

is minimal possible. In other words, $c \underset{\mathfrak{M}}{\downarrow} A$ implies $c \underset{A}{\downarrow} a'$, whence $c \underset{\mathfrak{M}}{\downarrow} Aa'$, and A dominates Aa' over $\mathfrak{M}$. $\square$

Now choose an uncountable Morley sequence A for q. Then there is an elementary superstructure $\mathfrak{N}$ of $\mathfrak{M}$ containing A and dominated by A over $\mathfrak{M}$. Since $\mathrm{tp}(A/\mathfrak{M})$ is orthogonal to p, there is no realization of p in $\mathfrak{N}$, so $p^{\mathfrak{N}} = p^{\mathfrak{M}}$ is countable. Therefore $\mathfrak{N}$ cannot be saturated. But there are saturated models of cardinality $|\mathfrak{N}|$, contradicting categoricity. So p cannot be orthogonal to q.

But now any two types p_1 and p_2 have realizations of non-forking extensions, say a_1 and a_2 over A, such that there is a realizing the non-forking extension of p with $a_1 \underset{A}{\not\downarrow} a$ and $a_2 \underset{A}{\not\downarrow} a$. Then $a \in \mathrm{acl}(Aa_1) \cap \mathrm{acl}(Aa_2) - \mathrm{acl}(A)$; it follows that p_1 and p_2 are non-orthogonal, and T is unidimensional. $\square$

Lemma 4.8.5 *Suppose P is a set of stationary types over $\emptyset$ and $\mathrm{tp}(a)$ is P-analysable. If T does not interpret an infinite group, then there is a finite tuple $\bar{b}$ of realizations of types in P such that $a \in \mathrm{acl}(\bar{b})$.*

Proof: First we prove:

Claim. If $\mathrm{tp}(a/A)$ is p-internal for some stationary type p, then there is a tuple $\bar{b}$ of realizations of p such that $a \in \mathrm{acl}(A\bar{b})$ (i.e. no parameters are needed to witness internality).

Proof of Claim: Consider the binding group of $\mathrm{tp}(a/\mathrm{acl}(A))$ over p. This is type-definable, and contained in a definable group by Theorem 2.1.17. As there is no definable infinite group, the binding group must be finite and a is algebraic over A together with finitely many realizations $\bar{b}$ of p. $\square$

Now consider a P-analysis $(a_i : i \leq \alpha)$ of $\mathrm{tp}(a)$. Then $\mathrm{tp}(a_i/a_j : j < i)$ is p_i-internal for some $p_i \in P$ for all $i \leq \alpha$, and by the claim there are realizations $\bar{b}_i$ of p_i for $i \leq \alpha$ such that $a_i \in \mathrm{acl}(\bar{b}_i a_j : j < i)$, whence $a_i \in \mathrm{acl}(\bar{b}_j : j \leq i)$. As $a \in \mathrm{acl}(a_\alpha) \subseteq \mathrm{acl}(\bar{b}_i : i \leq \alpha)$, there is a finite tuple $\bar{b}$ of realizations of types in P such that $a \in \mathrm{acl}(\bar{b})$. $\square$

Theorem 4.8.6 *A unidimensional theory T is superstable.*

Proof: As the fundamental order is closed under unions of chains, by Zorn's Lemma there is a minimal fundamental class which does not contain the formula $x = y$. If p is a type of that class, then p is non-algebraic, but any forking extension must represent the formula $x = y$ and hence be algebraic. It follows that $U(p) = 1$. We may assume that p is stationary and over $\emptyset$. The proof now splits into two cases.

Case 1: T does not interpret an infinite group.

We may assume that p is over some parameter set A. Consider any element a. Since p is non-orthogonal to every type, $\mathrm{tp}(a/A)$ must be p-analysable, and by Lemma 4.8.5 there is a finite tuple $\bar{b}$ of realizations of p such that $a \in \operatorname{acl}(A\bar{b})$. By the Lascar inequality $U(a/A)$ is bounded by the length of the tuple $\bar{b}$; in particular it is finite. Hence every type has finite Lascar rank and T is superstable.

Case 2: T interprets an infinite group G.

As $\operatorname{gen}(G)$ is non-orthogonal to p, by Theorem 3.1.1 there is a normal subgroup N such that the quotient G/N is p-internal, and must have finite Lascar rank by the Lascar inequality. Replacing G by G/N, we may assume that $U(G) < \omega$. Now every type is non-orthogonal to $\operatorname{gen}(G)$ and hence analysable in $\operatorname{gen}(G)$, and therefore in G. But G is a definable set, so by Lemma 3.0.8 every type q is analysable in G in finitely many steps; as $U(G)$ is finite, so is $U(q)$ by the Lascar inequality again, and T is superstable. $\square$

This theorem well underlines the importance of stable groups in the analysis of more general stable structures: although the statement does not mention groups at all, the only known proof makes essential use of them.

Theorem 4.8.7 *Suppose a countable theory T is uncountably categorical and does not interpret an infinite group. Then T is almost strongly minimal.*

Proof: By Proposition 4.8.4 the theory must be ω-stable and unidimensional, whence one-dimensional, and does not have the finite cover property, by Proposition 4.7.4. By Proposition 0.2.7 there is an atomic model $\mathfrak{M}$; as in the proof of Corollary 4.7.6 there is a minimal formula $\varphi(x, m)$ over $\mathfrak{M}$, which is strongly minimal by Lemma 4.7.5. Clearly $\operatorname{tp}(m)$ is isolated. By unidimensionality an arbitrary type $\operatorname{tp}(a/m)$ is $\varphi(x, m)$-analysable; since there is no infinite definable group, a is algebraic over m and a finite tuple $\bar{b}$ of realizations of φ by Lemma 4.8.5. Hence T is almost strongly minimal. $\square$

In a similar vein:

Proposition 4.8.8 *If T is dimensional and does not interpret an infinite group, then T is superstable.*

Proof: Let P be the collection of all regular superstable types, and Q the family of types carrying the dimensions. Then every type in Q has weight at most $|T|$, and can be non-orthogonal to at most $|T|$ pairwise orthogonal types in P. Hence there is a set P' of regular superstable types such that every superstable type is non-orthogonal to some type in P', and therefore P'-analysable. We add the parameters of the types in P' to the language.

Suppose that T is not superstable. Let a be an element and $(A_i : i < \alpha)$ an increasing sequence of parameter sets, such that $\operatorname{tp}(a/A_i)$ is not superstable for all $i < \alpha$. Then $\alpha < |T|^+$ by Corollary 0.2.25; we may hence assume that the sequence is maximal. Now consider $A := \bigcup_{i<\alpha} A_i$. By maximality of the sequence, $\operatorname{tp}(a/A)$ is superstable and therefore P'-analysable; by Lemma 4.8.5

there is a tuple $\bar{b}$ of realizations of types in P' such that $a \in \mathrm{acl}(A\bar{b})$. But then $a \in \mathrm{acl}(A_i\bar{b})$ for some $i < \alpha$ and already $\mathrm{tp}(a/A_i)$ would be superstable by Lascar's inequalities, contradicting our choice of the sequence $(A_i : i < \alpha)$. $\square$

4.9 Historical and Bibliographical Remarks

Locally modular regular types and groups were studied paradigmatically by Hrushovski in [52], after earlier work by Buechler [27, 28], Cherlin, Harrington and Lachlan [35], Pillay [84] and Zil'ber [132, 133]; the one-based group case had been dealt with by Hrushovski and Pillay [58]. Our exposition eliminates weight considerations; the quantifier elimination for abelian structures follows the lines of Prest [98] and Ziegler [128]. Neumann's Lemma comes from [81]. A different approach to Corollaries 4.1.14 and 4.1.17 can be found in [48].

The group configuration is due to Zil'ber [133], and was greatly elaborated on by Hrushovski [51]; the reader may also want to refer to Bouscaren [24]. The theorems on quasi-functions can be found in [114], those on endogenies in [118]. A precursor in this field is Loveys [68], who considers locally modular groups, and distinguishes those with a connected-by-finite quotient *(strongly regular like)* from those without *(weakly minimal like)*.

CM-triviality was introduced by Hrushovski in [57] and studied by Pillay in [87]. We follow the weightless line of [123].

Uncountable categoricity of a definably simple group or a field of finite Morley rank, Proposition 4.7.8, was shown by Zil'ber [130]; finite dimensionality of an arbitrary group of finite Morley rank, Corollary 4.7.6, by Lascar [66], and the decomposition of a group of finite Morley rank without an abelian normal subgroup into almost strongly minimal ones (Proposition 4.7.9) by Belegradek. Proposition 4.7.10 was proven by Poizat in [91] after previous work by Baldwin [5] and Belegradek [15]. Finally, the binding group was introduced by Zil'ber [131], who also proved Theorem 4.8.7; it was put to good use by Hrushovski in his thesis [51]. Theorem 4.8.6 can also be found in [54].

Chapter 5

Groups & Grades

In this chapter we shall analyse stable groups with the additional property $\mathfrak{R}$, a property common to small stable and superstable groups. We shall define it in section 0 and show that a small stable group is $\mathfrak{R}$; the proof also implies that the index of the image of an endomorphism of a small group is at most the cardinality of the kernel (as long as both are finite). We then derive some immediate consequences like the independence of the generic type of the group law. In section 1 we shall prove that an infinite $\mathfrak{R}$-group must always contain an infinite abelian subgroup, and look at $\mathfrak{R}$-fields: an $\mathfrak{R}$-ring without zero-divisors is a commutative, algebraically closed field, and any definable automorphism is $\emptyset$-definable and has finite fixed field.

In section 2 we shall prove an analysability result reminiscent of Remark 3.6.2, replacing regular types by abelian subgroups. In particular there exist abelian subgroups which are big, i.e. not co-foreign to the generic type of the whole group.

The next section deals with linear operations and the existence of definable fields. The problem here is one of comparing groups: if G operates on A as a group of automorphisms, in order to obtain a field, we have to take care that the sums $\sum_i (g_i A)$ remain bounded, for elements g_i from G. In a certain way this means that A and G have comparable size; technically it is achieved by requiring A to be G-minimal and $\text{gen}(A)$ not to be foreign to G. In the finite rank case of course, we could just use the Indecomposability Theorem.

In section 4 we shall employ the methods just developed to analyse soluble $\mathfrak{R}$-groups. We show that the ω-centre of the Φ-component of an $\mathfrak{R}$-group is relatively definable and hence equals some iterated centre. The commutator subgroup of a soluble Φ-component is nilpotent; if G^Φ is centreless, we get a structure theorem for the centre of $(G^\Phi)'$. This is used to prove that the radical of a Φ-component is definable. From this, together with the results from chapter 3 on groups without normal abelian subgroups, we derive a structure theorem for Φ-connected $\mathfrak{R}$-groups.

250

In section 5 we investigate nilpotency. We shall prove that if a normal subgroup N of a Φ-connected soluble ℜ-group G has a nilpotent quotient G/N, then there is a nilpotent supplement for N. We then develop the theory of Carter subgroups. Nilpotency also plays a major rôle in the next section where we study the Frattini subgroup. The body of the chapter finishes in section 7 with a proof that an ℜ-group has infinitely many conjugacy classes. To this end we first show that in a periodic ℜ-group the 2-Sylow subgroups are normal-by-finite; we may then reduce to the case of finite 2-Sylow subgroups, which is dealt with easily.

5.0   ℜ-Groups

We have seen in Example 2.1.1 that the generic type of a stable group need not be invariant under definable bijections. This is different in the case of a superstable group: as a 1-type is generic iff it has maximal rank, and as rank is invariant under definable bijections, so are the generic types. Furthermore, it follows from the Lascar inequalities that if b is algebraic over a, then $U(a)$ is not smaller than $U(b)$. In particular, if b is generic, so is a. This property, in fact, is common to a wider class of groups than the superstable ones, and has far-reaching consequences in the analysis of a stable group. The reason primarily lies in the fact that we can define a very weak rank-like notion as follows:

If p and q are two types, then $\mathrm{rk}_p(q) \geq \frac{n}{m}$ iff any n realizations of p are algebraic over m (suitably chosen) realizations of q.

If p is the generic type of some definable group, then property ℜ tells us that $\mathrm{rk}_p(p) = 1$. So we do have a notion for two groups G and H to be of comparable size: if p and q are the generic types of G and H, then both $\mathrm{rk}_p(q)$ and $\mathrm{rk}_q(p)$ should be finite. Technically, we shall work with the following definition:

Definition 5.0.1 A stable theory is ℜ (for "rank-like") iff for all definable transitive group actions, a generic element for this action can only be algebraic over another generic element. That is,

if G is any definable (with parameters) group acting definably and transitively on some definable set X, and x is a generic element of X (for that action) over some parameters A which is algebraic over $A \cup \{y\}$ for some element y of X, then y is generic over A as well.

An ℜ-group is a group whose theory is ℜ.

We immediately get

Lemma 5.0.1 *Let G be an $\Re$-group and φ an endomorphism of G with finite kernel. Then φ is surjective on G^0.*

Proof: The image of G^0 under φ lies again in G^0. A principal generic element a is algebraic over its image, so its image must be principal generic as well. But every element is the sum of two principal generic elements, and hence lies in the image of φ. □

Note that φ need not be surjective on G, as the example of $x \mapsto 2x$ on $\mathbb{Z}$ shows. We now aim to prove that a small theory is $\Re$.

Definition 5.0.2 A family of sets is *n-disjoint* if the intersection of any n-element subfamily is empty.

Definition 5.0.3 A *quasi-function of multiplicity n* is a (binary) relation F such that for any x there are at most n different y such that $F(x, y)$ holds.

Theorem 5.0.2 *Suppose $\{X_0, X_1, \ldots, X_{kl}\}$ is a collection of $(k+1)$-disjoint subsets of some set X, all of which are type-definable over A. Assume that for $i = 0, \ldots, kl$ there are A-definable quasi-functions φ_i of multiplicity l, which map X_i surjectively to (a superset of) X. Then there are uncountably many 1-types over A.*

Proof: Suppose there are only countably many 1-types r_i, for $i < \omega$. We shall construct a new one, r, which is not on the list. For this we shall construct a function $j \in (kl + 1)^\omega$ and a function $i \in \omega^\omega$ such that the set of conditions

$$P_j(n) \qquad \exists x_0, \ldots, x_n \quad x = x_0 \wedge \bigwedge_{m < n} [x_m \in X_{j(m)} \wedge \varphi_{j(m)}(x_m, x_{m+1})]$$

is consistent for $n < \omega$, but $P_j(i(n))$ contradicts each of $r_1, r_2, \ldots, r_n$.

First of all it is obvious that $P_j(n + 1)$ implies $P_j(n)$ for any j. On the other hand $P_j(n)$ is consistent: we choose x_n first and proceed backwards, using the surjectivity of φ_i at every step.

Suppose i has been constructed up to n_0 and j up to $i(n_0)$. For any $m < \omega$ there are $(kl + 1)^m$ choices for a continuation of j to $i(n_0) + m$, but at most $(kl)^{i(n_0)+m}$ of them can occur in r_{n_0+1}: if $x = x_0$ satisfies r_{n_0+1}, then x_0 belongs to at most k different X_i, so there are at most kl possibilities for $\text{tp}(x_1)$, and at most $(kl)^2$ possibilities for $\text{tp}(x_2)$, etc., and the $(kl)^{i(n_0)+m}$ possible $\text{tp}(x_{i(n_0)+m})$ represent all the paths consistent with r_{n_0+1}. Thus for $m = m_0$ big enough, there is some choice of $j\lceil_{i(n_0)+m_0}$ such that $P_j(n_0 + 1)$ contradicts r_{n_0+1}. So we can take $i(n_0 + 1) = i(n_0) + m_0$ and that particular choice of $j\lceil_{i(n_0+1)}$.

But now any completion of $\{P_j(n) : n < \omega\}$ is a type which does not occur on the list. Hence there are uncountably many 1-types over A. $\square$

For a small group we can thus improve Lemma 5.0.1.

Corollary 5.0.3 *Let φ be a definable endomorphism of the small group G. Then $|G : \varphi(G)| \leq |\ker(\varphi)|$, if the latter is finite. In particular, an injective endomorphism is surjective.*

Proof: If $|\ker(\varphi)| < |G : \varphi(G)|$, then the definable cosets $a_i\varphi(G)$ with the relation $a_i^{-1}x = \varphi(y)$ satisfy the requirements of Theorem 5.0.2 with $k = 1$ and $l = |\ker(\varphi)|$. $\square$

Theorem 5.0.4 *A small theory is ℜ.*

Proof: Let G be a definable group acting transitively on a definable set X, and b a generic element of X algebraic over some $a \in X$. Suppose for a contradiction that tp(a) is not generic. As every type is a translate of some extension of a generic type, we may assume that there is a finite set of parameters B such that b is generic over B, while a is generic over $\emptyset$ and dependent on B via $\models \psi(B, a)$, and that b is algebraic over B, a via $\varphi_1(a, B, b)$ with multiplicity l_1. Put

$$S = \{y \in X : \exists x\, \psi(B, x) \wedge \varphi_1(x, B, y)\}.$$

As $S \in \mathrm{tp}(b/B)$ is generic, there are elements $g_1, \ldots, g_m$ in G such that $\bigcup_{j=1}^{m} g_j S$ covers X. We incorporate these g_j into B and put

$$\varphi(x, B, y) = \bigvee_{j=1}^{m} \varphi_1(x, B, g_j^{-1}y).$$

Then φ is a surjective quasi-function from $\psi(B, x)$ to X of multiplicity $l = ml_1$. Now we consider a Morley sequence $(B_i : i < \omega)$, in stp(B), and put $X_i = \psi(B_i, x)$. If $\models \psi(B_i, d)$, then tp(B_i/d) forks over $\emptyset$. As the B_i form a Morley sequence, this can happen only finitely often for any fixed d by Theorem 0.2.31 and the X_i are $(k + 1)$-disjoint for some $k < \omega$. If we consider $X_0, \ldots, X_{kl}$ together with $\varphi_i(x, y) := \varphi(x, B_i, y)$, this satisfies the assumptions of Theorem 5.0.2 and we get the desired contradiction. $\square$

For our purposes it would be sufficient merely to require the property ℜ to hold for definable transitive actions of sections of the original group on definable sets, i.e. for interpretable group actions where the group is actually related to the one we are studying. In a more general line, one might ask whether the ℜ-property also holds for *type*-definable group actions in an ℜ-theory – or, for that matter, in a small stable group (it clearly does in a superstable group, as an easy rank calculation shows). We only have partial information in this respect:

Lemma 5.0.5 *Let G be a type-definable group in an $\Re$-theory which acts transitively on some type-definable set X. If x is a generic element of X with respect to that action and x is algebraic over some element y of X, then y is not contained in any non-generic definable subset of X.*

Note that we make no assertion about *relatively* definable subsets of X. In particular, this lemma does not imply that y is a generic element of X.

Proof: Let p be the principal generic type for X and suppose that the generic element x is algebraic over y via the formula $\models \varphi(y, x)$ and that y is contained in the non-generic definable subset ψ of X. Let $d_i x\, \psi(ux)$ be the finitely many $\psi(ux)$-definitions of generic types for X, let $\mathcal{G}$ be a definable supergroup of G, and consider the group

$$G_0 := \{g \in \mathcal{G} : \forall u \bigvee_i d_i x\, \psi(ux) \leftrightarrow \bigvee_i d_i x\, \psi(ugx)\}.$$

This is a supergroup of G which permutes the generic ψ-types of X. By remark 2.2.1 there are a descending sequence $(G_i : i < \omega)$ of subgroups of G_0 and a descending sequence of sets X_i with $G = \bigcap_{i<\omega} G_i$ and $X = \bigcap_{i<\omega} X_i$, such that G_i acts transitively on X_i (and the actions match up, of course).

Now $Y := \exists x\, \psi(x) \wedge \varphi(x, y)$ is generic for X and finitely many translates of Y by elements of G cover X. But the union of these translates is again definable, and must therefore cover X_i for all sufficiently big i by compactness. On the other hand, if ψ is generic for X_i, then X would be contained in $\bigcup_j g_j \psi$ for some finite number of elements g_j in G_i. Thus $g_{j_0} \psi$ would be in p for some j_0; since $g_{j_0} \in G_0$ permutes the generic ψ-types, ψ is generic for X, a contradiction. Hence ψ cannot be generic for X_i, but this contradicts the fact that the action of G_i on X_i is definable in an $\Re$-group. $\square$

Corollary 5.0.6 *Let G be a connected type-definable $\Re$-group and M a type-definable group of automorphisms of G. If φ is an endomorphism of G generated by elements of M such that φ has finite kernel, then φ is surjective.*

Proof: $G \rtimes M$ is the intersection of a descending sequence of groups X_i, containing subgroups G_i and M_i, such that $G = \bigcap G_i$ and $M = \bigcap M_i$. We intersect every G_i with all its $N_{M_0}(G)$-conjugates and replace M_i by $N_{M_i}(G_i)$. Then every M_i operates on G_i and so does φ. We may assume that (from some i onwards) φ acts on G_i with finite kernel. Now φ is surjective on each G_i^0 by Lemma 5.0.1, but then φ is surjective on G, as any element has only finitely many pre-images. $\square$

Note that due to Corollary 5.0.3, in a small theory we do not need connectivity:

Corollary 5.0.7 *Let G be a type-definable group and M a type-definable group of automorphisms of G, all in a small theory. If φ is an endomorphism of G generated by elements of M such that φ has finite kernel, then $|G : \varphi(G)| \leq |\mathrm{ker}(\varphi)|$.*

Proof: Consider M_i and G_i as above. We may assume that $|\mathrm{ker}(\varphi{\restriction}_{G_i})|$ is constant. Then by Corollary 5.0.3 we get $|G_i : \varphi(G_i)| \leq |\mathrm{ker}(\varphi{\restriction}_{G_i})| = |\mathrm{ker}(\varphi{\restriction}_G)|$. The corollary follows. $\square$

In particular, for abelian G we may take M to be the trivial group, and φ to be multiplication by some $n < \omega$.

Theorem 5.0.8 *The generic types of an $\Re$-group do not depend on the group law.*

Proof: Suppose G has two multiplications $*$ and $\circ$, and let x be generic for $*$ and y generic for $\circ$, with $x \downarrow y$. Consider $\mathrm{tp}(x*y/x)$. As y is generic for $\circ$ over x and algebraic over $(x*y), x$, $\mathrm{tp}(x*y/x)$ is generic for $\circ$ and does not fork over the empty set. So x is generic for $*$ over $x*y$ and algebraic over $y, (x*y)$. Therefore, y is generic for $*$. Similarly, x is generic for $\circ$. $\square$

5.1 Abelian $\Re$-groups and $\Re$-fields

It is a puzzling problem in the theory of stable groups whether they must always have an infinite abelian subgroup. By compactness, an (ω-saturated) stable group G without infinite abelian subgroup must have finite exponent; by Proposition 1.3.1 any locally finite subgroup of G must be nilpotent-by-finite and hence finite. A particularly bad group of this kind would be a stable Tarski monster; we have already encountered this possibility a number of times. $\Re$-groups, however, are well-behaved:

Theorem 5.1.1 *An $\Re$-group contains an infinite abelian subgroup.*

Proof: By the chain condition on centralizers, there is a smallest infinite intersection A of centralizers. We claim that A is abelian. So let x be a non-central element of A. By minimality, $C_A(x)$ is finite, hence a generic element g of A over x is algebraic over x^g, x. Therefore x^g must be generic over x. So every conjugacy class of a non-central element is generic, and there are only finitely many of them. It follows that $H := A^0/Z(A^0)$ is a definable group with a single non-trivial conjugacy class. So H is trivial by Theorem 1.0.3, and A^0 is central in A. But then A is abelian by minimality. $\square$

Proposition 5.1.2 *An infinite $\Re$-field k is algebraically closed.*

Proof: This is similar to the proof of the first claim in Theorem 3.7.2. If K is any finite extension of k, then K is interpretable in k and thus also an $\mathfrak{R}$-field. Now K is connected by Lemma 2.3.1, so every endomorphism of K with finite kernel is surjective by Lemma 5.0.1. In particular, $x \mapsto x^n$ and $x \mapsto x^p - x$ (where $p = \operatorname{char}(k)$) are both surjective. So k is perfect and no finite extension of k has a cyclic Galois extension, which implies that k is algebraically closed. $\square$

Lemma 5.1.3 *Let D be a division ring and K an additive or multiplicative subgroup co-foreign to the generic type $p := \operatorname{gen}(D)$. If $\varphi : D/K \to D$ is a definable injective map and the theory is $\mathfrak{R}$, then $\operatorname{im}(\varphi)$ is generic.*

Proof: Suppose that K is additive. Let a and b be two independent generic elements and suppose first that there are non-trivial c in K and b' in D with $a\varphi(b + K) = a\varphi(b' + K) + c$. As p is foreign to K and a is independent of $\varphi(b+K)$, it must be independent of $\varphi(b+K), c$, so $\varphi(b'+K) = \varphi(b+K) - a^{-1}c$ is generic for D.

Otherwise $x \mapsto a\varphi(x) + K$ is a generically injective map from D/K to itself. $\mathfrak{R}$ implies that the image of a generic element is generic, so for any generic element b' there are elements b in D and c in K such that $a\varphi(b+K) + c = b'$. So $\varphi(b + K) = a^{-1}(b' - c)$ is generic, since a, b' must be independent of c.

Now suppose that K is multiplicative. If a and b are two independent generic elements, we suppose first that for some non-trivial c in K and b' in D we have $\varphi(bK) + a = (\varphi(b'K) + a)c$. Then a is independent of $\varphi(bK), c$, so $\varphi(b'K) = \varphi(bK)c^{-1} + a(c^{-1} - 1)$ is generic for D.

Otherwise $x \mapsto \varphi(x) + a$ is a generically injective map from D/K to itself. Again, the image of a generic element is generic, so for any generic element b' there are elements b in D and c in K such that $(\varphi(bK) + a)c = b'$. So $\varphi(bK) = b'c^{-1} - a$ is generic, since a, b' are independent of c. $\square$

Theorem 5.1.4 *An infinite type-definable ring without zero-divisors in an $\mathfrak{R}$-theory is an algebraically closed field.*

Proof: By Theorem 2.3.3 and Proposition 2.3.2 the ring D is the intersection of definable division rings, and if those are all algebraically closed fields, then so is D. We may therefore assume that D is definable. Suppose that D is not commutative. Taking a minimal non-commutative intersection of centralizers, we may assume that the centralizer of any non-central element is commutative. Suppose the generic type p of D is foreign to all of them. By Lemma 5.1.3, taking conjugation for φ, all conjugacy classes of non-central elements are generic and we contradict Theorem 1.0.3. (Note that D is connected.)

So p is not foreign to some proper centralizer $C := C_D(x)$. Then D is C-internal by Corollary 3.1.2, and there is a surjective function $\varphi : C^n \to D$ for some n. If the C-vector space dimension of D were greater than n, then for linearly independent elements $u_1, \ldots, u_n$ of D/C, the sum $S = u_1 C \oplus u_2 C \oplus \cdots \oplus u_n C$ would be a non-generic subset of D, and $\varphi(\pi_1 y, \pi_2 y, \ldots, \pi_n y)$ would be a surjective function $S \to D$, contradicting the fact that D is $\Re$. (Here π_i denotes projection to the i-th component.)

Therefore $[D : C] \leq n$. But then $[D : Z(D)] \leq n^2$ by Lemma 3.7.3, so $Z(D)$ is infinite, hence algebraically closed. On the other hand, every element of K generates a finite field extension of $Z(D)$, which must be trivial. Hence $Z(D) = D$. $\square$

Corollary 5.1.5 *The generic type p of an infinite $\Re$-field D is foreign to all proper definable subfields C.*

Proof: This is just the second part of the above: if p is not foreign to C, then $[D : C]$ is finite; and as C is algebraically closed, $D = C$. $\square$

Theorem 5.1.6 *If K is an infinite $\Re$-field, π is a (partial) type and the generic type p of K is not foreign to π, then there are polynomials $P(X_1, X_2, \ldots, X_n)$ and $Q(X_1, X_2, \ldots, X_n)$ in $K[X_1, X_2, \ldots, X_n]$ (with parameters), such that*

$$K = \frac{P(\pi^K, \pi^K, \ldots, \pi^K)}{Q(\pi^K, \pi^K, \ldots, \pi^K)}.$$

Proof: Suppose p is not foreign to π, so K is π-internal, but that the conclusion does not hold. Then in particular the generic type is not the quotient of two such polynomials, as every element is the product of two (principal) generic elements and the product of two rational functions (in disjoint variables $\bar{X}$ and $\bar{Y}$) is obviously again rational. Therefore for any finite number of elements $a_1, a_2, \ldots, a_n$ in K, the field $F := \langle \bar{a}, \pi^K \rangle$ does not contain a generic element over $\bar{a}$ and hence must be a proper subfield of K (in a suitably saturated model). Thus there is some a in K such that F intersects $a\langle \pi^K \rangle$ trivially. But then for all $n < \omega$ there are elements $a_1, a_2, \ldots, a_n$ in K such that for any $x_1, x_2, \ldots, x_n$ in π^K, all the x_i are definable over $a_1, a_2, \ldots, a_n, (\sum_{i=1}^n a_i x_i)$. But K is π-internal, so there is $n_0 < \omega$ such that every generic element is definable over n_0 realizations of π. It follows from $\Re$ that $\sum_{i=1}^{n_0} a_i \pi^K$ is a generic subset of K, a contradiction. $\square$

Problem 5.1.1 If p is not foreign to π, is there a polynomial $P(\bar{X})$ such that $K = P(\pi^K, \ldots, \pi^K)$?

Theorem 5.1.7 *Let σ be a definable automorphism of an infinite $\Re$-field K. Then the fixed field K^σ is finite. In particular, σ has infinite order and is definable over $\mathrm{acl}(\emptyset)$.*

Proof: Let L be the fixed field of φ. Then $\psi(x) = \varphi(x)/x$ is a multiplicative endomorphism with kernel L. By Corollary 5.1.5, the generic type of K is foreign to L, so ψ is surjective by Lemma 5.1.3 and there is some x in K with $\psi(x) = -1$ (for char$(K) \neq 2$) or a cube root of unity (for char$(K) = 2$). Thus x is not in L, but x^2 lies in L (or x^3 does). Hence L is not algebraically closed, and cannot be infinite by Proposition 5.1.2. Now every automorphism of finite order of the algebraic numbers in K has an infinite fixed field, and therefore cannot be definable.

If Σ is a family of uniformly definable automorphisms, then the fixed field of $\sigma^{-1}\sigma'$ (for $\sigma, \sigma' \in \Sigma$) must have bounded cardinality by compactness. As a given finite field has only finitely many automorphisms, Σ must be finite, and any $\sigma \in \Sigma$ is determined by its action on the algebraic numbers and is therefore acl($\emptyset$)-definable. $\square$

In particular, an infinite $\Re$-field of characteristic zero has no definable automorphism.

5.2   A Decomposition Theorem for $\Re$-groups

We have already seen that an $\Re$-group must contain an infinite abelian subgroup. We shall show in this section that it must actually contain a big one, i.e. one which is not co-foreign to the generic type of the group. This will then be used to derive a decomposition theorem in terms of a normal series with quotients internal in abelian subgroups.

Lemma 5.2.1 *Let G_0 be a definable $\Re$-group and G a normal non-abelian type-definable subgroup of G_0. Then $p := \text{gen}(G)$ is not foreign to some centralizer (in G_0) of a non-central element of G.*

Proof: Suppose G is a counter-example. If G^0 is abelian and non-central in G, then p is not foreign to $C_{G_0}(x)$ for any non-central x in G^0, and if G^0 is central in G, then p is not foreign to $C_{G_0}(x)$ for any non-central x in G. Hence we may replace G by its connected component. We first prove the proposition for G centreless. Note that this implies that G has no abelian normal subgroups A, as $G/C_G(A)$ would be A-internal, hence $C_{G_0}(A)$-internal.

Let H be a minimal type-definable normal subgroup of G, which must be definably simple by Lemma 1.1.17. There are only finitely many G_0-conjugates of H, which are permuted by G_0 and stabilized by a definable subgroup G_1 of finite index in G_0 (necessarily containing G). Suppose that $\text{gen}(H)$ is foreign to all centralizers (in G_1) of its non-trivial elements.
Claim. Let $x \neq 1$ be an element of H, and g be generic for H over x. Put $C = C_{G_1}(x)$. If there are only finitely many elements c in C (note that such c are necessarily also in H) such that cx^g lies in x^H, then x^H is generic for H.

Proof of Claim: As Cg is interdefinable with x^g over x, it follows from the assumptions that Cg is algebraic over $\{Cx^g, x\}$. Now H is normal in G_1 and Cx^H is the intersection of definable sets contained in HC/C, so genericity of Cg for HC/C over x implies genericity of Cx^g for HC/C by Lemma 5.0.5. But now x^g is generic for H over x, since gen(H) is foreign to C. $\square$

Claim. H has a proper normal type-definable subgroup.

Proof of Claim: By Theorem 1.0.3 there must be a non-trivial element x in H such that x^H is not generic. Then by the above claim, there are infinitely many elements c in C such that $cx^g = x^{g'}$ for some g' in H. But Cg is algebraic over $\{x, c, Cg'\}$ and generic for HC/C over $\{x, c\}$, so property $\Re$ implies that Cg' is generic as well. Since gen(H) is foreign to C, this implies that g' is generic for H. Hence the stabilizer $C_1 := \mathrm{stab}_H(x^g/H)$ is an infinite subgroup of H; it is normal, as gh is also generic for H whenever $h \in H$ is independent from g, and

$$C_1^h = \mathrm{stab}_H(x^g/H)^h = \mathrm{stab}_H(x^{gh}/H) = \mathrm{stab}_H(x^g/H) = C_1.$$

Furthermore, as x^g is non-generic, C_1 must have infinite index in G by Proposition 2.1.24. $\square$

But this clearly contradicts the definable simplicity of H. Therefore there must be some non-trivial d in H such that gen(H) is not foreign to $C_{G_1}(d)$. Theorem 3.1.1 implies that H has a normal proper subgroup N such that H/N is $C_{G_0}(d)$-internal; N must be trivial by the definable simplicity of H. Furthermore, $G/C_G(h)$ is infinite and obviously H-internal for any non-trivial element h of H, and therefore $C_{G_0}(d)$-internal. It follows that gen(G) is not foreign to $C_{G_0}(d)$. This finishes the argument for centreless G.

Let now G have centre Z. Replacing G_0 by $C_{G_0}(Z)$, we may assume that Z is also central in G_0. If the generic type p of G is not foreign to Z, we are done. If there is some non-central element in the second centre of G, then $G/C_G(x)$ is infinite and Z-internal, and Z is not co-foreign to gen(G). So the only case left is $Z_2(G) = Z$ and G/Z is centreless. Let us assume that p is foreign to all centralizers of non-central elements.

As the lemma holds for centreless groups, there is some non-central element d in G such that the generic type of G/Z is not foreign to $C := C_{G_0/Z}(dZ)$. But commutation with d is a homomorphism from $C_{G_0}(d/Z)$ to Z whose kernel is contained in $C_{G_0}(d)$. Now gen(G) is foreign to both Z and $C_{G_0}(d)$ and thus also to $C_{G_0}(d/Z)$, and a fortiori to $C_{G_0/Z}(dZ) = C_{G_0}(d/Z)Z/Z$. However, this implies that this group is also co-foreign to gen(G/Z), a contradiction.

This completes the proof of Lemma 5.2.1. $\square$

Theorem 5.2.2 *Let G be a definable $\Re$-group which is not co-foreign to some type p. Then p is not foreign to some abelian subgroup A of G.*

Proof: Let C be a minimal intersection of centralizers in G such that p is not foreign to C. Suppose that C is non-abelian and let Σ be the set of all centralizers in C of non-central elements. By the remark after Definition 3.2.1 the generic type of C^Σ is foreign to Σ. If C^Σ were non-trivial, it could not be abelian, thus contradicting Lemma 5.2.1 applied to $C^\Sigma \trianglelefteq C$. Hence C is Σ-analysable. But p is foreign to Σ, so p is foreign to C, a contradiction. $\square$

We now easily get the decomposition theorem:

Theorem 5.2.3 *Let G be a definable $\Re$-group. Then there are a descending sequence*
$$G = G_0 \triangleright G_1 \triangleright \cdots \triangleright G_n = \{1\}$$
and abelian subgroups A_i of G for $i < n$ such that every quotient G_i/G_{i+1} is A_i-internal.

Proof: Let Σ be the set of all definable abelian subgroups of G. Then by Theorem 5.2.2 the generic type of any type-definable normal subgroup of G is not foreign to Σ; now Theorem 3.1.1 implies that G is Σ-analysable, and by Corollary 3.1.4 the analysis can be carried out in finitely many steps. $\square$

Corollary 5.2.4 *An $\Re$-group contains an abelian subgroup of the same cardinality.*

Proof: Immediate from Theorem 5.2.3. $\square$

5.3   Linear Operations

In this section, we shall prove the definability of a field from certain group actions. Recall that a group A acted upon by some group M of automorphisms is M-minimal if it is infinite, type-definable, M-invariant, and minimal with these properties.

Theorem 5.3.1 *Let A be a type-definable abelian group and M a definable infinite abelian group of automorphisms of A. Suppose A is M-minimal, the theory is $\Re$, and the generic type of A is not foreign to M. Then there are a definable field K, a definable isomorphism $K^+ \cong A$ and a definable embedding $M \hookrightarrow K^\times$.*

Proof: Any finite M-invariant subgroup of A must be centralized by the centralizer-connected component M^{cc}, and $C_A(M^{cc})$ is the maximal finite M-invariant subgroup of A. Let R be the ring of endomorphisms of A generated by M. For any $r \in R$ both $\ker(r)$ and $\mathrm{im}(r)$ are M-invariant; by minimality one of them is finite and the other the whole of A. But if $|\ker(r)| = n$, then by surjectivity of r, the cardinality of $\ker(r^k)$ is n^k. But this is bounded by $|C_A(M^{cc})|$, so $n = 1$ and every non-zero endomorphism in R is an automorphism. In particular $x \mapsto mx - x$ is injective for any $1 \neq m \in M$, and $C_A(M^{cc})$ is trivial.

Now R is an integral domain and gives rise to a field K of definable endomorphisms of A generated by M. Note that R is the union of an increasing chain of definable sets of endomorphisms, each of which consists of all those endomorphisms generated in a finite (increasing) number of ways by elements of M. Since such an endomorphism is zero iff it maps any non-zero element of A to zero, the set of invertible endomorphisms is also a union of an increasing chain of definable sets, and so is K.

The generic type of A is not foreign to M. By Theorem 3.1.1, there is an M-invariant subgroup A' of infinite index in A such that A/A' is M-internal. But A' can only be trivial, so actually A is M-internal, and there is a definable function from M^k to a superset of A, for some $k < \omega$.

For any $a \in A$ the orbit Ka is a subgroup of A (which is *a priori* undefinable). As K is a field, Ka is a minimal (truly minimal, without definability assumptions) K-invariant subgroup of A. Therefore for any elements $(a_i : i < n)$ of A such that $a_j \notin \sum_{i \neq j} Ka_i$ for all $j < n$, the sum

$$\sum_{i < n} Ka_i = \bigoplus_{i < n} Ka_i$$

must be direct. However, for any non-zero a in A the map $a \mapsto a^m$ defines a bijection between M and a^M. Thus M is a^M-internal.

Suppose $n > k$. Then over suitable $a_0, \ldots, a_k$ a generic element $(m_0, \ldots, m_k)$ of M^{k+1} is algebraic over $\sum_i m_i a_i$, which in turn is algebraic over k elements of M, i.e. an element of $M^k \times \{1\}$. But this is a non-generic subset of M^{k+1}, contradicting property $\mathfrak{R}$. We may thus consider A as a finite-dimensional vector space over K: $A = \sum_{i<n} Ka_i$ for some $n \leq k$. By compactness this means that there is a definable subset K' of K such that for any n elements $(a_i : i < n)$ of A with $a_j \notin \sum_{i \neq j} K'a_i$ for all $j < n$, already $A = \sum_{i<n} K'a_i$.

But then K is definable: there are at most n different automorphisms $(k_i : i < n)$ in K such that $k_j a \notin \sum_{i \neq j} K'(k_i a)$ for all $i < n$. Thus for every $k \in K$ there is $k' \in K' - \{0\}$ such that

$$k'ka \in \sum_{i < n} K'k_i a,$$

whence $k'k \in \sum_{i<n} K'k_i$ and $K = (K' - \{0\})^{-1} \sum_{i<n} K'k_i$ is definable. Therefore Ka is definable; as it is M-invariant, it must equal A.

So we get $A \cong K^+$ (and in particular A is definable), and obviously $M \hookrightarrow K^\times$. $\square$

Theorem 5.3.2 *Let G be an infinite definable centralizer-connected $\Re$-group acting as a group of automorphisms on the abelian G-minimal group B. Suppose that G has a definable normal abelian subgroup M, and that the generic type of B is not foreign to M. Then there is an infinite definable field K such that we can endow B definably with the structure of a finite-dimensional vector space over K; G acts K-linearly, and M is central in G.*

Proof: Let A be an M-minimal subgroup of B. Then $M/C_M(A)$ is A-internal, and so is M/N, where $N := \bigcap_{g \in G} C_M(A)^g$ (remember that M is normal in G). But now N is normal in G, so $C_B(N)$ is G-invariant and contains A. Minimality of B implies that $B = C_B(N)$, so N must be trivial. Hence M is A-internal, and also $(M/C_M(A))$-internal. Since $\text{gen}(B)$ is not foreign to M, and G-minimal, B is almost M-internal by Theorem 3.1.1, and therefore almost $(M/C_M(A))$-internal and almost A-internal, and the generic type of A is not foreign to $M/C_M(A)$.

Now we show that $B = \bigoplus_{i=1}^{n} g_i A$ for some $n < \omega$ and elements g_i of G. Indeed, by normality of M any intersection $A \cap \sum g_i A$ is M-invariant; it cannot be finite by Theorem 5.3.1, and if it is infinite, then it must be the whole of A by minimality. But as M is A-internal and B is almost M-internal, the length of any direct sum of A-conjugates in B is bounded by $\Re$. So $\langle GA \rangle = GA$ is infinitely definable as a finite direct sum of translates of A; since it is G-invariant, it must be equal to B. Similarly, $B_0 := G^0 A$ is type-definable as a finite direct sum of translates of A.

Let R denote the ring of endomorphisms of B_0 generated by M. By Theorem 5.3.1 the action of R on A is that of an integral domain R' embedding into a definable field K; there is a prime ideal $I = \text{ann}(A)$ with $R/I = R'$, and non-trivial endomorphisms in R' have trivial kernels on A. Now for n independent principal generic elements $g_1, \ldots, g_n$ of G^0, if the ideals $I^{g_1}, I^{g_2}, \ldots, I^{g_n}$ are pairwise distinct, then the corresponding $g_1^{-1}A, g_2^{-1}A, \ldots, g_n^{-1}A$ are linearly disjoint (by symmetry one ideal cannot be contained in another, so for any $i \leq n$ there is some element s_i in $\bigcap_{j \neq i} I^{g_j} - I^{g_i}$ which annihilates all $g_j^{-1}A$ for $j \neq i$ and has trivial kernel on $g_i^{-1}A$). Therefore n is bounded by $\Re$, as above. But G^0 acts on the I^{g_i} by inner automorphisms, this action is definable using the $g_i^{-1}A$; as G^0 is connected, it fixes I. Therefore $I = \text{ann}(A) = \text{ann}(gA)$ for all $g \in G^0$. But $\langle G^0 A \rangle = B_0$ and so $I = \text{ann}(B_0)$ is trivial. Thus $R = R'$ embeds into the definable field K, and G^0 acts by conjugation as a group of automorphisms on K.

By Theorem 5.1.7 the action is trivial (note that G^0 is connected), so $[M, G^0] \leq C_G(B_0) \leq C_G(A)$. But for any g in G, by normality of M the translate gA is also an M-minimal subgroup of B. Therefore $[M, G^0]$ is contained in $C_G(gA)$ for all $g \in G$. Since B is generated by (finitely many) gA, the commutator group $[M, G^0]$ must be contained in $C_G(B)$, which is trivial. By centralizer-connectivity, M must be central in G.

It follows that B is isomorphic to a vector space over K, on which G acts K-linearly and M scalarly. $\square$

We should note that the condition that $\mathrm{gen}(A)$ or $\mathrm{gen}(B)$ is not foreign to M in Theorems 5.3.1 and 5.3.2 is in particular satisfied if $U_P(A)$ (resp. $U_P(B)$) and $U_P(M)$ are comparable, i.e. the Cantor normal form begins with the same power ω^α. Indeed, $U_P(Ma) = U_P(M) \geq \omega^\alpha$ for all $a \in A$ and Ma is α_P-indecomposable with respect to all finite M-invariant subgroups. As there are no infinite ones, Ma, and hence also $Ma - a$, is α_P-indecomposable by Lemma 3.6.10 and by Theorem 3.6.11 generates an M-invariant P-connected group, which must be the whole of A by M-minimality. So A is M-internal. As any sum of G-translates of A is direct (after omitting superfluous summands) by M-minimality of A and normality of M in G, comparability of $U_P(A)$ (which is at least $U_P(M)$) and $U_P(B)$ implies that B is a finite direct sum of translates of A, and as such A- and therefore M-internal.

5.4 Solubility and Nilpotency

In this section we shall analyse the Φ-component of a soluble $\mathfrak{R}$-group. We shall show that its ω-centre is definable and that its commutator subgroup is nilpotent. This implies the definability of the soluble radical for any Φ-connected $\mathfrak{R}$-group. Finally we shall prove a decomposition theorem for the centre of the commutator subgroup of a centreless soluble Φ-connected $\mathfrak{R}$-group.

In order to simplify notation we shall simply write $[G, H]$ for the definable closure $\mathrm{dc}[G, H]$ of some commutator subgroup. Similarly, we write $G^{(n)}$ when we mean its definable closure.

Theorem 5.4.1 *Let G be a definable $\mathfrak{R}$-group and N a definable normal subgroup of G. Then every (iterated) commutator subgroup of G^Φ has a definable ω-centre modulo N, i.e. the ascending upper central series modulo N eventually becomes stationary.*

The main idea in this proof is a generalization of the vector space techniques used in the proof of Theorem 5.3.1: supposing we have an infinitely ascending central series, we take a minimal invariant type-definable group A not contained in the ω-centre Z and show that A/Z is a definable field. In particular Z must then be relatively definable and equal some iterated centre, contradicting our initial assumptions. The technical difficulty lies in manipulating the $(\bigwedge / \bigvee)$-definable quotient A/Z.

Proof: Suppose otherwise, and let $Z_1 < Z_2 < \cdots < \bigcup_{i<\omega} Z_i =: Z$ be the strictly ascending upper central series of $(G^\Phi)^{(n)}$ modulo N. We clearly may assume that G is saturated. In order to simplify notation, we may take N to be trivial, except when dealing with Frattini formulæ.

By the chain condition on centralizers, $C_G(Z) = C_G(Z_i)$ for some $i < \omega$, and $(G^\Phi)^{(n)}/C_{G^\Phi}(Z)$ is a group of automorphisms of Z_i which stabilizes a normal chain. So by Lemma 1.4.12 $(G^\Phi)^{(n)}/C_{G^\Phi}(Z)$ is nilpotent, of class c, say. So there is a maximal $k < n + c$ such that $H := (G^\Phi)^{(k)}$ is the (definable closure of the) last commutator subgroup of G^Φ such that Z is not contained in $C_G^j(H)$ for any $j < \omega$.

Z is locally nilpotent and hence soluble by Lemma 1.2.5; therefore Z is contained in a minimal G-invariant type-definable soluble group S. So there is a minimal s such that $Z^{(s)}$ is contained in $C_G^{j_s}(H)$ for some $j_s < \omega$; we replace N by $NC_G^{j_s}(H)$ (a definable normal subgroup of G). Then Z has derived length s, and the last non-trivial derived group $Z^{(s-1)}$ is not contained in $C_G^j(H)$ for any $j < \omega$. However, by maximality of k, it is contained in $C_G^i(H')$ for some $i < \omega$, so there is a maximal $t < \omega$ such that $C_G^t(H') \cap Z^{(s-1)}$ is contained in $C_G^{j_t}(H)$ for some $j_t < \omega$. We replace N by $NC_G^{j_t}(H)$, and consider $D := \mathrm{dc}(C_G(H') \cap Z^{(s-1)})$, a normal abelian group centralized by H' which is not contained in $C_G^j(H)$ for any $j < \omega$. (We are suppressing N again.)

Call a subgroup of D *tall* if it contains elements in the definable sets $C_G^{j+1}(H) - C_G^j(H)$ for all $j < \omega$. Note that D itself is tall, so by compactness there is a minimal type-definable G-invariant tall subgroup B of D. Furthermore, if X is an H-invariant subgroup of D and $x \in C_X^{j+1}(H) - C_X^j(H)$, then there is $h \in H$ with $[h, x] \in C_X^j(H) - C_X^{j-1}(H)$. So an H-invariant subgroup of D is tall iff it does not intersect $C_G^\omega(H)$ in a subgroup of $C_G^j(H)$ for any $j < \omega$. Note that a tall X must be connected, since $X^0 \cap C_G^\omega(H) \le C_G^i(H)$ for some $i < \omega$ would imply $X \cap C_G^\omega(H) \le C_G^{i+1}(H)$ by the connectivity of H. In particular, if A is a minimal H-invariant tall subgroup of B, then A and B are both connected.

Put $M := H/C_H(A)$. Since $H > C_H(A) \ge C_H(B) \ge H'$, this is an infinite connected type-definable abelian group of automorphisms of A. Let R be the ring of endomorphisms of A generated by M, and put $I = \{r \in R : rA < A\}$. Note that R is commutative, its elements are definable endomorphisms, and R is the union of an increasing chain of type-definable sets. However, what we really want to consider is the ring of M-induced endomorphisms of A modulo $A_\omega := C_A^\omega(H)$, and compare it with R.

Claim. $I = (0)$.

Proof of Claim: For any $r \in I$, since rA is still M-invariant, minimality of A implies that $rA \cap A_\omega \le C_G^i(H)$ for some $i < \omega$. Therefore $rA_\omega \le rA \cap A_\omega \le C_G^i(H)$, and $\{a \in A : ra \in C_G^i(H)\}$ is an M-invariant subgroup of A containing A_ω. So it is tall, and must equal A by minimality. In particular, if $r = (g - 1)$ for generic $g \in M$, then as every element of M is the product of two generic elements and $(gh - 1) = (g - 1)(h - 1) + (g - 1) + (h - 1)$, we get $A = C_A(H/C_G^i(H)) = C_G^{i+1}(H)$, a contradiction. Therefore $(g-1)A = A$ for all generic $g \in M$.

But now for arbitrary $r \in I$ and $i < \omega$ with $rA \leq C_G^i(H)$, we get $(g-1)^i rA = 0$ for any generic $g \in M$. As R is commutative and $(g-1)A = A$, this implies $rA = 0$. $\square$

In particular, any non-zero endomorphism in R is surjective, and R is an integral domain.

Claim. If A' is a proper relatively $\emptyset$-definable subgroup of A and $a \in A$ is generic over the parameters needed to define A', then $ra \in A'$ implies $r = 0$ for any $r \in R$. Furthermore the map $M \ni m \mapsto ma + A' \in A/A'$ is injective. Note that we do not suppose that a and r are independent.

Proof of Claim: Let X be the (type-definable) set of $\emptyset$-conjugates of r. Since $r \neq 0$, the set $\{a' \in A : ra' \in A'\}$ is not tall and intersects A_ω in $C_A^i(H)$ for some $i < \omega$, and the same is true for all $r' \in X$. However, a is generic and satisfies $\exists r \in X$ $(ra \in A')$. By minimality of A, it is clear that $A = \mathrm{dc}(A_\omega)$, so a definable generic property of A must be true on a generic subset of A_ω by Theorem 2.1.19; in particular it must be true generically for $C_A^i(H)$ and big $i < \omega$. By compactness, we find arbitrarily large $i < \omega$ and $a' \in C_A^{i+1}(H) - C_A^i(H)$, and $r' \in X$ with $r'a' \in A'$, a contradiction. Therefore $ra \notin A'$. But now for $m \neq n \in M$ clearly $m - n \neq 0$ in R, and hence $(m-n)a \notin A'$. Therefore $ma + A' \neq na + A'$ and the map $m \mapsto ma + A'$ is injective. $\square$

It follows that M is uniformly (A/A')-internal, independently of A'.

Claim. For any proper subgroup M_0 of M, the generic type of B is not foreign to M/M_0.

Proof of Claim: Suppose otherwise. As $\mathrm{gen}(H)$ is not foreign to M/M_0, Lemma 3.1.6 implies that $\mathrm{gen}(G^\Phi)$ is not foreign to M/M_0. Hence there is a definable subgroup D of G not contained in G^Φ such that $G^\Phi/(G^\Phi \cap D)$ is (M/M_0)-internal. We claim that G/D is a Frattini formula.

So let Σ be a set of formulæ such that G is $(\Sigma \cup \{G/D\})$-analysable. Then B is also $(\Sigma \cup \{G/D\})$-analysable; as its generic type is foreign to M/M_0 and hence to $G^\Phi/(G^\Phi \cap D)$, that cannot be foreign to $\Sigma \cup \{G/G^\Phi\}$, or to $\Sigma \cup \Phi(G)$. By Theorem 3.1.1 there is a relatively definable G-invariant proper subgroup B_0 of B such that B/B_0 is $(\Sigma \cup \Phi(G))$-internal, and by minimality, B_0 cannot be tall. So B_0 intersects A in a proper subgroup. Now M is $(A/(B_0 \cap A))$-internal, hence (B/B_0)-internal and thus $(\Sigma \cup \Phi(G))$-internal; and so is $G^\Phi/(G^\Phi \cap D)$ by M-internality. As G/G^Φ is $\Phi(G)$-analysable, this means that G/D is $(\Sigma \cup \Phi(G))$-analysable, whence G is $(\Sigma \cup \Phi(G))$-analysable and thus Σ-analysable.

But this clearly implies that D contains G^Φ, a contradiction. Note that the proof does not mention N, and is therefore unaffected by the substitutions at the beginning. □

Therefore the generic type of B is not foreign to M. By Theorem 3.1.1 there is a relatively definable G-invariant proper subgroup B_0 of B such that B/B_0 is M-internal. By minimality, B_0 is not tall, and $C_M(B/B_0)$ is a relatively definable subgroup of M, which must be proper, since $C^\omega_{B_0}(H) \leq C^i_B(H)$ for some $i < \omega$ and M cannot centralize $C^\omega_B(H)$ modulo $C^i_B(H)$. So by the last claim there is a proper relatively definable G-invariant subgroup B_1 of B such that B/B_1 is $(M/C_M(B/B_0))$-internal, and B_1 cannot be tall by minimality. Let $A_1 := B_1 \cap A$, and note that A_1 is M-invariant, whence $rA_1 \leq A_1$ for all $r \in R$.

The proof now proceeds similarly to that of Theorem 5.3.1. Put

$$\mathfrak{A} := \{\{a' \in A : ra' \in A_1\} : r \in R^\times\},$$

and $A_0 = \bigcup \mathfrak{A}$. As R is commutative and A_1 closed under R, this is a subgroup; by the second claim A_0 is proper in A and does not contain a generic element (generic over the parameters needed for A_1). By the first claim, R embeds into the field K of automorphisms of A/A_0 which are generated by M. Then $K\bar{a}$ (which is *a priori* a fairly undefinable object) is a minimal K-invariant subgroup of A/A_0 for any $\bar{a} \in A/A_0$. So for any $(\bar{a}_i : i \in I)$ in A/A_0, if $\bar{a}_j \notin \sum_{i \neq j} K\bar{a}_i$ for any $j < n$, then the sum is direct:

$$\sum_{i<n} K\bar{a}_i = \bigoplus_{i<n} K\bar{a}_i.$$

However, $G/C_G(B/B_0)$ is definable, (B/B_0)-internal and hence M-internal, and by the second claim M is (Ma/A_0)-internal for any generic $a \in A$, and hence $K\bar{a}$-internal. (By this we mean that $m \in \mathrm{dcl}(a, ma + A')$ for any $A' \in \mathfrak{A}$.) If the number n of direct summands in $\bigoplus_{i<n} K\bar{a}_i$ were not bounded, then (over suitable parameters $a_i \in A$) arbitrarily many elements $m_i \in M$ would be algebraic over a single element $a + A_0 = \sum m_i(a_i + A_0)$ of A/A_0, and the same would be true for arbitrarily many elements in $G/C_G(B/B_0)$. But $a + A_0 = [(a+B_1) \cap A] + A_0 \in \mathrm{dcl}(a+B_1)$, and $a + B_1$ is algebraic over a bounded number of elements $\bar{m} \in H/C_H(B/B_0)$ by the $(H/C_H(B/B_0))$-internality of B/B_1. This contradicts property $\mathfrak{R}$ for $G/C_G(B/B_0)$. So there is $n < \omega$ such that $A/A_0 = \sum_{i<n} K\bar{a}_i$ for all $a_0, \ldots, a_{n-1} \in A$ with $a_i + A_0 \notin \sum_{j \neq i} K\bar{a}_j$ for all $i < n$.

By compactness this means that there are a non-trivial type-definable subset $R' \subseteq R$ and some relatively definable M-invariant proper subgroup $A_2 \in \mathfrak{A}$ such that for any $a \in A$ and for any $a_0, \ldots, a_{n-1} \in A$ with $R'a_i \cap \sum_{j \neq i} R'a_j \leq A_2$ there is a non-zero $r' \in R'$ with $r'a = \sum_{i<n} R'a_i + A_2$. (It follows from the second claim that $r' = 0$ iff $r'a = 0$ for some generic element, so this is a definable property.) However, we can now replace the a_i by $r_i a$

for any generic element a in A and suitable endomorphisms r_i in R, to see that for any $q \in R$ there is a non-zero endomorphism r in R' satisfying

$$(\dagger) \qquad rqa \in A_2 + \sum_{i < n} R'(r_i a).$$

Put $R'' = \sum_{i=1}^{n} R' r_i$. Then $(\dagger)$ implies that for any q in R there are non-zero $r \in R'$ and $r' \in R''$ with $(rq - r')a \in A_2$. As a was generic over A_2, this means that $rq = r'$. However, $r \in R''$ is zero iff $ra = 0$, so by compactness there must be a $j < \omega$ such that for every non-zero $r \in R''$ the kernel of r intersects A_ω in $C_A^j(H)$. But if $a \in C_A^{j+1}(H) - C_A^j(H)$, then there is non-zero $q \in R$ with $qa = 0$. However, for some non-zero $r \in R'$ and $r' \in R''$ we have $rq = r'$, whence $r'a = 0$ and then $r' = 0$. Therefore $q = 0$, a contradiction. This finishes the proof. $\square$

Theorem 5.4.2 *Let G be a definable $\Re$-group and N a definable normal subgroup such that G^Φ/N is soluble. Then $(G^\Phi)'/N$ is nilpotent.*

Proof: Let $H = (G^\Phi)^{(n)}$ be the last (iterated) commutator subgroup which is not nilpotent modulo N. Assume for a contradiction that n is greater than zero. We shall show that $Z(H/N)$ is non-trivial; successively dividing out the iterated centres (i.e. replacing N by $C_G^i(H/N)$ for $i = 0, 1, 2, 3, \dots$) will then contradict Theorem 5.4.1, and we finish.

Let $A_0 \leq Z(H'/N)$ be minimal G^0-invariant. We may assume that $C_G(A_0)$ is maximal (varying A_0 among the minimal G^0-invariant subgroups of $Z(H'/N)$) and stabilized by some subgroup G_0 of finite index in G, which we may choose such that $G_0/C_{G_0}(A_0)$ is centralizer-connected. Let A be a minimal G_0-invariant subgroup of $C_{Z(H'/N)}(C_G(A_0))$. Then A contains a minimal G^0-invariant subgroup A', and clearly $C_G(A_0) \leq C_G(A) \leq C_G(A')$. So the maximality of $C_G(A_0)$ implies the equality of these centralizers. Hence $G_0/C_{G_0}(A)$ is a centralizer-connected group of automorphisms of A. If A is finite, then it is central and we are done. So suppose A is infinite.

Let M be some definable normal centralizer-connected abelian supergroup of $H/C_H(A)$ in $G_0/C_{G_0}(A)$. Suppose first that the generic type of A is not foreign to M. Theorem 5.3.2 implies that A is definably isomorphic to an n-dimensional vector space (for some $n < \omega$), on which $G_0/C_{G_0}(A)$ acts linearly and M scalarly. But, for any element m in $H/C_H(A)$, the determinant $\det(m) = 1$ (this is a commutator) and hence $m = \rho \cdot I$ for an n-th root of unity ρ. So $H/C_H(A)$ must be finite. By connectivity, H centralizes A.

Suppose next that the generic type of A is foreign to M. We claim that M is a Frattini formula. So let Σ be a set of formulæ such that G is $(\Sigma \cup \{M\})$-analysable. Then A is $(\Sigma \cup \{M\})$-analysable; as $\text{gen}(A)$ is foreign to M, it cannot be foreign to Σ. Now A is minimal G_0-invariant, so by Theorem 3.1.1

it is Σ-internal. Since $C_M(A)$ is trivial, M is A-internal, and therefore G is Σ-analysable, and M is a Frattini formula. By Lemma 3.1.6, the generic type of H is foreign to M. This can only happen if $C_H(A) = H$, i.e. A is central in H. $\square$

If G is definable and N a relatively definable normal subgroup of G^Φ, we can easily find a definable subgroup G_0 of G containing G^Φ and a definable normal subgroup N_0 of G_0 which intersects G^Φ in N. To wit, if $\bar{N}$ is a definable subgroup of G which intersects G^Φ in N, let N_0 be the intersection of all G^Φ-conjugates of $\bar{N}$, and G_0 the normalizer in G of N_0. We should note that $G_0^\Phi = G^\Phi$.

Theorem 5.4.3 *Let G be a definable $\Re$-group. Then the soluble radical $R_s(G^\Phi)$ is soluble and definable relative to G^Φ.*

Proof: Let S be a soluble normal subgroup of G. Then $(S^\Phi)'$ is nilpotent by Theorem 5.4.2, and soluble of derived length $(k-1)$ for some fixed $k < \omega$ independent of S by Lemma 1.2.5. By Theorem 1.1.12 the group generated by all soluble normal subgroups of derived length k is again soluble, and hence contained in a definable normal soluble subgroup N.

We now claim that S/N is a Frattini formula. So let Σ be a set of formulæ such that G is $(\Sigma \cup \{S/N\})$-analysable. Then in particular S is $(\Sigma \cup \{S/N\})$-analysable. N contains all normal soluble subgroups of derived length k, and in particular S^Φ, so S/N is a Frattini formula of S and S is already Σ-analysable. But then G is Σ-analysable.

Hence the generic type of G^Φ is foreign to S/N, so S/N is centralized by G^Φ/N. Since by Theorem 1.1.12 any normal soluble subgroup of G^Φ is contained in a definable normal soluble subgroup of G, we get $R_s(G^\Phi) = C_{G^\Phi}(G^\Phi/N)$. $\square$

We should note that $R_s(G^\Phi)$ is also the maximal locally soluble normal subgroup of G^Φ by Proposition 1.3.7. We can now iterate the process and consider $G_1 = C_G(G^\Phi/N)/R$, where R is any definable normal soluble subgroup of G containing $R_s(G^\Phi)$. If G satisfies the ωdcc^0, this eventually leads to the definability of $R_s(G)$. It is unknown whether the soluble radical is definable in an arbitrary $\Re$-group; similarly it is not known whether the locally soluble radical exists in an $\Re$-group.

As for the quotient group $G^\Phi/R_s(G^\Phi)$, Theorem 3.4.6 yields the following characterization:

Corollary 5.4.4 *Let G be a definable $\Re$-group. Then $G^\Phi/R_s(G^\Phi)$ decomposes as a direct product of finitely many $\Re$-connected subgroups.*

Proof: Let R be a definable normal soluble subgroup of G containing $R_s(G^\Phi)$. Then G^Φ/R has no abelian normal subgroups; by compactness there is a definable supergroup $G_1 \leq G$ of G^Φ such that G_1/R has no normal abelian subgroup. Note that $G_1^\Phi = G^\Phi$. Then G_1^Φ/R decomposes as a direct product of finitely many $\Re$-connected subgroups by Theorem 3.4.6, and clearly G_1^Φ/R is isomorphic to $G^\Phi/R_s(G^\Phi)$. $\square$

5.5 Complements and Carter subgroups

Again we shall write $[G, H]$ instead of $\mathrm{dc}[G, H]$.

Lemma 5.5.1 *Let G be an $\Re$-group and N a definable normal subgroup of G^Φ such that G^Φ/N is soluble. Let Z_ω be the ω-centre of G^Φ/N, and Z a normal abelian subgroup of G/NZ_ω centralized by $(G^\Phi)'$ and type-definable over $\emptyset$. Then for any $g \in G^\Phi$ the centralizer $C_Z(g)$ is relatively* $\mathrm{acl}(\emptyset)$-*definable.*

This implies in particular that a generic element of G^Φ acts without fixed points on $Z - \{0\}$, as by connectivity of G^Φ all generic elements must have the same centralizer, namely $C_Z(G^\Phi)$, which is trivial.

Proof: By Theorem 5.4.1 the ω-centre of G^Φ/N is relatively definable as a finitely iterated centre; replacing N by NZ_ω, we may assume that G^Φ/N is centreless. Furthermore $(G^\Phi)'$ is nilpotent by Theorem 5.4.2. Again we shall suppress N except when considering Frattini formulæ.

We may clearly assume that $C_Z(g)$ is non-trivial. As $(G^\Phi)'$ centralizes Z, the subgroup $C_Z(g)$ is normal in G^Φ. If M is an $\emptyset$-definable normal subgroup of G containing G^Φ and with abelian quotient $M/C_M(Z)$ (this exists since G^Φ is definably characteristic), then M also stabilizes $C_Z(g)$. Now let $(g_i : i < \omega)$ be a Morley sequence in $\mathrm{stp}(g)$ and put $I := \bigcap_{i<\omega} C_Z(g_i)$. Clearly I is $\mathrm{acl}(\emptyset)$-definable and M-invariant.

Suppose $C_Z(g)$ properly contains I. Let A be an M-minimal proper supergroup of I contained in $C_Z(g)$. Then A/I cannot be central in G^Φ/I, as otherwise $(g_i - 1)A$ would be central in G^Φ by commutativity for all i, hence trivial, and A would be contained in I. By Lemma 3.2.9 $\mathrm{gen}(A/I)$ is not foreign to $G^\Phi/C_{G^\Phi}(A/I)$ and *a fortiori* not to $M/C_M(A/I)$. Theorem 5.3.1 implies that $M/C_M(A/I)$ acts on A/I as a multiplicative subgroup of a field acting on the additive group; in particular there are no fixed points.

Let $(g_i, A_i : i < \omega)$ be a Morley sequence in $\mathrm{stp}(g, A)$. Put $C := \bigcap_{i<\omega} C_M(A_i/I)$; this is a supergroup of $C_M(Z/I)$, so M/C is abelian and acts on each A_i/I. For $i \neq j$ the element g_i operates without fixed points on A_j/I, hence $g_i - 1$ is zero on A_i/I and injective on all other A_j/I, and the sum of the A_i/I must be direct. Let $\bar{A}$ be a minimal type-definable M-invariant subgroup of Z such that $\bar{A}$ contains A_i for generic A_i (i.e. the parameters needed for the definition of A_i are independent of those needed for $\bar{A}$). The centralizer modulo I of all the A_i equals that of any big subset of them, in particular of a generic subset. Hence $C_M(\bar{A}/I) = C$.

Claim. $\text{gen}(\bar{A}/I)$ is not foreign to M/C.

Proof of Claim: Suppose that $\text{gen}(\bar{A}/I)$ is foreign to M/C. We shall derive that M/C is a Frattini formula. But this implies that C contains G^Φ, whence A is central in G^Φ modulo I, a contradiction.

So consider any set Σ of formulæ such that G is $(\Sigma \cup \{M/C\})$-analysable. Then $\bar{A}/I$ is $(\Sigma \cup \{M/C\})$-analysable; as $\text{gen}(\bar{A}/I)$ is foreign to M/C, it cannot be foreign to Σ. By Theorem 3.1.1 and the minimality of $\bar{A}$ there is an M-invariant relatively definable subgroup $B \geq I$ of $\bar{A}$ intersecting generic A_i in I such that $\bar{A}/B$ is Σ-internal. Now if J is any finite set of generic indices (i.e. for $j \in J$ the parameters needed for A_j are generic over those needed for $\bar{A}$ and B), then $B + \sum_{j \in J} A_j$ must be direct modulo I, since for any $i \in J$ multiplication by $\prod_{j \neq i}(g_j - 1)$ maps an element $x = b + \sum_{j \in J} a_j$ with $a_i \neq 0$ to $\prod_{j \neq i}(g_i - 1)b + \prod_{j \neq i}(g_i - 1)a_i$, where the second summand is non-trivial modulo I. As I, B and A_i are all M-invariant and $B \cap A_i = I$, this implies $x \notin I$.

Therefore since $\bar{A}/B$ is Σ-internal, so is $\sum_{j \in J}(A_j/I) = \bigoplus_{j \in J}(A_j/I)$. On the other hand, $C = C_M(A_i/I : i < \omega) = C_M(\sum_{j \in J}(A_j/I))$ for sufficiently large generic J, and so M/C is $(\bigoplus_{j \in J}(A_j/I))$-internal. Therefore G is Σ-analysable, and M/C is Frattini. $\square$

Hence $\text{gen}(\bar{A}/I)$ is not foreign to M/C. But then there is B as above such that $\bar{A}/B$, and thus also $\bigoplus_J(A_i/I)$, is (M/C)-internal. As M/C is internal in any sufficiently big partial sum, this contradicts $\mathfrak{R}$: there is a definable surjection $(M/C)^n \to \bigoplus_J(A_i/I)$ with n independent of J, and one from $\bigoplus_J(A_i/I) \to (M/C)^{2n}$ for sufficiently big generic J, so a generic element of $(M/C)^{2n}$ could be algebraic over a non-generic one.

It follows that I equals $C_Z(g)$, and thus is $\text{acl}(\emptyset)$-definable. $\square$

Corollary 5.5.2 *Let G be an $\mathfrak{R}$-group with a definable normal subgroup N such that G^Φ/N is soluble and centreless. Suppose Z/N is a definable normal abelian subgroup of $C_{G/N}((G^\Phi)')$. Then $C_Z(g)$ is G^Φ-invariant for all $g \in G^\Phi$.*

Proof: Let $h \in G^\Phi$ be generic over g. Then $C_Z(g^h)$ is $\text{acl}(g)$-definable and does not depend on h. In particular, it is G^Φ-invariant, as for any $h' \in G^\Phi$ independent of h the product hh' is again principal generic for G^Φ. But then $C_Z(g^h) = C_Z(g)^h = C_Z(g)$. $\square$

Proposition 5.5.3 *Let G be a type-definable $\mathfrak{R}$-group and N a normal abelian type-definable subgroup of G with nilpotent quotient G/N. Suppose some $g \in G^0$ acts without fixed points on N. Then there is a nilpotent complement H in G for N (i.e. $G = NH$ and $N \cap H = \{1\}$), any nilpotent* subgroup L with $G = NL$ is a complement, and any two complements are conjugate.

Proof: We claim first that N is connected. Indeed, $(g - 1)$ acts surjectively on N^0 by Corollary 5.0.6, and by Lemma 1.1.7 the group $[G^0, N]$ is connected, whence $N^0 = [g, N^0] \leq [G^0, N] \leq N^0$. So for any n in N there is some m in N^0 with $[g, n] = [g, m]$, whence $g^n = g^m$ and nm^{-1} lies in $C_N(g)$, which is trivial.

We now use induction on the nilpotency class of G/N. If it is one, then $C_G(g)$ is a complement: By assumption $C_N(g)$ is trivial, so we only have to show that $G = NC_G(g)$. But for any h in G the commutator $[g, h]$ lies in N, and by surjectivity there is some n in N with $[g, h] = [g, n]$, whence hn^{-1} centralizes g. Thus $G = N \rtimes C_G(g)$.

For the induction step we consider the last element K of the descending central series of G which is not contained in N. By surjectivity of $(g - 1)$ on N, for any k in K there is some n in N with $[g, k] = [g, n] \in N$, so kn^{-1} lies in $C_{KN}(g) =: C$. As $C_N(g)$ is trivial, $NK = N \rtimes C$. Now for any h in $N_G(C)$ and any c in C the commutator $[h, c]$ lies in $C \cap N = \{1\}$, so h centralizes C and $N_G(C) = C_G(C)$.

Claim. $NG^i = NC_{NG^i}(C)$.

Proof: We have just seen that this is true for $G^i = K$, and we shall use induction up the descending central series. So suppose the claim holds for G^i and consider any h in NG^{i-1}. There are some n in N and c in $C_{NG^i}(C^h)$ with $[g, h] = nc \in NG^i$ (the induction hypothesis also holds for C^h in place of C), and there is some m in N with $n = [g, m]$. Hence

$$
\begin{aligned}
C^h &= C_{NK}(g^h) = C_{NK}(g[g, h]) = C_{NK}(gnc) \\
&\leq C_{NK}(gn) = C_{NK}(g[g, m]) = C_{NK}(g^m) \\
&= C^m;
\end{aligned}
$$

as NK is normal and C is relatively definable, Lemma 1.0.9 implies $C^h = C^m$. Hence hm^{-1} lies in $N_{NG^{i-1}}(C) = C_{NG^{i-1}}(C)$. This proves the claim. $\square$

If we put $G_0 := C_G(C)$ and $N_0 := N \cap G_0$, then $G = NG_0$ implies $G^0 = NG_0^0$. So there are g_0 in G_0^0 and n in N with $g = ng_0$. As C intersects N trivially, the centralizer $C_{N_0}(g/C) = C_{N_0}(g_0/C)$ must be trivial as well. As $(G_0/C)/(N_0/C)$ is nilpotent of smaller class, we can apply the induction hypothesis and $G_0/C = (N_0C/C) \rtimes (H/C)$ for some nilpotent complement H/C. But then H must be a nilpotent complement for N_0 in G_0, as C is central and $C \cap N_0$ trivial. Now $G = NG_0$ implies $G = N \rtimes H$. Note that both N and H are relatively definable, as they must equal relatively definable supergroups with trivial intersection.

Let now L be nilpotent with $G = NL$. First note that N intersects $Z(G)$ trivially, as $C_N(g) = \{1\}$. But if $N \cap L$ were non-trivial, it would be normal in

L and contain a central element h. Then h would also be central in $NL = G$,
a contradiction. So $N \cap L = \{1\}$, and L is a complement.

Finally, let L be any complement for N in G. If $g = kn$ for some n in N
and k in L, then $C_N(k) = C_N(g)$ is trivial and $(k - 1)$ acts surjectively on
N. Hence there is m in N with $n = [k, m]$, that is $g = k^m$. We conjugate by
m^{-1} and assume $g = k$.

We shall again use induction on the nilpotency class of G/N, and consider
the last element K of the descending central series of G which is not contained
in N. As L is nilpotent and $L \cap N$ is trivial, $D := L \cap NK$ must be central
in L. Hence $D = C_D(g)$ is contained in $C = C_{NK}(g)$. On the other hand,
$N \rtimes C = NK = N \rtimes D$ (intersect $G = N \rtimes L$ with NK), so $C = D$ follows.
As, obviously, H contains C, this proves the case of nilpotency class one. But
inductively we may assume that L/C and H/C are conjugate in $C_G(C)/C$,
whence L and H are conjugate in G. $\square$

Note that we only needed g to be from the connected component G^0 in order
to get N connected and $g - 1$ to act surjectively on N. However, for a small
group G we may use Corollary 5.0.7 instead of Corollary 5.0.6 and derive

Corollary 5.5.4 *Let G be a type-definable small group and N a type-
definable normal abelian subgroup of G with nilpotent quotient G/N. Suppose
some g in G acts without fixed points on N. Then there is a nilpotent complement H in G for N, any nilpotent subgroup L with $G = NL$ is a complement,
and any two complements are conjugate.*

Proof: Immediate. $\square$

Theorem 5.5.5 *Let G be an $\Re$-group and N a definable normal subgroup
with soluble quotient G^Φ/N. Suppose K is a type-definable normal subgroup
of G^Φ containing N such that G^Φ/K is nilpotent. Then there is a nilpotent
connected supplement H/N for K/N in G^Φ/N (i.e. $G^\Phi = KH$).*

Proof: As $G^\Phi/(G^\Phi)'$ is abelian, $G^\Phi/((G^\Phi)' \cap K)$ is nilpotent by Lemma
0.1.3. We may thus assume that K is contained in the commutator subgroup
$(G^\Phi)'$, so K/N is itself nilpotent by Theorem 5.4.2. By Theorem 5.4.1 we
may assume that G^Φ/N is centreless, replacing N by NZ_ω, where Z_ω is the
ω-centre of G^Φ/N. Now by Lemma 5.5.1 a generic element g of G^Φ operates
on $Z((G^\Phi)'/N)$ without fixed points. We use induction on the nilpotency
class of $(G^\Phi)'/N$. If it is one, the theorem follows from Proposition 5.5.3.

For the induction step, we consider a definable abelian normal subgroup
Z/N of G/N centralizing $(G^\Phi)'$ modulo N which contains the centre of $(G^\Phi)'$
modulo N, such that $C_Z(g/N) = N$. By the inductive hypothesis there is a
connected subgroup L with nilpotent quotient LZ/Z such that $G^\Phi = KL$.

As G^{Φ}/K is nilpotent, so is $L/(L \cap K)$, whence $L/(L \cap K \cap Z)$ is nilpotent as well. Now if h is an element in L and n in K with $g = hn$, then the centralizer in $L \cap K \cap Z$ of g/N equals that of h/N, so h operates on $(L \cap K \cap Z)/N$ without fixed points. By Proposition 5.5.3 there is a nilpotent complement H/N for $(L \cap K \cap Z)/N$ in L/N, and $G^{\Phi} = KL = K(L \cap K \cap Z)H = KH$. $\square$

We now prove a structure theorem for the centre of the commutator subgroup of a Φ-connected centreless soluble $\mathfrak{R}$-group.

Theorem 5.5.6 *Let G be an $\mathfrak{R}$-group and N an $\emptyset$-definable normal subgroup of G with soluble quotient G/N. Suppose G^{Φ}/N is centreless, and Z is a normal abelian subgroup of G/N centralized by G' and type-definable over $\emptyset$. Then there is a finite sequence of $\emptyset$-definable G-invariant groups $N = Z_0 < Z_1 < \cdots < Z_n = Z$, such that every quotient Z_i/Z_{i-1} decomposes as a finite direct sum of definable G-minimal groups Z_i^j/Z_{i-1} (for $j < n_i$). Every summand Z_i^j/Z_{i-1} is also G^{Φ}-minimal, and definably isomorphic to the additive group of an algebraically closed field; $G/C_G(Z_i^j/Z_{i-1})$ embeds into and generates the multiplicative group. Finally, there are only finitely many centralizers $C_G(Z_i^j/Z_{i-1})$, all of which occur on the first level.*

Proof: We shall suppress N. Firstly, we note that by Lemma 5.5.1 a generic $g \in G^{\Phi}$ acts without fixed points on Z. Hence if A is a type-definable G^{Φ}-invariant subgroup of Z, then by Corollary 5.0.6 the operation of $(g - 1)$ on A^0 is surjective, so $[g, A^0] = A^0 = [G^{\Phi}, A] = [g, A]$. Hence for any $a \in A$ there is $a' \in A^0$ with $[g, a] = [g, a']$, that is, $g^a = g^{a'}$. So $a'a^{-1}$ centralizes g. As $C_Z(g)$ is trivial, $a = a'$ and A is connected.

Secondly, a generic element g in G^{Φ} acts without fixed points on Z/A: if $(g - 1)z$ lies in A for some z in Z, then by surjectivity there is an element a in A with $(g-1)a = (g-1)z$, whence $z - a$ commutes with g and $z = a \in A$.

Thirdly, if A is proper and relatively definable, and if B/A is a G-minimal subgroup of Z/A, then B/A is definable and definably isomorphic to the additive group of an algebraically closed field, and $G/C_G(B/A)$ embeds into and generates the multiplicative group. Indeed, by Lemma 3.2.9, as B/A is not centralized by G^{Φ}, its generic type is not foreign to $G^{\Phi}/C_{G^{\Phi}}(B/A)$ and hence not foreign to $G/C_G(B/A)$. The assertion now follows from Theorem 5.3.1. In particular, if R denotes the ring of endomorphisms of Z generated by G, then $\text{ann}(B/A)$, the annihilator of B modulo A, is a prime ideal and all elements of R not in $\text{ann}(B/A)$ act automorphically on B/A.

Furthermore, B/A is also G^{Φ}-minimal. For otherwise there would be a proper subgroup H of G containing G^{Φ}, and an H-minimal subgroup C/A of B/A, such that C/A is the additive group of an algebraically closed field into whose multiplicative group $H/C_H(C/A)$ embeds. Note that $C_H(C/A) =$

$C_H(B/A)$, and in both cases field multiplication is derived from the multiplicative group action. So after choosing an appropriate common multiplicative unit in C/A, it naturally becomes a subfield of B/A. By Corollary 5.1.5 the generic type of B/A is foreign to C/A and hence to $G^\Phi/C_{G^\Phi}(C/A)$, a contradiction.

Claim. There is a maximal direct sum of definable G-minimal subgroups of Z.

Note that any sum of G-minimal subgroups of Z is automatically direct (after omission of unnecessary summands) by G-minimality.

Proof of Claim: Suppose $\bigoplus_{i<\omega} A_i$ were an infinite direct sum. By the chain condition on centralizers, we may assume that $C := \bigcap_{i<\omega} C_G(A_i)$ is maximal possible among all such infinite sums. Hence for any infinite subset I of ω there are finitely many indices i_j in I with $C = \bigcap_j C_G(A_{i_j})$. As usual, G/C is $(\bigoplus_j A_{i_j})$-internal.

Let A be a minimal intersection of relatively definable G-invariant subgroups of Z such that for any finite subintersection X there are finitely many $A_{i_0}, \ldots, A_{i_k}$ such that $X \oplus \bigoplus_{j \le k} A_{i_j}$ contains an infinite subsum of $\bigoplus_{i<\omega} A_i$. Suppose gen(A) were foreign to G/C. We claim that then G/C is a Frattini formula, whence C contains G^Φ, a contradiction.

Indeed, consider any set Σ of formulæ such that G is analysable in $\Sigma \cup \{G/C\}$. Then gen(A) cannot be foreign to Σ, so there is a relatively definable proper subgroup B of A such that A/B is Σ-internal. Hence there are G-invariant definable $A_0 \ge A$ and $B_0 < A_0$, with $B_0 \cap A$ properly contained in A, such that A_0/B_0 is Σ-internal. By minimality of A (reducing A_0 and thus B_0 if necessary), B_0 plus any finite subsum of $\bigoplus_{i<\omega} A_i$ contains only finitely many A_i, but A_0 plus a finite subsum S contains infinitely many A_i. So there is an infinite subset $I \subseteq \omega$ with

$$\bigoplus_{i \in I} A_i \le A_0 + S \quad \text{but} \quad \bigoplus_{i \in I} A_i \cap (B + S) = \{0\}$$

(we first choose A_{i_0} with $A_0 + S \ge A_{i_0} \nleq B_0 + S$, then A_{i_1} with $A_0 + S \ge A_{i_1} \nleq B + S + A_{i_0}$, etc.), whence for any finite $I_0 \subset I$ the sum $\bigoplus_{i \in I_0} A_i$ is Σ-internal (note that $(A_0 + S)/(B_0 + S)$ is (A_0/B_0)-internal). As G/C is $(\bigoplus_{i \in I_0} A_i)$-internal for a sufficiently big finite subset I_0 of I, it is also Σ-internal, and G is Σ-analysable: G/C is a Frattini formula.

Therefore gen(A) is not foreign to G/C and we can repeat the argument with G/C in place of Σ. So we get A_0, B_0, S, I as above such that $(B_0 + S) \oplus \bigoplus_{i \in I} A_i \le A_0 + S$ and $(A_0 + S)/(B_0 + S)$ is (G/C)-internal. This contradicts $\Re$ for some sufficiently big (finite) subset of I, as there are infinitely many disjoint finite subsets $I_j \subset I$ such that G/C is $(\bigoplus_{i \in I_j} A_i)$-internal, but for some fixed $k < \omega$ there is a surjective function $(G/C)^k$ to (a superset of) $(A_0 + S)/(B_0 + S)$, which contains definably a copy of any finite subsum of

$\bigoplus_{j<\omega}\bigoplus_{i\in I_j}A_i$.

This finishes the proof of the claim. $\square$

Thus any direct sum must be finite, hence definable. If $Z_1 = Z_1^0\oplus\cdots\oplus Z_1^{n_1-1}$ is a maximal such sum and A is another G-minimal subgroup, then $A \leq Z_1$, since otherwise $A\cap Z_1 = \{0\}$ by G-minimality, and $A + Z_1$ is a longer direct sum, contradicting the maximality of Z_1. Therefore Z_1 is $\emptyset$-definable. Note, however, that the decomposition of Z_1 need not be unique and we might need parameters to define the Z_1^j. (Although the Z_1^j are G-minimal, they need not be minimal in the class of all G-invariant subgroups.)

Claim. The set of annihilators $\mathrm{ann}(Z_1^j)$ does not depend on the particular decomposition of Z_1.

Proof of Claim: Suppose A were a G-minimal subgroup of Z whose annihilator $\mathrm{ann}(A)$ is different from all $\mathrm{ann}(Z_1^j)$. Then $A \leq Z_1$, and for any non-zero $a\in A$ there are $z_j \in Z_1^j$ for $j < n_1$ with $a = \sum_{j<n_1} z_j$. We choose an a such that the sum has a minimal number of non-zero terms. If $i < n_1$ is such that there is $r \in \mathrm{ann}(Z_1^i)-\mathrm{ann}(A)$, then $0 \neq ra \in A$ and $ra = \sum_{j<n_1} rz_j$, so by minimality $z_i = 0$. Hence there must be some $i < n_1$ such that there is $r \in \mathrm{ann}(A) - \mathrm{ann}(Z_1^i)$ and $z_i \neq 0$, and then $0 = \sum_{j<n_1} rz_j$ and $rz_i \neq 0$, contradicting the directness of $\sum_{j<n_1} Z_1^j$. $\square$

We divide out by Z_1 and iterate the process, getting an ascending sequence of subgroups $\{0\} = Z_0 < Z_1 < Z_2 < \cdots$, such that every Z_{i+1}/Z_i decomposes as a maximal direct sum of G-minimal groups Z_{i+1}^j/Z_i, for $j < n_i$.

Claim. Any annihilator $\mathrm{ann}(Z_{i+1}^j/Z_i)$ equals one on the first level.

Proof of Claim: Suppose on some level k there is an annihilator $\mathrm{ann}(A/Z_k)$ for some G-minimal group A/Z_k which does not occur on the preceding level as $\mathrm{ann}(Z_k^j/Z_{k-1})$ for some $j < n_k$. We may assume that $k = 1$. Let J be the set of all indices j such that there is $r_j \in \mathrm{ann}(Z_1^j) - \mathrm{ann}(A/Z_1)$, and $J' = n_1 - J$. Then $r := \prod_{j\in J} r_j \in \bigcap_{j\in J}\mathrm{ann}(Z_1^j) - \mathrm{ann}(A/Z_1)$. On the other hand, an ideal in an infinite integral domain cannot be finite (it must contain a principal ideal, which is already infinite), so $\mathrm{ann}(Z_1^j)$ has infinite index in $\mathrm{ann}(A/Z_1)$ for all $j \in J'$. By Neumann's Lemma there is $r' \in \mathrm{ann}(A/Z_1) - \bigcup_{j\in J'}\mathrm{ann}(Z_1^j)$. Then $rZ_1 = \bigoplus_{j\in J'} Z_1^j$, and both r and r' act automorphically on it. For any a in A the image $rr'a$ lies in rZ_1, so by surjectivity there is z in rZ_1 with $r'rz = r'ra$, i.e. $r(a - z)$ lies in $A' := \ker(r') \cap rA$.

Now $ra' \in Z_1$ implies $a' \in Z_1$ for $a' \in A$, since r is an automorphism of A/Z_1. Therefore $rA \cap Z_1 = rZ_1 \leq \sum_{j\in J'} Z_1^j$. On the other hand, $\ker(r') \cap \sum_{j\in J'} Z_1^j = \{0\}$. Therefore $A' \cap Z_1 = \{0\}$ and A' contains a G-minimal subgroup disjoint from Z_1, contradicting maximality. $\square$

Now suppose the sequence $(Z_i : i < \omega)$ does not reach Z. Let A be a minimal type-definable G-invariant subgroup of Z not contained in $Z_\omega := \bigcup_{i<\omega} Z_i$.

By minimality $\mathrm{ann}(A/Z_\omega) =: I$ is a prime ideal, any $r \in R - I$ induces an automorphism of A/Z_ω, and R/I is an integral domain with fraction field K.
Claim. I is not equal to $\mathrm{ann}(Z_1^j)$ for any $j < n_1$.
Proof of Claim: Suppose otherwise. Then K is actually definable (and additively isomorphic to Z_1^j), and of the form $\bar{R}^{-1}\bar{R} \cup \{0\}$ for some definable subset $\bar{R}$ of $R/I - \{0\}$. Let $a \in A$ be generic. By compactness there is some $i < \omega$ such that for any $k \in K$ there is a well-defined action of k mapping a to some element $ka + Z_i$ of A/Z_i, and such that for $k, k' \in K$ we have $(k + k')a = ka + ka'$. But this means that Ka/Z_i is a G-minimal group. Therefore it must be contained in Z_{i+1}, whence $a = 1a \in Z_{i+1}$, a contradiction. $\square$

Let J be the set of $j < n_1$ with $\mathrm{ann}(Z_1^j) \not\subseteq I$, and $J' = n_1 - J$. Then as above there are some $r \in \bigcap_{j \in J} \mathrm{ann}(Z_1^j) - I$ and some $r' \in I - \bigcup_{j \in J'} \mathrm{ann}(Z_1^j)$. By compactness, there is $k < \omega$ such that $r'A$ is contained in Z_k. Replacing Z by $\ker(r'/Z_k)$ we may assume that $r' = 0$, whence $J' = \emptyset$. Now if $q \in I$, then $qA \leq Z_i$ for some $i < \omega$. However, r annihilates all quotients Z_j/Z_{j+1}, whence $r^i qA = 0$. On the other hand, $rA = A$ by minimality, whence $qA = 0$. It follows that $I = \mathrm{ann}(A)$; by minimality of A in fact $I = \mathrm{ann}(a)$ for any $a \in A - Z_\omega$.
Claim. The generic type of A is not foreign to $G/C_G(A)$.
Proof of Claim: Suppose otherwise. We claim that then $G/C_G(A)$ is a Frattini formula, contradicting $C_G(A) \not\geq G^\Phi$. So let Σ be a set of formulæ such that G is $(\Sigma \cup \{G/C_G(A)\})$-analysable. Then A is analysable in $\Sigma \cup \{G/C_G(A)\}$, and $\mathrm{gen}(A)$ cannot be foreign to it. As it is foreign to $G/C_G(A)$, it is not foreign to Σ and Theorem 3.1.1 yields a proper G-invariant subgroup B such that A/B is Σ-internal. But then $G/C_G(A)$ is (A/B)-internal, hence Σ-internal, and G is Σ-analysable. $\square$

By Theorem 3.1.1 there is a relatively definable G-invariant proper subgroup $B < A$ such that A/B is $(G/C_G(A))$-internal. However, $B \leq Z_i$ for some $i < \omega$, and r^i maps A/B onto A. So A is $(G/C_G(A))$-internal and there is $k < \omega$ such that any element in A is definable over k elements of $G/C_G(A)$. Note that $C_G(A) = C_G(a)$ for any $a \in A - Z_\omega$.

For any $(a_i : i < n) \in A$ with $a_i + Z_\omega \notin \sum_{j \neq i} K(a_j + Z_\omega)$ for all $j < n$, the sum $\sum_{i < n} K(a_i + Z_\omega)$ must be direct. It must therefore have bounded length $n \leq k$, as otherwise, after adding parameters $(a_i : i < n)$, arbitrary $g_0 C_G(A), \ldots, g_n C_G(A)$ in $G/C_G(A)$ would be definable over $\sum_{i \leq k} g_i a_i$, and this element is definable over only k elements of $G/C_G(A)$, thus contradicting $\Re$. By compactness there are a definable subset $R' \subset R - I$ and a G-invariant relatively definable proper subgroup $B < A$ such that for any $(a_i : i < k) \in A$ with

$$R'a_i + B \cap \sum_{j \neq i} R'a_j + B = B$$

(for $j < k$) and for any $a \in A$ there is $r \in R' - I$ with $r'a \in \sum_{i<k} R'a_i + B$. (Remember that $r \in I$ iff $ra = 0$ for any $a \in A - Z_\omega$, so this is a definable property.) But now there are at most k elements $(r_i : i < k)$ in R such that for generic $a \in A$ and any $r' \in R$ there is $q \in R' - I$ with

$$qr'a \in \sum_{i<k} R'r_ia + B.$$

Hence there is $q' \in \sum_{i<k} R'r_i =: R''$ with $qr' - q' = 0$. But by compactness there is some $j < \omega$ such that $\ker(q) \leq Z_j$ for all $q \in R'' - I$. On the other hand, for arbitrarily big i there is some $a \in Z_i - Z_{i-1}$, and $r^ia = 0$; putting $r' = r^i$ yields $q'a = 0$, whence $q = 0$ and $r = 0$, a contradiction. This finishes the proof. $\square$

Corollary 5.5.7 *Suppose G is a soluble $\mathfrak{R}$-group. Then there is a finite series $G^\Phi = G_0 \rhd G_1 \rhd \cdots \rhd G_n = \{1\}$ of relatively definable G-invariant subgroups of G^Φ such that every quotient G_i/G_{i+1} either is central in G_0/G_{i+1} or decomposes as a direct sum of definable G-minimal groups, each of which is definably isomorphic to the additive group of a definable algebraically closed field.*

Proof: The commutator subgroup $(G^\Phi)' =: H$ is nilpotent by Theorem 5.4.2; by Theorem 5.4.1 in every relatively definable characteristic quotient $M_i := Z_{i+1}(H)/Z_i(H)$ the centralizer $C^\omega_{M_i}(G^\Phi)$ is equal to some finitely iterated centralizer $C^{n_i}_{M_i}(G^\Phi) =: N_i$, and finally by Theorem 5.5.6 for every M_i/N_i there is a finite series of definable G-invariant subgroups such that the quotients of successive elements in the series decompose as a direct sum of additive groups of definable algebraically closed fields. $\square$

Corollary 5.5.8 *Let G be an $\mathfrak{R}$-group and N be a relatively definable normal subgroup of G^Φ with soluble quotient G^Φ/N. Suppose G^Φ/N is centreless, and X is a normal subgroup of G^Φ/N centralized by $(G^\Phi)'$ and type-definable over $\emptyset$. Then there is a finite sequence of $\emptyset$-definable groups $N = X_0 < X_1 < \cdots < X_n = X$, such that every X_i/X_{i-1} decomposes as a finite direct sum of definable G^Φ-minimal groups X_i^j/X_{i-1}. Every X_i^j/X_{i-1} is definably isomorphic to the additive group of an algebraically closed field; $G^\Phi/C_{G^\Phi}(X_i^j/X_{i-1})$ embeds into the multiplicative group. In particular, X is definable.*

If we knew that X was actually G-invariant, then we could use an immediate adaptation of Theorem 5.5.6. But for type-definable X its normalizer need only be type-definable and *a priori* we need not be able to find a definable supergroup of G^Φ normalizing X. However, once we know that X is definable, of course $G \geq N_G(X) \geq G^\Phi$ would do.

Proof: First let $\bar{N}$ be a definable subgroup of G which intersects G^Φ in N. We may assume $\bar{N}$ to be G^Φ-invariant; then we may replace N by $\bar{N}$ and G by $N_G(\bar{N})$ to get N definable and normal. Reducing G even further, we may assume that G/N is soluble and $C_{G^\Phi}(G'/N) = C_{G^\Phi}((G^\Phi)'/N)$.

We use induction on the minimal number $n = \sum n_i$ of summands of a decomposition of $C_{G^\Phi}(G'/N)$ as given by Theorem 5.5.6. So let Z_i^j be a decomposition with n summands and consider a minimal i such that Z_i intersects X non-trivially. Then there is a maximal proper subset J of n_i such that X intersects $\sum_{j\in J} Z_i^j$ trivially. As for $j_0 \notin J$ the group $Z_i^{j_0}/Z_{i-1}$ is also G^Φ-minimal, $Z_i^{j_0}$ is contained in $X \oplus \sum_{j\in J} Z_i^j$. But then for any definable G^Φ-invariant supergroup $\bar{X}$ of X intersecting $\sum_{j\in J} Z_i^j$ trivially, the intersection $X^{j_0} := \bar{X} \cap (Z_i^{j_0} + \sum_{j\in J} Z_i^j)$ must be a G^Φ-minimal definable group; in particular it may not vary with $\bar{X}$. Therefore $X^{j_0} = X \cap (Z_i^{j_0} + \sum_{j\in J} Z_i^j)$. Clearly $\sum_{j\notin J} X^j$ is direct and $\emptyset$-definable. We may divide out and inductively obtain a decomposition in G^Φ-minimal groups. $\square$

Corollary 5.5.9 *Let G be an $\mathfrak{R}$-group, and N a relatively definable normal subgroup of G^Φ with soluble centreless quotient G^Φ/N. Then there is no infinitely ascending sequence of G^Φ-invariant subgroups of $Z((G^\Phi)'/N)$.*

Proof: For definable G^Φ-invariant subgroups K and L of $Z((G^\Phi)'/N)$ such that K is contained in L, the minimal number of summands of a decomposition of $Z((G^\Phi)'/L)$ is strictly smaller than that of a decomposition of $Z((G^\Phi)'/K)$, and we get the desired bound. $\square$

Recall that a Carter subgroup is a nilpotent self-normalizing subgroup. If C is a Carter subgroup of G and H is a subgroup of G containing C, then obviously C is also a Carter subgroup of H. In particular C is a maximal nilpotent subgroup, as nilpotent groups have the normalizer property. Thus any Carter subgroup is relatively definable.

Theorem 5.5.10 *Let G be an $\mathfrak{R}$-group and N a relatively definable normal subgroup of G^Φ with soluble quotient $H = G^\Phi/N$. Then H has a Carter subgroup C. If K is a type-definable subgroup of H containing C^0, then $C \cap K$ is a Carter subgroup of K. If D is a nilpotent subgroup of K of finite index in the normalizer of its connected component D^0, then D^0 is conjugate to C^0 in K. Furthermore $K = K'(C \cap K)$ and $C = N_H(C^0)$.*

Proof: We shall use induction on the nilpotency class of H'. If it is one, then H is abelian and its own Carter subgroup. For the induction step, we may first divide out by the ω-centre, which is definable by Theorem 5.4.1, and suppose that H is centreless. Now consider $Z := Z(H')$; this group is definable by Corollary 5.5.8. By the inductive hypothesis, H/Z has a Carter subgroup

D/Z with $H = H'D = H'D^0$ and $D = N_H(D^0)$. By Lemma 5.5.1 a generic element $g \in H$ acts without fixed points on Z; then $(g-1)$ acts surjectively on Z^0 by Corollary 5.0.6, and $Z = Z^0 = [g, Z]$ must be connected as usual. If $g = dh$ with $d \in D^0$ and $h \in H'$, then the centralizer in Z of d equals that of g and must be trivial, so by Proposition 5.5.3 there is a nilpotent complement C for Z in D, i.e. $D = Z \rtimes C$. Hence $H = H'D^0 = H'ZC^0 = H'C^0$. But $X := N_H(C^0) = C^0 N_{H'}(C^0)$ is a product of two nilpotent subgroups normalizing one another and therefore must be itself nilpotent. As X contains C and is contained in $N_H(C^0Z) = N_H(D^0) = N_H(D/Z) = D$, we get $C \leq X \leq D = Z \rtimes C$. However, if X intersected Z non-trivially, then we could find a non-trivial element of $X \cap Z$ central in X, hence in $XH' = H$, a contradiction. Therefore $C = N_H(C^0) = X$ and $C \leq N_H(C) \leq N_H(C^0) = C$: it follows that C is a Carter subgroup.

Let now K be a type-definable subgroup of H containing C^0. Then $C_1 = N_K(C^0)$ is obviously nilpotent and self-normalizing in K, i.e. a Carter subgroup. Consider a nilpotent subgroup D of K of finite index in the normalizer of its connected component. Let Z_1 be a definable G^Φ-minimal subgroup of Z as given by Theorem 5.5.6. If D^0 centralizes Z_1, then Z_1 normalizes D^0, so $K \cap Z_1$ must be contained in $N_K(D^0)$. But $N_K(D^0 Z_1) = N_K(D^0(K \cap Z_1)) \leq N_K(N_K(D^0))$, and the latter equals $N_K(D^0)$, as D has finite index in it. Therefore DZ_1/Z_1 is nilpotent and has finite index in the normalizer (in KZ_1/Z_1) of its connected component.

If D^0 does not centralize Z_1, then there is $d \in D^0$ with $C_{Z_1}(d) < Z_1$; by Corollary 5.5.2 this centralizer is G^Φ-invariant and must be trivial by the minimality of Z_1. Therefore d acts without fixed points on $Z_1 \cap K$. Suppose x is an element of K normalizing $D^0 Z_1$. By Proposition 5.5.3, as $(D^0)^x$ is also a complement for $Z_1 \cap K$ in $(Z_1 \cap K)D^0$, there is some $y \in (Z_1 \cap K)D^0$ with $(D^0)^x = (D^0)^y$, whence xy^{-1} normalizes D^0. So $N_K(D^0 Z_1) = N_K(D^0)(Z_1 \cap K)$, and again DZ_1/Z_1 is nilpotent and of finite index in the normalizer of its connected component.

On the other hand, if the generic element g of H equals hc with $h \in H'$ and $c \in C^0$, then c acts without fixed points on Z_1 and we see as above that $(C \cap K)Z_1/Z_1$ is also a Carter subgroup of KZ_1/Z_1. Hence we may use induction on the nilpotency class of H' and the number of decomposition factors of $Z(H')$, and assume that $C^0 Z_1$ and $D^0 Z_1$ are conjugate by some element kz of KZ_1, hence also by $k \in K$. We may conjugate by k and suppose $C^0 Z_1 = D^0 Z_1$. But now c lies in $C^0(Z_1 \cap K)$ and acts without fixed points on $Z_1 \cap K$. So by Proposition 5.5.3 again C^0 and D^0 must be conjugate complements.

It follows that $N_H(D^0)$ is also a Carter subgroup of H and conjugate to C by some element in K. Furthermore the first case cannot occur: Z_1 must intersect D trivially.

Finally, we may inductively assume $KZ = K'N_{KZ}(C^0Z)$. Putting $X = Z \cap K$ we get $K = K'N_K(C^0Z) = K'N_K(C^0X) = K'N_K(C^0X^0)$. But for $k \in N_K(C^0X^0)$ the conjugate $(C^0)^k$ is also a Carter subgroup of C^0X^0; in particular it is connected, nilpotent and self-normalizing. So there is $l \in C^0X^0$ with $(C^0)^k = (C^0)^l$ and $kl^{-1} \in N_K(C^0) = C \cap K$. Hence $K = K'X^0(C \cap K)$. But the map $x \mapsto [x, c]$ is surjective on X^0. Therefore $X^0 \leq K'$ and $K = K'(C \cap K)$. $\square$

Corollary 5.5.11 *Let H and C be as above, and K be a subgroup of H containing C. Then K is self-normalizing, and for any relatively definable normal subgroup L of K with nilpotent quotient K/L the product LC equals K.*

Proof: For any g normalizing K the conjugate C^g is again contained in K and nilpotent of finite index in the normalizer of its connected component. Hence there is some h in K with $(C^0)^g = (C^0)^h$. But as $C = N_H(C^0)$, this implies $C^g = C^h$, and gh^{-1} normalizes C. Therefore gh^{-1} lies in C and g lies in $CK = K$, whence $N_H(K) = K$.

Now suppose L is a relatively definable normal subgroup of K with nilpotent quotient K/L. As LC is self-normalizing in K, LC/L is self-normalizing in K/L. By nilpotency, $LC/L = K/L$, that is $LC = K$. $\square$

Note that we could not prove conjugacy of all Carter subgroups, as it might conceivably happen that although D is nilpotent and self-normalizing, D^0 is contained in some other nilpotent group as a subgroup of infinite index. However, this cannot occur in a small stable group:

Theorem 5.5.12 *Let G be a small stable group and N a relatively definable normal subgroup of G^Φ with soluble quotient $H = G^\Phi/N$. Suppose C is a Carter subgroup of H. If K is a type-definable subgroup of H containing C^0, then all Carter subgroups of K are conjugate.*

Proof: We use the notation from the proof of Theorem 5.5.10. We have to show that if D is a Carter subgroup of K, then D is conjugate to $C \cap K$ in K, and again we consider a minimal G-invariant subgroup Z_1 of $Z = Z(H')$.

If Z_1 is centralized by the whole of D, then $Z_1 \cap K$ must be contained in D, as D is self-normalizing in K. Hence DZ_1 is self-normalizing in KZ_1 and DZ_1/Z_1 is a Carter subgroup of Z_1K/Z_1.

If some element d of D acts on Z_1 without fixed points, then we use Corollary 5.5.4 to see that DZ_1 is self-normalizing in KZ_1, whence again DZ_1/Z_1 is a Carter subgroup of KZ_1/Z_1. Similarly, $(C \cap K)Z_1/Z_1$ is a Carter subgroup of KZ_1/Z_1, and by induction on the nilpotency class of H' and the number of decomposition factors of $Z(H')$ we may assume that $DZ_1 =$

$(C \cap K)Z_1$. But then by Corollary 5.5.4 again D and $C \cap K$ are conjugate complements of $Z_1 \cap K$ in $D(Z_1 \cap K) = (C \cap K)(Z_1 \cap K)$. $\square$

5.6 The Frattini Subgroup

In chapter 4, given an abelian group, we defined $\mathfrak{A}$ to be the set of its subgroups without supplement. These subgroups reappear in connection with a generalization of the Frattini subgroup.

Definition 5.6.1 Let G be a group. A *non-supplement* in G is a relatively definable normal subgroup N of G which has no relatively definable proper supplement in G. We define $\mathfrak{A}(G)$ to be the set of all non-supplements in G, and $\mathfrak{A}_s(G)$ to be the set of all soluble non-supplements.

Note that both $\mathfrak{A}(G)$ and $\mathfrak{A}_s(G)$ are closed under G-invariant subgroups and products. Thus $\bigcup \mathfrak{A}(G)$ and $\bigcup \mathfrak{A}_s(G)$ are normal subgroups; and the latter is contained in the soluble radical of G. If G is finite, they are both equal to the Frattini subgroup $\Phi(G)$. To see this, we first note that $\Phi(G)$ is itself a nilpotent non-supplement. But if there were a non-supplement F properly containing $\Phi(G)$, then we could find some maximal subgroup M of G not containing F. By maximality, $MF = G$, contradicting $F \in \mathfrak{A}(G)$. In particular, $\bigcup \mathfrak{A}(G)$ is nilpotent and $\mathfrak{A}(G) = \mathfrak{A}_s(G)$ for any finite group G.

Lemma 5.6.1 *Let $\mathfrak{N}$ be a family of uniformly definable non-supplements of a stable group G. Then there is a non-supplement N such that $\langle \mathfrak{N} \rangle / N$ is nilpotent. In particular, if a saturated group G is contained in the union of the non-supplements, then there is a non-supplement N such that G/N is nilpotent.*

Proof: Suppose the groups in $\mathfrak{N}$ are defined by some formula $\varphi(x,b)$, for some parameters b. It is a partial type on b to say that φ is a non-supplement, call it π. We may then assume that $\mathfrak{N} = \{\varphi(x,b) : b \models \pi\}$. Let p_i $(i \in I)$ be the types over $\mathrm{acl}(\emptyset)$ in π, take a Morley sequence $(b_j^i : j \in J)$ in p_i and put $N_i = \bigcap_{j \in J} \varphi(G, b_j^i)$. This is a definable non-supplement. If B is an infinite independent set of realizations of p_i, then modulo N_i the family $\{\varphi(x,b) : b \in B\}$ has empty intersection, so by Theorem 1.1.13 $\varphi(x,b^i)$ is nilpotent of class n_i modulo N_i for any b^i realizing p_i. By compactness there is a finite $J \subset \omega$ with the property that for any $i \in I$ there is $j \in J$ such that $\varphi(x,b^i)$ is nilpotent of class n_j modulo N_j. If $N = \prod_{j \in J} N_j$ and $n = \max\{n_j : j \in J\}$, then N is a non-supplement and modulo N any element of $\mathfrak{N}$ is nilpotent of class n. Hence Theorem 1.1.12 implies that $\mathfrak{N}$ generates a nilpotent group modulo N.

Now suppose that G is saturated and contained in the union of all non-supplements. Then G is connected, as, if H is a subgroup of finite index in G, there is a non-supplement F with $G = HF = H$.

For any (principal) generic element g of G there is a non-supplement N_g containing g. These N_g form a uniformly definable family, which generates a nilpotent group modulo their (relatively definable) intersection N. But this group contains all the generic elements, so it equals G. $\square$

Theorem 5.6.2 *Let G be a saturated $\mathfrak{R}$-group and $G^\Phi / \bigcup \mathfrak{A}_s(G^\Phi)$ nilpotent. Then G^Φ is nilpotent.*

Proof: By Theorem 5.4.3 the soluble radical R of G^Φ is relatively definable and soluble; since it contains $\bigcup \mathfrak{A}_s(G^\Phi)$, it follows from the assumptions that G^Φ / R is nilpotent. Hence G^Φ is soluble. We suppose, by way of contradiction, that G^Φ is not nilpotent. Let N be a nilpotent relatively definable normal subgroup of G^Φ containing $(G^\Phi)'$; then the ascending central series of N is normal and relatively definable. Let k be maximal such that $G^\Phi / Z_k(N)$ is not nilpotent, and let Z be the ω-centre of G^Φ modulo $Z_k(N)$ (which is definable). Then G^Φ / Z is centreless; by Corollary 5.5.9 the intersection $N := (\bigcup \mathfrak{A}_s(G^\Phi) \cdot Z) \cap Z_{k+1}(N)$ is definable. By our choice of k the quotient $G^\Phi / Z_{k+1}(N)$ is nilpotent, therefore the nilpotency of $G^\Phi / \bigcup \mathfrak{A}_s(G^\Phi)$ implies the nilpotency of G^Φ / N. By Theorem 5.5.5 there is a nilpotent H with $G^\Phi = NH$. However, N is the product of a non-supplement with Z; therefore $G^\Phi = N(ZH)$ implies $G^\Phi = ZH$. So G^Φ / Z, and even $G^\Phi / Z_k(N)$, is nilpotent, contradicting our choice of k. $\square$

5.7 Involutions and Conjugacy Classes

We have already investigated involutions in periodic substable groups in chapter 1, and in stable groups in chapter 3. We shall now analyse the structure of the 2-Sylow subgroups in an $\mathfrak{R}$-group.

Theorem 5.7.1 *Let G be an $\mathfrak{R}$-group of finite exponent. Then there is a normal nilpotent 2-group I such that G/I has finite 2-Sylow subgroups.*

Proof: Consider a 2-Sylow subgroup S and suppose that the locally connected component S^{lo} is not normal. Then there is a conjugate T of S with $S^{lo} \neq T^{lo}$ and maximal intersection $I = S \cap T$, and by Theorem 1.5.8 $N = N_G(S^{lo}) \cap N_G(T^{lo}) \cap N_G(I)$ acts transitively on the infinite abelian group $A = (N_S(I)/I)[2]$. Furthermore $N/C_N(A)$ has no infinite abelian subgroup.

Since A is infinite, so is $N/C_N(A)$. But then $N/C_N(A)$ must have an infinite abelian subgroup by Theorem 5.1.1, a contradiction. $\square$

We shall now successively weaken the periodicity condition on G.

Theorem 5.7.2 *Let N be a definable group of automorphisms of the definable abelian group A of exponent p such that A is N-analysable. If the structure satisfies $\mathfrak{R}$ and definable abelian subgroups of N have p-divisible connected components, then an algebraically closed field is definable.*

Proof: By Theorem 5.2.2 there are finitely many definable abelian subgroups N_i, for $i < k$, of N such that N (and hence also A) is $\{N_i : i < k\}$-analysable. By taking k minimal and successively replacing N_i by its $\{N_j : j > i\}$-connected component, we may assume that the generic types of the N_i are pairwise foreign. Note that the new N_i are possibly only type-definable; in any case they are connected. Put $M := N_0^\Phi$.

Let $Z = C_A(M)$, a proper subgroup of A. Then if an element a of A is centralized by some element n of M modulo Z, we get

$$0 = (n-1)^2 a = (n-1)^p a = (n^p - 1)a,$$

so a is centralized by n^p. However, the assumption implies that type-definable abelian subgroups of N also have p-divisible connected components, whence $M^p = M$. Therefore no point in A/Z is stabilized by the whole of M. We now choose an N-minimal subgroup B of A/Z.

We claim that the generic type of B is not foreign to N_0. Indeed, as A is analysable in N and thereby in $\{N_i : i < k\}$, so is B. Hence if $\mathrm{gen}(B)$ were foreign to N_0, by minimality it would be almost N_i-internal for some $i \neq 0$. On the other hand $N_0/C_{N_0}(B)$ is an infinite B-internal quotient, as already $M/C_M(B)$ is infinite; together this implies that $N_0/C_{N_0}(B)$ is almost N_i-internal and $\mathrm{gen}(N_0)$ cannot be foreign to N_i, a contradiction.

Next we claim that $\mathrm{gen}(B)$ is not foreign to M. For otherwise, as it is not foreign to N_0, there is a Frattini formula $\varphi \in \Phi(N_0)$ such that $\mathrm{gen}(B)$ is not foreign to φ. By minimality B is almost φ-internal; as $M/C_M(B)$ is infinite and B-internal, $\mathrm{gen}(M)$ cannot be foreign to φ, a contradiction. By minimality again, B must be M-analysable.

Let M_0 be a definable abelian supergroup of M with $C_A(M_0) = C_A(M)$, and let B_0 be an M_0-minimal subgroup of B. Finally, we show that $\mathrm{gen}(B_0)$ is not foreign to $M_0/C_{M_0}(B_0)$. So suppose otherwise, and consider a set Σ of formulæ such that M (whence also B and B_0) is $(\Sigma \cup \{M_0/C_{M_0}(B_0)\})$-analysable. So $\mathrm{gen}(B_0)$ cannot be foreign to Σ; by minimality B_0 is almost Σ-internal. But clearly $M_0/C_{M_0}(B_0)$ is B_0-internal, hence almost Σ-internal, and M must be Σ-analysable. Therefore $M_0/C_{M_0}(B_0)$ is a Frattini formula for M and $C_{M_0}(B_0)$ contains M, a contradiction.

Therefore $\text{gen}(B_0)$ is not foreign to $M_0/C_{M_0}(B_0)$. By Theorem 5.3.1 there is a definable algebraically closed field K with B_0 isomorphic to K^+ and $M_0/C_{M_0}(B_0)$ embedding into $K^\times$. $\square$

Note that by Corollary 5.0.6 the divisibility condition is in particular satisfied if N has only finitely many elements of order p.

Proposition 5.7.3 *Let G be a small stable group, and S and T be two infinite definable 2-Sylow subgroups such that $I = S \cap T$ is maximal subject to having infinite index in S. Then $N = N_G(I) \cap N_G(S^{co}) \cap N_G(T^{co})$ acts transitively on the infinite abelian group $A = (N_S(I)/I)[2]$. N/I does not contain involutions, and induces a definable field structure on A (addition is the group law on A, and multiplication is derived from conjugation).*

Note that the 2-Sylow subgroups are definable iff the 2-exponent of G is finite, i.e. the order of any 2-element of G is bounded.

Proof: We should remark first that if I has infinite index in S, then it must have infinite index in T as well. Indeed, since S and T are conjugate by Theorem 1.5.4, so are the locally connected components S^{co} and T^{co}. But if I had finite index in T, then it would contain T^{co}, and in fact T^{co} would be properly contained in S^{co}. So a definable set would be conjugate to a proper subset, contradicting Lemma 1.0.2.

By Corollary 1.2.23 the normalizers of I both in S and in T are proper supergroups of I. So we can fix an involution jI in $N_T(I)/I$ and consider any involution iI in $N_S(I)/I$. Then i and j invert ij modulo I, so there is a definable abelian subgroup X of $N_G(I)/I$ which is inverted by i and by j. By Corollary 4.4.4 it is divisible-plus-bounded. Suppose X contains an involution k. Then k commutes with both i and j, giving rise to 2-Sylow subgroups $S' \supset I \cup \{i,k\}$ and $T' \supset I \cup \{k,j\}$ whose intersections properly contain I. But maximality of I implies $S^0 = (S')^0 = (T')^0 = T^0$, a contradiction. Therefore X is 2-divisible and there is some element r, inverted both by i and by j, with $r^2 = ij$. Then modulo I we get $i^r = r^{-1}ir = irr = i(ij) = j$ and $j^r = r^{-1}jr = jr^2 = jij = i^j$. So there is an element $n_i := rj$ in $N_G(I)$ with $(iI)^{n_i} = jI$ and $(jI)^{n_i} = iI$, whence by maximality of I we have $(S^{n_i})^{co} = T^{co}$ and $(T^{n_i})^{co} = S^{co}$. If i' is another involution, then $n = n_i n_{i'}^{-1}$ takes i to i' (modulo I) and normalizes both S^{co} and T^{co}. As there is a central involution in $N_S(I)/I$, all involutions are central and $(N_S(I)/I)[2]$ forms a definable abelian group A of exponent 2. It must be infinite, since $N_S(I)/I$ is an infinite nilpotent 2-group of finite exponent. Therefore A must intersect S^{co}/I non-trivially; since S^{co}/I is normalized by N, it contains the whole of A. Hence N acts transitively on $A - \{0\}$.

Now if N were to contain an involution k, then we could find 2-Sylow subgroups $S' \supset S^{co} \cup I \cup \{k\}$ and $T' \supset T^{co} \cup I \cup \{k\}$, and by the maximality of I we have $S^{co} = (S')^{co} = (T')^{co} = T^{co}$, a contradiction. Indeed, even

$N/C_N(A)$ does not contain an involution: If $n \in N$ with $n^2 \in C_N(A)$, then $Z(C_N(n))$ is a definable abelian subgroup of N. By Corollary 4.4.4 it is 2-divisible (since it contains no involutions), so there is $z \in Z(C_N(n))$ with $z^2 = n^2$, and $(z^{-1}n)^2 = 1$. Hence $n = z \in C_N(A)$.

Finally we apply Theorem 5.7.2 to $A \rtimes N/C_N(A)$. As N acts transitively, A must be $(N/C_N(A))$-analysable and since $N/C_N(A)$ has no involutions, any definable abelian subgroup $N/C_N(A)$ is 2-divisible. Obviously, the resulting field will have characteristic 2. $\square$

If the 2-Sylow subgroups are not definable, this proposition need not hold: Let K be an algebraically closed field of characteristic different from 2, and $G = K^+ \rtimes K^\times$ the group of affine transformations. Then G is small and stable (even of Morley rank 2) and the 2-Sylow subgroups are not normal-by-finite (they are the 2-subgroups of $K^\times$ and isomorphic to the Prüfer group C_{2^∞}), but there is no definable field of characteristic 2, as any interpretable field must be isomorphic to K. Since we may take K to be locally finite, the proposition fails even for locally finite groups.

Recall that the Prüfer rank of a group is the supremum of the minimal numbers of generators of its finitely generated subgroups.

Lemma 5.7.4 *Let G be a locally nilpotent group with normal subgroups M and N. Then the Prüfer rank of G is finite iff the Prüfer rank of G/N and the Prüfer rank of N are finite. The Prüfer rank of $G/(M \cap N)$ is finite iff the Prüfer ranks of G/M and G/N are both finite.*

Proof: The first assertion follows from the fact that a subgroup of a finitely generated nilpotent group is finitely generated (Lemma 1.2.2), and the second assertion follows from the first. $\square$

Remark 5.7.1 The Prüfer rank of a locally nilpotent $\mathfrak{M}_c$-p-group G is finite iff G has a normal abelian subgroup of finite index which is the direct sum of finitely many Prüfer groups.

Proof: G is nilpotent-by-finite by Theorem 1.2.15. Since any subgroup of finite index in a finitely generated group is finitely generated, G has finite Prüfer rank iff any subgroup of finite index has finite Prüfer rank. So we may assume that G is nilpotent. By Lemma 1.2.14 the exponent of $G/Z(G)$ is finite, and by the preceding lemma G has finite Prüfer rank iff both $G/Z(G)$ and $Z(G)$ have finite Prüfer rank. However, the Prüfer rank of a nilpotent p-group of finite exponent is finite iff the group is finite (as the group is uniformly locally finite), and the Prüfer rank of an abelian p-group is finite iff the abelian group is the direct sum of finitely many Prüfer groups C_{p^∞} with a finite group. It follows that G has finite Prüfer rank iff the direct sum $\bigoplus C_{p^\infty}$ of those Prüfer groups has finite index in G. $\square$

Theorem 5.7.5 *Let G be a small stable group whose 2-Sylow subgroups are not normal-by-(finite Prüfer rank). Then there is a definable algebraically closed field of characteristic 2.*

Owing to the fact that neither 2-Sylow subgroups nor finite Prüfer rank are definable in general, the proof of this theorem is slightly complicated and splits into two halves: first we obtain some kind of minimal counter-example where any two 2-Sylow subgroups have either finite intersection or finite Prüfer rank over their intersection; from this we then construct the field.
Proof: Let S be a 2-Sylow subgroup. Then S is nilpotent-by-finite, so there is a definable nilpotent group N with $S \cap N$ as 2-Sylow subgroup and such that the index $|S : S \cap N|$ is finite. Let $\bar{S}$ be the intersection of all conjugates of N which intersect S in a subgroup of finite index, and put $S^c = S \cap \bar{S}$. Then S^c is a relatively definable nilpotent subgroup of finite index in S whose normalizer is maximal among all such subgroups. Furthermore S^c is the set of 2-elements of $\bar{S}$, and $N_G(\bar{S}) = N_G(S^c)$.
Claim. Let N be nilpotent and substable. If $Z(N)$ has finite Prüfer rank, then so does N. If I is a relatively definable subgroup of N such that $N/\bigcap_{n\in N} I^n$ has infinite Prüfer rank, then $N_N(I)/I$ has infinite Prüfer rank.
Proof of Claim: Let i be maximal such that $Z_i(N)$ has finite Prüfer rank, and assume $i > 0$. We claim that $Z_{i+1}(N)/Z_i(N)$ has finite Prüfer rank. Indeed, $Z_i(N)$ is the centralizer modulo $Z_{i-1}(N)$ of finitely many elements in N, but for every $n \in N$ the quotient $Z_{i+1}(N)/C_{Z_{i+1}(N)}(n/Z_{i-1}(N))$ is isomorphic to $[Z_{i+1}(N), n]/Z_{i-1}(N)$, which is contained in $Z_i(N)/Z_{i-1}(N)$ and has finite Prüfer rank. Hence $Z_{i+1}(N)/Z_i(N)$ has finite Prüfer rank by Lemma 5.7.4, and so has $Z_{i+1}(N)$. This implies that N has finite Prüfer rank, proving the first assertion.
   For the second assertion, we divide out by $\bigcap_{n\in N} I^n$ and assume this is trivial. Then $Z(N)$ must have infinite Prüfer rank, and obviously normalizes I. Now if $Z(N)/(Z(N)\cap I)$ had finite Prüfer rank, so would $Z(N)/(Z(N)\cap I^n)$ for every $n \in N$; as $\bigcap_{n\in N} I^n$ is trivial and equals a finite subintersection, this would contradict Lemma 5.7.4. This proves the claim. $\square$

Put $S_0 := \bigcap\{T^c : T \text{ a 2-Sylow subgroup such that } S^c/(S^c \cap \bigcap_{s\in S^c}(T^c)^s)$ has finite Prüfer rank$\}$. Then S_0 is a normal subgroup of S^c and S^c/S_0 has finite Prüfer rank. Suppose for every 2-Sylow subgroup T the Prüfer rank of $S^c/(S^c \cap \bigcap_{s\in S^c}(T^c)^s)$ is finite. As the intersection $I = \bigcap\{T^c : T \text{ a 2-Sylow subgroup}\}$ equals a finite subintersection, this implies that S/I has finite Prüfer rank by Lemma 5.7.4; since I is normal in G, this contradicts our assumptions. If T is another 2-Sylow subgroup, let $I(S,T)$ be the intersection of $S^c \cap T^c$ with all U^c (where U runs over all other 2-Sylow subgroups) such that $U^c \cap S^c \cap T^c$ has finite index in $S^c \cap T^c$. As S^c is uniformly definable relative to the set of all 2-elements, so is $I(S,T)$, and by the above and the icc we can fix some 2-Sylow subgroups S and T such that $I = I(S,T)$ is maximal subject to $S_0 \neq T_0$. Note that I is relatively definable as the 2-subgroup of some nilpotent definable group N with $N_G(N) = N_G(I)$.

Claim. The Prüfer rank of $N_{S^c}(I)/I$ is infinite.

Proof of Claim: Suppose it were finite. Note that $N_{S^c}(I) \geq N_{S_c}(S^c \cap T^c)$, so $N_{S^c}(S^c \cap T^c)/(S^c \cap T^c)$ also has finite Prüfer rank. By the last claim $T^c \geq S_0$; since T^c/T_0 has finite Prüfer rank, so does $S_0/(S_0 \cap T_0)$, and $S_0 \leq T_0$. By Lemma 1.0.9 we have equality, contradicting $S_0 \neq T_0$. $\square$

Similarly, $N_{T^c}(I)/I$ has infinite Prüfer rank. Now let S' and T' be two 2-Sylow subgroups of $N_G(I)$ and extend them to 2-Sylow subgroups $\bar{S}$ and $\bar{T}$ of G.

Claim. If $(S' \cap T')/I$ is infinite, then $\bar{S}_0 = \bar{T}_0$.

Proof of Claim: If $(S' \cap T')/I$ is infinite, then I has infinite index in $\bar{S} \cap \bar{T}$, and therefore also in $I(\bar{S}, \bar{T})$. By maximality of I we get $\bar{S}_0 = \bar{T}_0$. $\square$

In particular, if S' extends $N_{S^c}(I)$ and T' extends $N_{T^c}(I)$, then $(S' \cap T')/I$ is finite, and both S'/I and T'/I have infinite Prüfer rank. Thus the 2-Sylow subgroups of $N_G(I)$ are not normal-by-(finite Prüfer rank), and we may replace G by $N_G(I)$. We may thus assume that any two 2-Sylow subgroups S and T either intersect in a finite subgroup modulo I or satisfy $S_0 = T_0$.

Any 2-Sylow subgroup T is contained in a uniformly definable soluble subgroup $\bar{T}$ of G such that T^c is normal in $\bar{T}$. Then $(S \cap T)/I$ is finite iff $(S \cap \bar{T})/I$ is finite; by uniformity there is a maximal finite (modulo I) intersection $S \cap \bar{T}$, which leads to a maximal intersection $J := S \cap T)$ of two 2-Sylow subgroups S and T which is a finite extension of I.

Claim. $N_{S_0}(J)[2]/J$ is infinite.

Proof of Claim: As S_0/I has infinite Prüfer rank, so does $Z(S_0)/I$ by the first claim, and $Z(S_0)[2]/I$ is an infinite abelian 2-group. We prove inductively that if a finite 2-group M acts on an infinite elementary abelian 2-group A, then it stabilizes an infinite subgroup. So suppose M is a minimal counter-example. Then M is nilpotent and has a normal subgroup of index 2; by the inductive hypothesis we may assume that it stabilizes A. So we are reduced to the action of an involution i on an infinite elementary abelian 2-group. But $(i + 1)^2 A = (i^2 + 1)A = 0$, and i does indeed stabilize an infinite subgroup $\ker(i + 1)$ of A.

Taking $M = J/I$ and $A = Z(S_0)[2]/I$ we obtain a subgroup $A' \leq C_A(J)$ of finite index in A. Clearly $A'I \leq N_{S_0}(J)$; as J/I is finite, the result follows. $\square$

Similarly, $N_{T_0}(J)/J$ contains infinitely many involutions. We now fix an involution j in $Z(N_{T_0}(J)/J)$ and consider any involution $i \in N_S(J)/J$. Let J_0 be a finite subgroup of J with $J = IJ_0$. But I is the increasing union of definable characteristic normal subgroups by Lemma 1.2.20; since i and j

both normalize J modulo I, there is a definable characteristic subgroup $I_0 \leq I$ such that J_0^i and J_0^j are contained in $I_0 J_0$. Hence i and j both normalize $I_0 J_0$; we may further assume that i^2 and j^2 are both contained in $I_0 J_0$. We may now consider a definable abelian subgroup A of $N_G(I_0 J_0)/I_0 J_0$ containing ij and inverted by both i and j. If A contains an involution k, then there are 2-Sylow subgroups $S' \supset I \cup J_0 \cup \{i, k\}$ and $T' \supset I \cup J_0 \cup \{k, j\}$; then $S \cap S'$, $S' \cap T'$ and $T' \cap T$ all properly contain J and must be infinite modulo I by maximality of J. This implies $S_0 = S_0' = T_0' = T_0$, a contradiction. So A does not contain an involution; it must then be the sum of a group of odd exponent with a divisible group, and thus 2-divisible: there is $r \in A$ with $r^2 = ij$. Putting $n_i = rj$, we have $i^{n_i} = jr^{-1}irj = jir^2j = j$ and $j^{n_i} = jr^{-1}jrj = r^2j = i$. Therefore $S^{n_i} \cap T \supseteq J \cup \{j\}$ and $T^{n_i} \cap S \supset J \cup \{i\}$; the maximality of J now yields $S_0^{n_i} = T_0$ and $T_0^{n_i} = S_0$.

It follows that i must be a central involution of S_0/J. Since $n_i n_{i'}^{-1}$ maps i to i' for any two involutions $i, i' \in N_S(J)/J$, the group $N := N_G(S_0) \cap N_G(T_0)$ acts transitively on $(N_S(J)/J)[2]$ and $(N_T(J)/J)[2]$; if N contained an involution k, this would give rise to 2-Sylow subgroups $S' \supset S_0 \cup \{k\}$ and $T' \supset T_0 \cup \{k\}$, implying $S_0 = S_0' = T_0' = T_0$ by the maximality of J, a contradiction.

Let $A := (N_S(J)/J)[2]$. We can then apply Theorem 5.7.2 to $A \rtimes (N/C_N(A))$. As N acts transitively, A must be $(N/C_N(A))$-analysable, and since $N/C_N(A)$ has no involutions, any definable abelian subgroup $N/C_N(A)$ is 2-divisible. This will yield a definable field of characteristic 2. $\square$

We should note that one may easily obtain Proposition 5.7.3 as a corollary to a more precise formulation of this theorem, since a soluble 2-group of finite exponent has finite Prüfer rank iff it is finite.

In chapter 1 we proved that a substable 2-group is locally finite, and in chapter 2 we considered the question of groups of generic exponent 2 or 3. Here we shall, in a way, put these results together.

Theorem 5.7.6 *Let G be an $\Re$-group of exponent $3 \cdot 2^n$. Then G is nilpotent-by-finite.*

Proof: We suppose, by way of contradiction, that G is an $\Re$-group of exponent $3 \cdot 2^n$ which is not nilpotent-by-finite, and work in a saturated model. By Proposition 1.3.1 it is enough to show that G is soluble-by-finite, and by Theorem 5.7.1 we may assume that the 2-Sylow subgroups of G are finite. Furthermore, by Proposition 1.2.21 we may divide out by the soluble radical.

Consider a locally nilpotent subgroup S of G. By Lemma 1.2.5 it is soluble of bounded derived length; since the exponent is finite, it is uniformly locally nilpotent. By Corollary 1.2.10 it is nilpotent, and compactness implies a bound on the nilpotency class. Therefore all maximal locally nilpotent subgroups are uniformly definable and nilpotent.

Since G is not nilpotent-by-finite, by uniform definability and the icc there are two maximal nilpotent subgroups S and T with maximal intersection $I = S \cap T$, such that I has infinite index in both S and T. For instance, by Theorem 5.1.1 there is an infinite maximal nilpotent subgroup S of G, which cannot intersect all its G-conjugates in a subgroup of finite index, since otherwise its conjugacy-connected component would be a non-trivial normal soluble subgroup. But the conjugacy-connected components of S and one of its conjugates S^g are themselves conjugate, so I must actually have infinite index in both S and S^g by Lemma 1.0.2. By Lemma 1.1.9 the indices of I in $N_S(I)$ and $N_T(I)$ are both infinite.

Consider two maximal nilpotent subgroups S'/I and T'/I of $N :=$ $N_G(I)/I$ whose intersection has infinite index in both. Then S' is locally finite, hence nilpotent-by-finite. But since I is normal in S', the maximal normal subgroup of finite index S'_0 must contain I. Similarly, $T'_0 \geq I$. But S'_0 and T'_0 can be extended to maximal locally nilpotent subgroups S_0 and T_0 of G whose intersection must have infinite index in both S_0 and T_0, so maximality of I implies that $S_0 \cap T_0 = I$. It follows that if for any maximal nilpotent subgroup S of N we denote by S^n the intersection of all maximal nilpotent subgroups of $N_G(I)$ intersecting the pre-image of S in a subgroup of finite index, then S^n is a locally connected nilpotent subgroup of finite index in S, and any two maximal such groups S^n and T^n either are equal or intersect trivially. In particular, N must have finite centre and $N^{cc}/Z(N^{cc})$ is centreless. (Note that it might conceivably happen that S^n is properly contained in T^n for some maximal nilpotent subgroups S and T of N. However, these groups are uniformly definable and maximal ones do exist.)

By the chain condition on centralizers we may replace G by a minimal intersection of centralizers which is not nilpotent-by-finite; in particular G is now centralizer-connected. Note that maximal nilpotent subgroups S of G remain uniformly definable, and so are their components S^n, any two of which still intersect trivially (if they are not equal). In particular, maximal connected nilpotent subgroups are disjoint. Hence $Z(G)$ must be finite and $Z_2(G) = Z(G)$; as $|C_G(A/Z(G)) : C_G(A)|$ is then finite, we may divide out by $Z(G)$ and assume that G is centreless (and retain all the other properties).

If G^0 contains no involutions, we are done by Theorem 2.4.3. So let i be an involution in the connected component. By Theorem 3.3.1 it must have an infinite centralizer $C_G(i)$, which is a proper subgroup of G and hence nilpotent-by-finite. So there is a unique maximal nilpotent 3-subgroup $B(i)$ containing $C_G(i)^0$ with $B(i) = B(i)^n$. (Remember that the 2-Sylow subgroups are finite, so the 2-part of a nilpotent group lies outside the connected component.) If $N_B(B(i))/B(i)$ were infinite, it would contain an infinite abelian subgroup $A/B(i)$ by Theorem 5.1.1 and A would have a normal nilpotent subgroup in which $B(i)$ has infinite index, in contradiction to the maximality of $B(i)$. Therefore $B(i)$ has finite index in its normalizer.

Claim. If j is an involution not normalizing $B(i)$, then $B(i)j$ contains exactly one involution (namely j).

Proof of Claim: Suppose k were a second involution in $B(i)j$. Then jk would equal some element b in $B(i)$, and $b^j = b^{-1}$. Hence b would lie in $B(i) \cap B(i)^j$, whence $B(i) = B(i)^j$, and j does normalize $B(i)$, a contradiction. $\square$

Claim. $C_G(i) \cdot i^G$ is generic over i.

Proof of Claim: Let g be a principal generic element of G over i. If i^g normalizes $B(i)$, then i^G is contained in $N_G(B(i))$ and generates a normal locally finite subgroup F of G. If F is finite, it is central; if F is infinite, F^0 is contained in all connected maximal nilpotent subgroups; either case contradicts our assumptions on G.

Therefore i^g does not normalize $B(i)$ and is definable over the canonical parameter of $B(i)i^g$ as the unique involution in that set; and $C_G(i)g$ is definable over i and i^g. But $B(i)$ intersects $C_G(i)$ in a subgroup of finite index, so $C_G(i)g$ is algebraic over $C_G(i)i^g$ (and i). Since the former is a generic element of $G/C_G(i)$ over i, so is the latter by $\Re$. Furthermore it is principal generic, as i^G is contained in the connected component of G. $\square$

It follows that $C_G(i)^0 i^G$ is principal generic in G. Hence for a principal generic element g over i there are $k = i^h$ in i^G and b in $C_G(i)^0$ with $g = bk$. Now k cannot normalize $B(i)$ since $g \notin N_G(B(i))$, so k is the only involution in $B(i)k$. This implies that k and b are unique, given g. Furthermore, if there were $c \in B(i)^0 - C_G(i)$ (independent of g), then cg would also be a principal generic element over $\emptyset$ which lies in $B(i)k$, and hence $cg = b'k$ for some b' in $C_G(i)^0$. So $cg = cbk = b'k$, whence $c = b'b^{-1}$ lies in $C_G(i)^0$, a contradiction. Therefore $B(i)^0 = C_G(i)^0$.

Claim. i is algebraic over $B(i)$ (that is, the canonical parameter needed for its definition).

Proof: If C is the intersection with $B(i)$ of all centralizers of involutions containing $B(i)^0$, and if i and j are two involutions centralizing C, then i and j normalize $B(i)$. So there are only finitely many possibilities for their cosets modulo C. But if $Ci = Cj$, then $ij \in C \leq C_G(i)$ and conjugation by i both inverts and fixes ij. Since $ij \in C \leq B(i)$, it is a 3-element and must be trivial. So $i = j$. $\square$

If g is principal generic over i with $g = bk$ as above, it follows from $\Re$ that b^k is generic: first $B(i^k)$ is definable over b^k as the unique conjugate of $B(i)$ containing b^k, and i^k is algebraic over $B(i^k)$ by the last claim. Therefore $C_G(i)k$ is algebraic over b^k, and so is $B(i)k$. Now k is the unique involution in $B(i)k$. Hence b is algebraic over b^k, and so is the generic element $g = bk$.

Thus there is a generic element of order 3, and G is nilpotent-by-finite by Theorem 2.4.6. $\square$

This certainly was much ado about almost nothing. However, the proof illustrates both the reduction process used to obtain a suitably minimal counterexample, and the use of $\Re$ to obtain a generic element of finite order. For comparison, we shall indicate a proof in the case where the group has finite U-rank:

A counter-example G of minimal U-rank must have disjoint maximal connected proper subgroups, which are also nilpotent and have finite index in their normalizer. Hence if H is such a maximal connected proper subgroup, then $U(H) = U(N_G(H))$ and $U(G) = U(G/H) + U(H)$. But the family of conjugates of $H - \{1\}$ is a family of rank $U(G/N_G(H))$ consisting of disjoint sets of rank $U(H)$; so $\bigcup_{g \in G} H^g$ has rank $U(H) + U(G/N_G(H)) = U(G)$ and is generic. But it consists entirely of elements of order 3, and we get a generic element of order 3.

In our context, the minimization process is slightly more complicated (we actually have to say with regard to what we minimize, rather than just taking care of everything by a minimal rank assumption), but the main difficulty arises from the fact that we cannot deduce the genericity of the set of conjugates of H (or $B(i)$, in our proof) simply from a rank equality. In order to use property $\Re$, we have to define a transversal i^G to $B(i)$, and then use algebraization.

We have already seen that a stable group must have more than one nontrivial conjugacy class. We shall now show that an $\Re$-group must in fact have infinitely many of them.

Theorem 5.7.7 *A stable $\Re$-group has infinitely many conjugacy classes.*

Proof: Consider a counter-example G. This group has only finitely many normal subgroups, so its connected component is definable and has finite index. Hence G^0 has only finitely many conjugacy classes as well, and we may assume that G is actually connected. Note that the torsion of G must be bounded by some finite $t < \omega$.

Claim. G has finite exponent.

Proof: Let A be a minimal type-definable subgroup of G of infinite exponent. Then A is abelian, since it is contained in $Z(C_G(a))$ for any element $a \in A$ of infinite order, and torsion-free, as $A = t!A$. Furthermore, if $a^g \in A$ for some $g \in G$, then $A \cap A^g$ is infinite and must equal A. Therefore $N_G(A)$ acts on A with finitely many orbits. By Proposition 1.3.3 this gives rise to a type-definable field K, with $A = K^+$ and $N_G(A)/C_G(A)$ isomorphic to $K^\times$. The torsion of $K^\times$ is non-trivial, as it contains -1. On the other hand the multiplicative torsion is bounded, as otherwise a G-conjugacy class in $N_G(A)$ would contain elements of arbitrarily large finite order modulo $C_G(A)$, i.e. some element x would have arbitrarily large finite order modulo conjugates of $C_G(A)$, and the sequence of $C_G(x^{i!})$ would be infinitely ascending. This already contradicts Proposition 5.1.2, but we shall indicate an argument which avoids the use of $\Re$ at this stage.

If the torsion T of $K^\times$ is bounded by n, then $M := K^{n!}$ is torsion-free. Let p be prime and $M_0/C_G(A)$ a definable abelian p-torsion-free supergroup of M contained in $N_G(C_G(A))/C_G(A)$. Consider some $x \in M_0$ and some $m < \omega$ such that x^{p^m} lies in a maximal set of G-conjugates of M_0. (This set is increasing with m, so by stability there is such a maximum.) As there are only finitely many conjugacy classes, there are some j and i with $j > i + m \geq 2m$ such that x^{p^j} is conjugate to x^{p^i}. Then x^{p^i} and x^{p^j} lie in the same conjugates of M_0 as x^{p^m}; as x^{p^j} is p^{j-m}-divisible in those conjugates, so is x^{p^i}. As $M_0/C_G(A)$ is p-torsion-free, this implies that $xC_G(A)$ is p-divisible. Therefore any p-torsion-free definable supergroup of M, and hence M itself, is p-divisible; as p was arbitrary, M is divisible and $K^\times = M \times T$. But T is finite and non-trivial, contradicting the connectivity of K. □

By Theorem 5.7.1 we may divide out by a definable normal subgroup and assume that the 2-Sylow subgroups are finite. We may further assume by the chain condition on centralizers that no definable subgroup of finite index in some proper centralizer of G has only finitely many conjugacy classes. We now take the connected component, divide out by a maximal normal 2-subgroup, take a minimal subgroup of finite index in a centralizer with finitely many conjugacy classes, and repeat the whole process. As the 2-Sylow subgroups shrink every time, the process stops eventually.

Let a be a principal generic element of G. Then a^{-1} is again principal generic, hence conjugate to A by some element g. But a cannot be an involution since G is not abelian, so the order of g must be divisible by 2 and G contains involutions.

By Theorem 3.3.1 an involution has an infinite centralizer. Let A be a maximal 2-group with infinite centralizer. As A is finite, $|N_G(A) : C_G(A)|$ is finite. Clearly, $C_G(A)$ is a proper subgroup of G.

Suppose that A is not a maximal 2-subgroup, so there is some element b normalizing A with $b^2 \in A$. By the maximality of A the centralizer $C_G(A,b)$ is finite, so b acts on $C_G(A)$ as an involution with finitely many fixed points. By Theorem 3.3.1 it inverts an abelian subgroup C_b of finite index in $C_G(A)$, and the locally connected 2'-component $C := (C_b^{lo})_{2'}$ is a definable subgroup of finite index in $C_G(A)$ without involutions.

Claim. For any $c \in C$ the group A is a 2-Sylow subgroup of $C_G(c)$.

Proof of Claim: If $b \in C_G(c)$ normalizes A with $b^2 \in A$, then b inverts C, whence c, and c must be an involution, giving a contradiction. $\square$

If A is a maximal 2-subgroup, we just put $C := C_G(A)$. In either case, for any two elements c_1 and c_2 of C which are conjugate by g in G, both A and A^g are 2-Sylow subgroups of $C_G(c_2)$, and conjugate by Theorem 1.5.4. So there is h in $C_G(c_2)$ with $A^{gh} = A$, and $c_1^{gh} = c_2$. Thus $N_G(A)$ has as many orbits on C as G does; as the index of C in $N_G(A)$ is finite, C has only finitely many conjugacy classes, contradicting the minimality of G. $\square$

We should note that the only point where we have used $\mathfrak{R}$ is where we applied Theorem 5.7.1.

Problem 5.7.1 Is there a stable group with finitely many conjugacy classes?

5.8 Historical and Bibliographical Remarks

Sections 0–3 are taken from [114]; we have already seen some of the super-stable analogues in the earlier chapters. Section 4 on solubility and nilpotency comes from [116]; sections 5 and 6 from [120]; both sections 4 and 5 generalize work of Nesin [77, 78] in the case of finite Morley rank. Theorems 5.7.1 and 5.7.6 can be found in [119], as well as a preliminary version of Proposition 5.7.3 and Theorem 5.7.5. Theorem 5.7.7 is a generalization of Aguzarov, Farey and Goode [1], taken from [112] and simplified further.

Groups & Glory

References

[1] I. Aguzarov, R. E. Farey, and John B. Goode. An infinite superstable group has infinitely many conjugacy classes. *Journal of Symbolic Logic*, 56:618–623, 1991.

[2] Friedrich Bachmann. *Aufbau der Geometrie aus dem Spiegelungsbegriff.* Springer-Verlag, Berlin, 1959.

[3] John Baldwin. *Fundamentals of Stability Theory.* Springer-Verlag, Berlin, 1988.

[4] John Baldwin and Jan Saxl. Logical stability in group theory. *Journal of the Australian Mathematical Society*, 21:267–276, 1976.

[5] John T. Baldwin. α_T is finite for $\aleph_1$-categorical T. *Transactions of the American Mathematical Society*, 181:37–51, 1973.

[6] John T. Baldwin, editor. *Classification Theory: Proceedings of the US-Israel Binational Workshop on Model Theory in Mathematical Logic, Chicago, '85* Lecture Notes in Mathematics 1292. Springer-Verlag, Berlin, 1987.

[7] John T. Baldwin and Anand Pillay. Semisimple stable and superstable groups. *Annals of Pure and Applied Logic*, 45:105–127, 1989.

[8] Andreas Baudisch. Decidability and stability of free nilpotent Lie algebras and free nilpotent p-groups of finite exponent. *Annals of Mathematical Logic*, 23:1–25, 1982.

[9] Andreas Baudisch. On superstable groups. *Journal of the London Mathematical Society*, 42:452–464, 1990.

[10] Andreas Baudisch. Another stable group. *Annals of Pure and Applied Logic*, 80:109–138, 1996.

[11] Andreas Baudisch. A new uncountably categorical group. *Transactions of the American Mathematical Society*, 348:3889–3940, 1996.

[12] Andreas Baudisch and John S. Wilson. Stable actions of torsion groups and stable soluble groups. *Journal of Algebra*, 153:453–457, 1991.

[13] Walter Baur. Elimination of quantifiers for modules. *Israel Journal of Mathematics*, 25:64–70, 1976.

[14] Walter Baur, Gregory Cherlin, and Angus Macintyre. Totally categorical groups and rings. *Journal of Algebra*, 57:407–440, 1979.

[15] Oleg V. Belegradek. On almost categorical theories (in Russian). *Sibirskiĭ Matematicheskiĭ Zhurnal*, 14:277–288, 1973.

[16] Chantal Berline. Superstable groups; a partial answer to conjectures of Cherlin and Zil'ber. *Annals of Pure and Applied Logic*, 30:45–61, 1986.

[17] Chantal Berline and Daniel Lascar. Superstable groups. *Annals of Pure and Applied Logic*, 30:1–43, 1986.

[18] Chantal Berline and Daniel Lascar. Des gros sous-groupes abéliens pour les groupes stables. *Comptes Rendus de l'Académie des Sciences à Paris*, 305:639–641, 1987.

[19] Maurice Boffa and Françoise Point. Identités de Engel généralisées. *Comptes Rendus de l'Académie des Sciences à Paris*, 313, I:909–911, 1991.

[20] Alexandre V. Borovik and Ali Nesin. *Groups of Finite Morley Rank*. Oxford University Press, Oxford, 1994.

[21] Alexandre V. Borovik and Bruno P. Poizat. Simple groups of finite Morley rank without non-nilpotent connected subgroups (in Russian). Preprint mimeographed by VINITI, N2062-B, 1990.

[22] Alexandre V. Borovik and Bruno P. Poizat. Tores et p-groupes. *Journal of Symbolic Logic*, 55:478–491, 1991.

[23] Alexandre V. Borovik and Simon R. Thomas. On generic normal subgroups. In Richard Kaye and Dugald Macpherson, editors, *Automorphisms of First-Order Structures*, pages 319–324. Oxford University Press, Oxford, 1994.

[24] Elisabeth Bouscaren. The group configuration — after E. Hrushovski. In Nesin and Pillay [80], pages 199–209.

[25] Roger M. Bryant. Groups with minimal condition on centralizers. *Journal of Algebra*, 60:371–383, 1979.

[26] Roger M. Bryant and Brian Hartley. Periodic locally soluble groups with the minimal condition on centralizers. *Journal of Algebra*, 61:328–334, 1979.

[27] Steven Buechler. The geometry of weakly minimal types. *Journal of Symbolic Logic*, 50:1044–1053, 1985.

[28] Steven Buechler. Locally modular theories of finite rank. *Annals of Pure and Applied Logic*, 30:83–94, 1986.

[29] Steven Buechler. Vaught's conjecture for superstable theories of finite rank. Preprint, 1992.

[30] Chen C. Chang and Jerome H. Keisler. *Model Theory*. North-Holland, Amsterdam, 1973.

[31] Olivier Chapuis. Universal theory of certain solvable groups and bounded Ore group-rings. *Journal of Algebra*, 176:368–391, 1995.

[32] Zoë Chatzidakis and Ehud Hrushovski. Algebraically closed fields with an automorphism. Preprint, 1995.

[33] Gregory Cherlin. Superstable division rings. In Angus Macintyre, Leszek Pacholski, and Jeff Paris, editors, *Logic Colloquium '77*, pages 99–111. North-Holland, Amsterdam, 1978.

[34] Gregory Cherlin. Groups of small Morley rank. *Annals of Mathematical Logic*, 17:1–28, 1979.

[35] Gregory Cherlin, Leo Harrington, and Alistair Lachlan. $\aleph_0$-categorical $\aleph_0$-stable structures. *Annals of Pure and Applied Logic*, 28:103–135, 1985.

[36] Gregory Cherlin and Ehud Hrushovski. Permutation groups with few orbits on 5-tuples, and their infinite limits. Preprint, 1995.

[37] Gregory Cherlin and Saharon Shelah. Superstable fields and groups. *Annals of Mathematical Logic*, 18:227–270, 1980.

[38] Luis Jaime Corredor. Bad groups of finite Morley rank. *Journal of Symbolic Logic*, 54:768–773, 1989.

[39] Jamshid Derakhshan and Frank O. Wagner. Nilpotency in groups with chain conditions. *Oxford Quarterly Journal of Mathematics*, to appear.

[40] Juri Ershov. Fields with solvable theory (in Russian). *Doklady Akademii Nauk SSSR*, 174:19–20, 1967.

[41] Ulrich Felgner. $\aleph_0$-categorical stable groups. *Mathematische Zeitschrift*, 160:27–49, 1978.

[42] K. W. Gruenberg. The Engel elements of a soluble group. *Illinois Journal of Mathematics*, 3:151–168, 1959.

[43] Claus Grünenwald and Frieder Haug. On stable groups in some soluble group classes. In Martin Weese and Helmut Wolter, editors, *Proceedings of the 10th Easter Conference on Model Theory*, pages 46–59. Humboldt Universität, Berlin, 1993.

[44] Claus Grünenwald and Frieder Haug. On stable torsion-free nilpotent groups. *Archive for Mathematical Logic*, 32:451–462, 1993.

[45] Philip Hall. Some sufficient conditions for a group to be nilpotent. *Illinois Journal of Mathematics*, 2:787–801, 1958.

[46] Hermann Heineken. Endomorphismenringe und Engelsche Elemente. *Archiv der Mathematik*, 13:29–37, 1962.

[47] Ward Henson. Countable homogeneous relational structures and $\aleph_0$-categorical theories. *Journal of Symbolic Logic*, 37:494–500, 1972.

[48] Bernhard Herwig, James G. Loveys, Anand Pillay, Predrag Tanović, and Frank O. Wagner. Stable theories without dense forking chains. *Archive for Mathematical Logic*, 31:297–303, 1992.

[49] K. A. Hirsch. Über lokal nilpotente Gruppen. *Mathematische Zeitschrift*, 63:290–294, 1955.

[50] Wilfrid Hodges. *Model Theory*. Cambridge University Press, Cambridge, 1993.

[51] Ehud Hrushovski. *Contributions to Stable Model Theory*. PhD thesis, University of California at Berkeley, 1986.

[52] Ehud Hrushovski. Locally modular regular types. In Baldwin [6], pages 132–164.

[53] Ehud Hrushovski. On superstable fields with automorphisms. In Nesin and Pillay [80], pages 186–191.

[54] Ehud Hrushovski. Unidimensional theories are superstable. *Annals of Pure and Applied Logic*, 50:117–138, 1990.

[55] Ehud Hrushovski. Pseudo-finite fields and related structures. Preprint, 1991.

[56] Ehud Hrushovski. The Mordell-Lang conjecture for function fields. Preprint, 1993.

[57] Ehud Hrushovski. A new strongly minimal set. *Annals of Pure and Applied Logic*, 62:147–166, 1993.

[58] Ehud Hrushovski and Anand Pillay. Weakly normal groups. In Ch. Berline, E. Bouscaren, M Dickmann, J.-L. Krivine, D. Lascar, M. Parigot, E. Pelz, and G. Sabbagh, editors, *Logic Colloquium '85*, pages 233–244. North-Holland, Amsterdam, 1987.

[59] Ehud Hrushovski and Boris Zil'ber. Zariski geometries. *Bulletin of the American Mathematical Society*, 28:315–323, 1993.

[60] Bertram Huppert. *Endliche Gruppen I*. Springer-Verlag, Berlin, 1967.

[61] S. V. Ivanov and A. Yu. Ol'shanskii. Some applications of graded diagrams in combinatorial group theory. In C. M. Campbell and E. F. Robertson, editors, *Groups St Andrews 1989, Volume 2*, London Mathematical Society Lecture Notes Series 160, pages 258–308. Cambridge University Press, Cambridge, 1992.

[62] Otto Kegel. Four lectures on Sylow theory in locally finite groups. In Kai Nah Cheng and Yu Kaiang Lerong, editors, *Group Theory (Singapore 1987)*, pages 3–27. Walter de Gruyter, Berlin, 1989.

[63] E. I. Khukhro. On locally nilpotent groups admitting a splitting automorphism of prime order (in Russian). *Matematicheskiĭ Sbornik*, 130:120–127, 1986.

[64] Byunghan Kim. Forking in simple unstable theories. *Journal of the London Mathematical Society*, to appear.

[65] Byunghan Kim and Anand Pillay. Simple theories. *Annals of Pure and Applied Logic*, to appear.

[66] Daniel Lascar. Les groupes ω-stables de rang fini. *Transactions of the American Mathematical Society*, 292:451–462, 1985.

[67] Daniel Lascar. *Stability in Model Theory*. Longman, New York, 1987.

[68] James G. Loveys. Abelian groups with modular generic. *Journal of Symbolic Logic*, 56:250–259, 1991.

[69] Angus Macintyre. On ω_1-categorical theories of abelian groups. *Fundamenta Mathematicae*, 70:253–270, 1971.

[70] Angus Macintyre. On ω_1-categorical theories of fields. *Fundamenta Mathematicae*, 71:1–25, 1971.

[71] A. I. Mal'cev. A correspondence between groups and rings. In *The Metamathematics of Algebraic Systems. Collected Papers: 1936-1967*, pages 124–137. North-Holland, Amsterdam, 1971.

[72] Alan Mekler. Stability of nilpotent groups of class 2 and prime exponent. *Journal of Symbolic Logic*, 46:781–788, 1981.

[73] Margit Messmer. *Groups and Fields Interpretable in Separably Closed Fields*. PhD thesis, University of Illinois at Chicago, 1992.

[74] L. Monk. *Elementary-Recursive Decision Procedures*. PhD thesis, University of California at Berkeley, 1975.

[75] Michael Morley. Categoricity in power. *Transactions of the American Mathematical Society*, 114:514–538, 1965.

[76] Ali Nesin. Nonsolvable groups of Morley rank 3. *Journal of Algebra*, 124:199–218, 1989.

[77] Ali Nesin. Solvable groups of finite Morley rank. *Journal of Algebra*, 121:26–39, 1989.

[78] Ali Nesin. On solvable groups of finite Morley rank. *Transactions of the American Mathematical Society*, 321:659–690, 1990.

[79] Ali Nesin. Generalized Fitting subgroup of a group of finite Morley rank. *Journal of Symbolic Logic*, 56:1391–1399, 1991.

[80] Ali Nesin and Anand Pillay, editors. *The Model Theory of Groups*. University of Notre Dame Press, Notre Dame, Indiana, 1989.

[81] Bernhard H. Neumann. Groups covered by permutable subsets. *Journal of the London Mathematical Society*, 29:236–248, 1954.

[82] Ludomir Newelski. On type definable subgroups of a stable group. *Notre Dame Journal of Formal Logic*, 32:173–187, 1991.

[83] Anand Pillay. *An Introduction to Stability Theory*. Clarendon Press, Oxford, 1983.

[84] Anand Pillay. Simple superstable theories. In Baldwin [6], pages 247–263.

[85] Anand Pillay. Groups and fields definable in an o-minimal structure. *Journal of Pure and Applied Algebra*, 53:239–255, 1988.

[86] Anand Pillay. Some foundational questions concerning differential algebraic groups. Preprint, 1994.

[87] Anand Pillay. The geometry of forking and groups of finite Morley rank. *Journal of Symbolic Logic*, 60:1251–1259, 1995.

[88] Anand Pillay. *Geometrical Stability Theory*. Oxford University Press, Oxford, 1996.

[89] B. I. Plotkin. On the nilradical of a group (in Russian). *Doklady Akademii Nauk SSSR (N.S.)*, 98:341–343, 1954.

[90] Françoise Point. Conditions of quasi-nilpotency in certain varieties of groups. *Communications in Algebra*, 22:365–390, 1994.

[91] Bruno P. Poizat. Une preuve par la théorie de la déviation d'un théorème de John Baldwin. *Comptes Rendus de l'Académie des Sciences à Paris*, 287:589–591, 1978.

[92] Bruno P. Poizat. Sous-groupes définissables d'un groupe stable. *Journal of Symbolic Logic*, 46:137–146, 1981.

[93] Bruno P. Poizat. Groupes stables, avec types génériques réguliers. *Journal of Symbolic Logic*, 48:339–355, 1983.

[94] Bruno P. Poizat. *Cours de théorie des modèles.* Nur Al-Mantiq Wal-Ma'rifah, Villeurbanne, France, 1985.

[95] Bruno P. Poizat. *Groupes Stables.* Nur Al-Mantiq Wal-Ma'rifah, Villeurbanne, France, 1987.

[96] Bruno P. Poizat. Équations génériques. In B. Dahn and H. Wolter, editors, *Proceedings of the 8th Easter Conference on Model Theory.* Humboldt Universität, Berlin, 1990.

[97] Bruno P. Poizat and Frank O. Wagner. Sous-groupes périodiques d'un groupe stable. *Journal of Symbolic Logic*, 58:385–400, 1993.

[98] Mike Prest. *Model Theory of Modules.* Cambridge University Press, Cambridge, 1988.

[99] Joachim Reineke. Minimale Gruppen. *Zeitschrift für Mathematische Logik*, 21:357–359, 1975.

[100] Derek J. S. Robinson. *Infinite Soluble and Nilpotent Groups.* Queen Mary College Mathematics Notes, London, 1968.

[101] A. Seidenberg. An elimination theory for differential algebra. *University of California Mathematical Publications*, 3:31–65, 1956.

[102] Saharon Shelah. *Categoricity of Classes of Models.* PhD thesis, The Hebrew University, Jerusalem, 1969.

[103] Saharon Shelah. Stable theories. *Israel Journal of Mathematics*, 7:187–202, 1969.

[104] Saharon Shelah. Differentially closed fields. *Israel Journal of Mathematics*, 16:314–328, 1973.

[105] Saharon Shelah. *Classification Theory.* North-Holland, Amsterdam, 1978.

[106] Saharon Shelah. Simple unstable theories. *Annals of Pure and Applied Logic*, 19:177–203, 1980.

[107] Michio Suzuki. *Group Theory II.* Springer-Verlag, Berlin, 1986.

[108] Wanda Szmielew. Elementary properties of abelian groups. *Fundamenta Mathematicae*, 41:203–271, 1955.

[109] Alfred Tarski. *A Decision Method for Elementary Algebra and Geometry.* University of California Press, Berkeley, California, 1951.

[110] Lou van den Dries. Definable groups in characteristic 0 are algebraic groups. *Abstracts of the American Mathematical Society*, 3:142, 1982.

[111] Lou van den Dries, Angus Macintyre, and Dave Marker. The elementary theory of restricted analytic fields with exponentiation. *Annals of Mathematics*, 140:183–205, 1994.

[112] Frank O. Wagner. *Stable Groups and Generic Types.* PhD thesis, Oxford, 1990.

[113] Frank O. Wagner. Subgroups of stable groups. *Journal of Symbolic Logic*, 55:151–156, 1990.

[114] Frank O. Wagner. Small stable groups and generics. *Journal of Symbolic Logic*, 56:1026–1037, 1991.

[115] Frank O. Wagner. À propos d'équations génériques. *Journal of Symbolic Logic*, 57:548–554, 1992.

[116] Frank O. Wagner. More on $\mathfrak{R}$. *Notre Dame Journal of Formal Logic*, 33:159–174, 1992.

[117] Frank O. Wagner. Commutator conditions and splitting automorphisms for stable groups. *Archive for Mathematical Logic*, 32:223–228, 1993.

[118] Frank O. Wagner. Quasi-endomorphisms in small stable groups. *Journal of Symbolic Logic*, 58:1044–1051, 1993.

[119] Frank O. Wagner. Stable groups, mostly of finite exponent. *Notre Dame Journal of Formal Logic*, 34:183–192, 1993.

[120] Frank O. Wagner. Nilpotent complements and Carter subgroups in stable $\mathfrak{R}$-groups. *Archive for Mathematical Logic*, 33:23–34, 1994.

[121] Frank O. Wagner. A note on defining groups in stable structures. *Journal of Symbolic Logic*, 59:575–578, 1994.

[122] Frank O. Wagner. The Fitting subgroup of a stable group. *Journal of Algebra*, 174:599–609, 1995.

[123] Frank O. Wagner. CM-triviality and stable groups. submitted to *Journal of Symbolic Logic*, 1996.

[124] Frank O. Wagner. Hyperstable theories. In Charles Steinhorn Wilfrid Hodges, Martin Hyland and John Truss, editors, *Logic: from Foundations to Applications (European Logic Colloquium 1993)*, pages 483–514. Oxford University Press, Oxford, 1996.

[125] André Weil. On algebraic groups of transformations. *American Journal of Mathematics*, 77:302–271, 1955.

[126] Alex Wilkie. Model completeness results for expansions of the ordered field of real numbers by restricted Pfaffian functions and the exponential function. *Journal of the American Mathematical Society*, 9:1051–1094, 1996.

[127] Carol Wood. Notes on the stability of separably closed fields. *Journal of Symbolic Logic*, 44:412–416, 1979.

[128] Martin Ziegler. Model theory of modules. *Annals of Pure and Applied Logic*, 26:149–213, 1984.

[129] Martin Ziegler. Stabilitätstheorie. Vorlesungsskript WS 88/89, Freiburg, 1989.

[130] Boris Zil'ber. Groups and rings whose theory is categorical (in Russian). *Fundamenta Mathematicae*, 95:173–188, 1977.

[131] Boris Zil'ber. Totally categorical theories; structural properties and the non-finite axiomatizability. In L. Pacholski, J. Wierzejewski, and A. Wilkie, editors, *Model Theory of Algebra and Arithmetic*, Lecture Notes in Mathematics 834, pages 381–410. Springer-Verlag, Heidelberg, 1980.

[132] Boris Zil'ber. Strongly minimal countably categorical theories II (in Russian). *Sibirskiĭ Matematicheskiĭ Zhurnal*, 25:71–88, 1984.

[133] Boris Zil'ber. Strongly minimal countably categorical theories III (in Russian). *Sibirskiĭ Matematicheskiĭ Zhurnal*, 25:63–77, 1984.

[134] Boris Zil'ber. The structure of models of uncountably categorical theories. In Zbigniew Ciesielski and Czeslaw Olech, editors, *Proceedings of the International Congress of Mathematicians, August 16–24, 1983, Warszawa*, pages 359–368. North-Holland, Amsterdam, 1984.

[135] Boris Zil'ber. ω-stability of the field of complex numbers with a predicate distinguishing the roots of unity. Preprint, 1994.

Groups & Gobbledegook

Index

Printed in the United States
By Bookmasters